Useful Astronomical Data

(1 AU = mean distance from earth to sun = 1.50×10^{11} m)

Object	Mass	Radius	Mean Orbital Radius	Orbital Period	Orbital Eccentricity
Sun	1.99×10^{30} kg	696,000 km	—	—	—
Moon	7.36×10^{22} kg	1,740 km	384,000 km = 1.28 s	27.3 days	0.055
Mercury	$0.0558 M_E$	2,439 km	0.387 AU = 3.23 min	0.241 y	0.206
Venus	$0.815 M_E$	6,060 km	0.723 AU = 6.03 min	0.615 y	0.007
Earth	5.98×10^{24} kg $\equiv M_E$	6,380 km	1.000 AU = 8.33 min	1.000 y	0.017
Mars	$0.107 M_E$	3,370 km	1.524 AU = 12.7 min	1.88 y	0.093
Jupiter	$318 M_E$	69,900 km	5.203 AU = 43.4 min	11.9 y	0.048
Saturn	$95.1 M_E$	58,500 km	9.539 AU = 1.32 h	29.5 y	0.056
Uranus	$14.5 M_E$	23,300 km	19.182 AU = 2.66 h	84.0 y	0.047
Neptune	$17.2 M_E$	22,100 km	30.058 AU = 4.17 h	165 y	0.009
Pluto/Charon	$0.0025 M_E$	3500/1800 km	39.785 AU = 5.53 h	248 y	0.254

Based mostly on data in D. Halliday and R. Resnick, *Fundamentals of Physics*, 3d ed., New York: Wiley, p. A6.

1 min = 18×10^6 km
1 h = 1.08 Tm = 1.08×10^9 km
1 day = 2.59×10^{13} m = 25.9×10^9 km

Six Ideas That Shaped

Unit R: The Laws of Physics Are Frame-Independent

Physics

Second Edition

Thomas A. Moore

Boston   Burr Ridge, IL   Dubuque, IA   Madison, WI   New York   San Francisco   St. Louis
Bangkok   Bogotá   Caracas   Kuala Lumpur   Lisbon   London   Madrid   Mexico City
Milan   Montreal   New Delhi   Santiago   Seoul   Singapore   Sydney   Taipei   Toronto

McGraw-Hill Higher Education

*A Division of The **McGraw-Hill** Companies*

SIX IDEAS THAT SHAPED PHYSICS, UNIT R: THE LAWS OF PHYSICS
ARE FRAME-INDEPENDENT, SECOND EDITION

2 3 4 5 6 7 8 9 0 QPD/QPD 0 9 8 7 6 5

ISBN   0-07-239714-4

Publisher: *Kent A. Peterson*
Sponsoring editor: *Daryl Bruflodt*
Developmental editor: *Spencer J. Cotkin, Ph.D.*
Marketing manager: *Debra B. Hash*
Senior project manager: *Susan J. Brusch*
Lead production supervisor: *Sandy Ludovissy*
Media project manager: *Sandra M. Schnee*
Lead media technology producer: *Judi David*
Designer: *David W. Hash*
Cover/interior designer: *Rokusek Design*
Cover image: © *CERN*
Senior photo research coordinator: *Lori Hancock*
Photo research: *Chris Hammond/PhotoFind LLC*
Supplement producer: *Brenda A. Ernzen*
Compositor: *Interactive Composition Corporation*
Typeface: *10/12 Palatino*
Printer: *Quebecor World Dubuque, IA*

Figure R2.2: *From A. Brillet and J.L. Hall, "Improved Laser Test of the Isotropy Space," Physical
Review Letters, Vol. 45, No. 9, February 26, 1979, pp. 549–552. Copyright © 1979 by the American
Physical Society.* Page 9 figure: *"Used with permission of the Stanford Linear Accelerator Center,
Stanford University."*

Credit List: **Chapter 1** Page 4: © Owen Franken/Stock Boston; p. 16: National Oceanic
and Atmospheric Administration/Department of Commerce. **Chapter 2** R2.12: © Syracuse
Newspapers/The Image Works. **Chapter 3** R3.8: U.S. Naval Observatory. **Chapter 5** R5.4:
Courtesy of Brookhaven National Laboratory; p. 91: © Corbis/Vol. 103. **Chapter 7** Page 129:
NASA. **Chapter 8** Page 149: © Ernest H. Robl; p. 150: NASA. **Chapter 9** Page 169:
© PhotoDisc/Vol. 39. **Chapter 10** R10.12: Fermi National Accelerator Laboratory.

Library of Congress Cataloging-in-Publication Data

Moore, Thomas A. (Thomas Andrew)
    Six ideas that shaped physics. Unit R, The laws of physics are frame-independent /
    Thomas A. Moore. — 2nd ed.
      p.   cm.
    Includes index.
    ISBN 0-07-239714-4 (acid-free paper)
    1. Special relativity (Physics). I. Title.: Laws of physics are frame-independent. II. Title.
QC173.65 .M657   2003
530.11—dc21   2002024404
                                               CIP

www.mhhe.com

Dedication

For My Parents, Stanley and Elizabeth,
who taught me the joy of wondering.

Table of Contents for
Six Ideas That Shaped Physics

Contents: Unit R
The Laws of Physics Are Frame-Independent

About the Author

Introduction

Thomas A. Moore graduated from Carleton College (magna cum laude with Distinction in Physics) in 1976. He won a Danforth Fellowship that year that supported his graduate education at Yale University, where he earned a Ph.D. in 1981. He taught at Carleton College and Luther College before taking his current position at Pomona College in 1987, where he won a Wig Award for Distinguished Teaching in 1991. He served as an active member of the steering committee for the national Introductory University Physics Project (IUPP) from 1987 through 1995. This textbook grew out of a model curriculum that he developed for that project in 1989, which was one of only four selected for further development and testing by IUPP.

He has published a number of articles about astrophysical sources of gravitational waves, detection of gravitational waves, and new approaches to teaching physics, as well as a book on special relativity entitled "A Traveler's Guide to Spacetime" (McGraw-Hill, 1995). He has also served as a reviewer and an associate editor for American Journal of Physics. He currently lives in Claremont, California, with his wife Joyce and two college-aged daughters. When he is not teaching, doing research in relativistic astrophysics, or writing, he enjoys reading, hiking, scuba diving, teaching adult church-school classes on the Hebrew Bible, calling contradances, and playing traditional Irish fiddle music.

Preface

Introduction

This volume is one of six that together comprise the text materials for *Six Ideas That Shaped Physics,* a fundamentally new approach to the two- or three-semester calculus-based introductory physics course. *Six Ideas That Shaped Physics* was created in response to a call for innovative curricula offered by the Introductory University Physics Project (IUPP), which subsequently supported its early development. In its present form, the course represents the culmination of more than a decade of development, testing, and evaluation at a number of colleges and universities nationwide.

Opening comments about Six Ideas That Shaped Physics

This course is based on the premise that innovative approaches to the presentation of topics and to classroom activities can help students learn more effectively. I have completely rethought, from the ground up, the presentation of every topic, taking advantage of research into physics education wherever possible, and have done nothing just because "that is the way it has always been done." Recognizing that physics education research has consistently emphasized the importance of active learning, I have also provided tools supporting multiple opportunities for active learning both inside and outside the classroom. This text also strongly emphasizes the process of building and critiquing physical models and using them in realistic settings. Finally, I have sought to emphasize contemporary physics and to view even classical topics from a thoroughly contemporary perspective.

I have not sought to "dumb down" the course to make it more accessible. Rather, my goal has been to help students become *smarter.* I intentionally set higher-than-usual standards for sophistication in physical thinking, and I then used a range of innovative approaches and classroom structures to help even average students reach this standard. I don't believe that the mathematical level required by these books is significantly different from that in most university physics texts, but I do ask students to step beyond rote thinking patterns to develop flexible, powerful conceptual reasoning and model-building skills. My experience and that of other users are that normal students in a wide range of institutional settings can, with appropriate support and practice, meet these standards.

Six volumes comprise the complete *Six Ideas* course:

The six volumes of the *Six Ideas* text

Unit C (**C**onservation laws):	Conservation Laws Constrain Interactions
Unit N (**N**ewtonian mechanics):	The Laws of Physics Are Universal
Unit R (**R**elativity):	The Laws of Physics Are Frame-Independent
Unit E (**E**lectricity and magnetism):	Electric and Magnetic Fields Are Unified
Unit Q (**Q**uantum physics):	Particles Behave Like Waves
Unit T (**T**hermal physics):	Some Processes Are Irreversible

I have listed these units in the order that I recommend they be taught, although other orderings are possible. At Pomona, we teach the first three units during the first semester and the last three during the second semester of a year-long course, but one can easily teach the six units in three quarters or even over three semesters if one wants a slower pace. The chapters of all these texts have been designed to correspond to what one might realistically discuss in a single 50-minute class session at the *highest possible pace*. A reasonable course syllabus will therefore set an average pace of not more than one chapter per 50 minutes of class time.

For more information than I can include in this short preface about the goals of the *Six Ideas* course, its organizational structure (and the rationale behind that structure), the evidence for its success, and guidance on how to cut and/or rearrange material, as well as many other resources for both teachers and students, please visit the *Six Ideas* website (see the next section).

Important Resources

Instructions about how to use this text

I have summarized important information about how to read and use this text in an *Introduction for Students* immediately preceding the first chapter. Please look this over, particularly if you have not seen other volumes of this text.

The *Six Ideas* web site

The *Six Ideas* website contains a wealth of up-to-date information about the course that I think both instructors and students will find very useful. The URL is

www.physics.pomona.edu/sixideas/

Essential computer programs

One of the most important resources available at this site is a number of computer applets that illustrate important concepts and aid in difficult calculations. Past experience indicates that students learn the ideas much more effectively when these programs are used both in the classroom and for homework. These applets are freeware and are available for both the Mac (Classic) and Windows operating systems.

Some Notes Specifically About Unit R

Why spend so much time on special relativity?

Unit R is a relatively short unit that focuses on developing the theory of special relativity as a logical consequence of the principle of relativity. Typically, little time in a traditional introductory physics course is spent exploring relativity, and as a result, few students understand or appreciate the beauties it has to offer. The experience of those of us who have used preliminary versions of this text is that if two to four weeks of class time are devoted to the study of relativity using the approach outlined in this unit, students at almost any level can develop a robust and satisfying understanding of the logic and meaning of relativity, and many will become genuinely excited about really being able to *understand* such a well-known but counterintuitive topic in physics (the intensity of this excitement actually surprised some of our early users).

There is also perhaps no better or more accessible example in all of physics illustrating how carefully thinking through consequences can uncover unexpected truths beyond the realm of daily experience. Therefore, studying special relativity can provide students with a glimpse of both the process and the rewards of theoretical physics, as well as help them take an important first step into the world of contemporary physics, where reaching beyond the level of our daily experience requires an increasing reliance on logical reasoning and abstract models.

This text emphasizes relativity's logical structure

This text therefore emphasizes the logical structure of relativity, clearly showing how well-known and bizarre relativistic effects such as length contraction and time dilation are the *inevitable* consequences of the principle of relativity. If students come away from this unit feeling that the universe not only *is* in fact as described by relativity, but indeed almost *has* to be, then this unit has been successful. I urge instructors to tailor their efforts toward this goal.

This unit should follow a treatment of newtonian mechanics that includes nonrelativistic kinematics, Newton's second law, conservation of momentum and energy, and some study of reference frames. This book can be used as a supplement to a traditional introductory text anytime after these topics are covered. In a *Six Ideas* course, this unit should definitely follow units C and N.

How this unit is related to the other units

On the other hand, I think that it is good to go against history and schedule unit R before unit E for several reasons. First, knowing some relativity can actually make certain aspects of electricity and magnetism simpler, and unit E takes some advantage of the relativistic perspective in general and the ideas in chapter R8 in particular (see the preface to unit E for a fuller description of exactly how it depends on unit R). Second, I think that it is good for students to get a taste of some exciting contemporary physics between the many weeks of classical physics represented by units C, N, and E. This is especially true if this is the last unit in the first semester: ending the first semester with unit R means that many students will leave the course excited and intrigued about physics and (perhaps) more eager to continue their studies in the second semester.

Unit Q uses relativity only in a couple of places, once in chapter Q4 (where the relativistic de Broglie equation is mentioned) and then in the section on nuclear physics (where the relationship between mass and energy is needed). These ideas could be summarized for students if one really wanted to study unit Q before unit R for some reason. Unit T does not draw on relativity at all.

The shortest possible treatment of relativity using this book would be to omit chapters R5 and R8 through R10. This would yield a six-session introduction to basic relativistic kinematics (with no dynamics or $E = mc^2$). Adding chapter R5 and/or R8 would provide a richer introduction to pure kinematics.

How to make cuts if absolutely necessary

The shortest introduction that includes dynamics would be to omit chapters R5, R7, and R8 and add a single class session devoted to sections R5.1 through R5.4 and section R8.4 (and possibly section R8.1 if there is time). This would get everything that is useful for units E and Q within eight class sessions.

However, students find the material in chapters R5 and R7 some of the most interesting in the book, and chapter R7 is also where they really test their understanding of relativistic kinematics in the context of tough paradoxes. Therefore, I really recommend doing the whole unit if you have time.

Appendix RB on the Doppler shift can be covered (if desired) anytime after section R5.2, and can displace some of the latter sections of chapter R5, displace some of the middle sections of chapter R8, or be added to chapter R7 (which, though challenging, involves fewer new ideas than the other chapters).

Please see the instructor's manual for more detailed comments about this unit and suggestions about how to teach it effectively.

Appreciation

A project of this magnitude cannot be accomplished alone. I would like to thank first the others who served on the IUPP development team for this project: Edwin Taylor, Dan Schroeder, Randy Knight, John Mallinckrodt,

Thanks!

Alma Zook, Bob Hilborn, and Don Holcomb. I'd like to thank John Rigden and other members of the IUPP steering committee for their support of the project in its early stages, which came ultimately from an NSF grant and the special efforts of Duncan McBride. Users of the texts—especially Bill Titus, Richard Noer, Woods Halley, Paul Ellis, Doreen Weinberger, Nalini Easwar, Brian Watson, Jon Eggert, Catherine Mader, Paul De Young, Alma Zook, Dan Schroeder, David Tanenbaum, Alfred Kwok, and Dave Dobson—have offered invaluable feedback and encouragement. I'd also like to thank Alan Macdonald, Roseanne Di Stefano, Ruth Chabay, Bruce Sherwood, and Tony French for ideas, support, and useful suggestions. Thanks also to Robs Muir for helping with several of the indexes. My editors Jim Smith, Denise Schanck, Jack Shira, Karen Allanson, Lloyd Black, J. P. Lenney, and Daryl Bruflodt as well as Spencer Cotkin, Donata Dettbarn, David Dietz, Larry Goldberg, Sheila Frank, Jonathan Alpert, Zanae Roderigo, Mary Haas, Janice Hancock, Lisa Gottschalk, Debra Hash, David Hash, Patti Scott, Chris Hammond, Brittney Corrigan-McElroy, Rick Hecker, and Susan Brusch have all worked very hard to make this text happen, and I deeply appreciate their efforts. I'd like to thank all the reviewers, including Edwin Carlson, David Dobson, Irene Nunes, Miles Dressler, O. Romulo Ochoa, Qichang Su, Brian Watson, and Laurent Hodges for taking the time to do a careful reading of various units and offering valuable suggestions. Thanks to Connie Wilson, Hilda Dinolfo, and Connie Inman, and special student assistants Michael Wanke, Paul Feng, Mara Harrell, Jennifer Lauer, Tony Galuhn, and Eric Pan, and all the Physics 51 mentors for supporting (in various ways) the development and teaching of this course at Pomona College. Thanks also to my Physics 51 students, and especially Win Yin, Peter Leth, Eddie Abarca, Boyer Naito, Arvin Tseng, Rebecca Washenfelder, Mary Donovan, Austin Ferris, Laura Siegfried, and Miriam Krause, who have offered many suggestions and have together found many hundreds of typos and other errors. Eric and Brian Daub were indispensable in helping me put this edition together. Finally, very special thanks to my wife, Joyce, and to my daughters, Brittany and Allison, who contributed with their support and patience during this long and demanding project. Heartfelt thanks to all!

Thomas A. Moore
Claremont, California

Introduction for Students

Introduction

Welcome to *Six Ideas That Shaped Physics!* This text has been designed using insights from recent research into physics learning to help you learn physics as effectively as possible. It thus has many features that may be different from science texts you have probably encountered. This section discusses these features and how to use them effectively.

Why Is This Text Different?

Research consistently shows that people learn physics most effectively if they participate in *activities* that help them *practice* applying physical reasoning in realistic situations. This is because physics is not a collection of facts to absorb, but rather is a set of *thinking skills* requiring practice to master. You cannot learn such skills by going to factual lectures any more than you can learn to play the piano by going to concerts!

This text is designed, therefore, to support *active learning* both inside and outside the classroom by providing (1) resources for various kinds of learning activities, (2) features that encourage active reading, and (3) features that make it easier for the text (as opposed to lectures) to serve as the primary source of information, so that more class time is available for active learning.

The Text as Primary Source

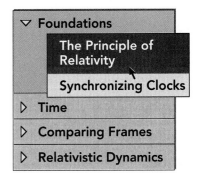

To serve the last goal, I have adopted a conversational style that I hope will be easy to read, and I tried to be concise without being so terse that you need a lecture to fill in the gaps. There are also many text features designed to help you keep track of the big picture. The unit's **central idea** is summarized on the front cover where you can see it daily. Each chapter is designed to correspond to one 50-minute class session, so that each session is a logically complete unit. The two-page **chapter overview** at the beginning of each chapter provides a compact summary of that chapter's contents to consider before you are submerged by the details (it also provides a useful summary when you review for exams). An accompanying **chapter-location diagram** uses a computer menu metaphor to display how the current chapter fits into the unit (see the example at the upper right). Major unit subdivisions appear as gray boxes, with the current subdivision highlighted in color. Chapters in the current subdivision appear in a submenu with the current chapter highlighted in black and indicated by an arrow.

All technical terms are highlighted using a **bold** type when they first appear, and a **glossary** at the end of the text summarizes their definitions. Please also note the tables of useful information, including definitions of common symbols, that appear inside the front cover.

A physics *formula* is both a mathematical equation and a *context* that gives the equation meaning. Every important formula in this text appears in a **formula box.** Each contains the equation, a **Purpose** (describing the formula's

Features that help the text serve as the primary source of information meaning and utility), a definition of the **Symbols** used in the equation, a description of any **Limitations** on the formula's applicability, and possibly some other useful **Notes.** Treat everything in such a box as an *indivisible unit* to be remembered and used together.

Active Reading

Like passively listening to a lecture, passively scanning a text does not really help you learn. *Active* reading is a crucial study skill for effectively learning from this text (and other types of technical literature as well). An active reader stops frequently to pose internal questions such as these: *Does this make sense? Is this consistent with my experience? Am I following the logic here? Do I see how I might use this idea in realistic situations?* This text provides two important tools to make this easier.

Use the **wide margins** to (1) record *questions* that occur to you as you read (so that you can remember to get them answered), (2) record *answers* when you receive them, (3) flag important passages, (4) fill in missing mathematics steps, and (5) record insights. Doing these things helps keep you actively engaged as you read, and your marginal comments are generally helpful as you review. Note that I have provided some marginal notes that summarize the points of crucial paragraphs and help you find things quickly.

The **in-text exercises** help you develop the habits of (1) filling in missing mathematics steps and (2) posing questions that help you *practice* using the chapter's ideas. Also, although this text has many examples of worked problems similar to homework or exam problems, *some* of these appear in the form of in-text exercises (as you are more likely to *learn* from an example if you work on it a bit yourself instead of just scanning someone else's solution). Answers to *all* exercises appear at the end of each chapter so you can get immediate feedback on how you are doing. Doing at least some of the exercises as you read is probably the *single most important thing you can do* to become an active reader.

Active reading does take effort. *Scanning* the 5200 words of a typical chapter might take 45 minutes, but active reading could take several times as long. I personally tend to "blow a fuse" in my head after about 20 minutes of active reading, so I take short breaks to do something else to keep alert. Pausing to fill in missing math also helps me to stay focused longer.

Class Activities and Homework

The problems appearing at the end of each chapter are organized into categories that reflect somewhat different active-learning purposes. **Two-minute problems** are short, concept-oriented, multiple-choice problems that are primarily meant to be used *in* class as a way of practicing the ideas and/or exposing conceptual problems for further discussion. (The letters on the back cover make it possible to display responses to your instructor.) The other types of problems are primarily meant for use as homework *outside* class. **Basic** problems are simple drill-type problems that help you practice in straightforward applications of a single formula or technique. **Synthetic** problems are more challenging and realistic questions that require you to bring together multiple formulas and/or techniques (maybe from different chapters) and to think carefully about physical principles. These problems define the level of sophistication that you should strive to achieve. **Rich-context** problems are yet more challenging problems that are often written in a narrative format and ask you to answer a practical, real-life question rather than explicitly asking for a numerical result. Like situations you will encounter in real life, many provide too little information and/or too much information, requiring you to make estimates and/or discard irrelevant data (this is true of some *synthetic* problems as well). Rich-context problems are generally too difficult for most students to solve alone; they are designed for *group* problem-solving sessions. **Advanced** problems are very sophisticated problems that provide supplemental discussion of subtle or advanced issues related to the material discussed in the chapter. These problems are for instructors and truly exceptional students.

Read the Text *Before* Class!

You will be able to participate in the kinds of activities that promote real learning *only* if you come to each class having already read and thought about the assigned chapter. This is likely to be *much* more important in a class using this text than in science courses you may have taken before! Class time can also (*if* you are prepared) provide a great opportunity to get your *particular* questions about the material answered.

Class time works best if you are prepared

The Principle of Relativity

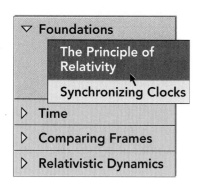
Chapter Overview

Introduction

In units C and N, we have explored the newtonian model of mechanics. In this unit, we will explore a *different* model, called the *special theory of relativity,* that better explains the behavior of objects, especially objects moving at close to the speed of light. This chapter lays the foundations for that exploration by describing the core idea of the theory and linking it to newtonian mechanics.

Section R1.1: Introduction to the Principle

We can informally state the great idea of this unit, the **principle of relativity,** as follows:

> The laws of physics are the same inside a laboratory moving at a constant velocity as they are inside a laboratory at rest.

The theory of special relativity essentially spells out the logical consequences of this idea.

This unit is divided into four subdivisions. The first (chapters R1 and R2) discusses the principle itself and develops important tools for future use. The second (chapters R3, R4, and R5) explores the relativistic concept of time. The third (chapters R6, R7, and R8) discusses how observers in different reference frames will view a sequence of events. Finally, the fourth (chapters R9 and R10) examines the consequences for the laws of mechanics.

Section R1.2: Events and Spacetime Coordinates

What exactly does the principle mean by a "laboratory"? The first step to understanding this better is to describe *operationally* how we measure a particle's motion. An **event** is something that happens at a well-defined place and time. An event's **spacetime coordinates** are a set of four numbers that locate the event in space and time. A particle's motion is a series of events.

Section R1.3: Reference Frames

A **reference frame** is a tool for assigning spacetime coordinates to events. We can visualize a reference frame as being a cubical lattice with a clock at every intersection. This ensures that there is a clock present at every event, but it also implies that we must synchronize the clocks somehow. An **observer** is a person who *interprets* results obtained in a reference frame to reconstruct the motions of particles. A real reference frame does not actually consist of a cubical lattice of clocks, but must be functionally equivalent.

Section R1.4: Inertial Reference Frames

An **inertial reference frame** is a frame in which an isolated object is *always* and *everywhere* observed to move at a constant velocity. We can check whether a frame is inertial by distributing **first-law detectors** around the frame to test for violations of Newton's first law.

A consequence of this definition is that two inertial frames in the same region of space must move at a constant velocity relative to each other. Conversely, if a given frame moves at a constant velocity relative to another inertial frame in the same region of space, the first must be inertial also.

Section R1.5: The Final Principle of Relativity

Note that in our original statement of the principle of relativity, the "laboratory moving at a constant velocity" and the "laboratory at rest" are *both* inertial frames. Moreover, the principle itself implies that there is no physical way to distinguish a frame in motion from one at rest: only the relative velocity between two inertial reference frames is measurable. Our final, polished statement of the principle therefore expresses the core issue without referring to "moving" or being at "rest":

The laws of physics are the same in all inertial reference frames.

Section R1.6: Newtonian Relativity

What does the phrase "the laws of physics are the same" mean? We can examine this issue in the context of newtonian physics if we temporarily embrace Newton's hypothesis about time, which is that *time is universal and absolute* and thus independent of reference frame. Consider two inertial frames that have constant relative velocity $\vec{\beta}$ and which are in **standard orientation** relative to each other; that is, the axes of both point in the same directions in space and the Other (primed) Frame moves in the $+x$ direction relative to the Home (unprimed) Frame. The concept of universal time implies that time measured in both frames is the same ($t = t'$); this, together with some simple vector addition and some calculus, implies that

$$\vec{r}\,'(t') = \vec{r}(t) - \vec{\beta}t \qquad (R1.1)$$

$$\vec{v}\,'(t') = \vec{v}(t) - \vec{\beta} \qquad (R1.3)$$

$$\vec{a}\,'(t') = \vec{a}(t) \qquad (R1.4)$$

Purpose: These equations describe how to compute an object's position $\vec{r}\,'$, velocity $\vec{v}\,'$, and acceleration $\vec{a}\,'$ at any given time t' in the Other Frame, given the object's position, position $\vec{r}$, velocity $\vec{v}$, and acceleration $\vec{a}$ at the same time t in the Home Frame.

Symbols: $\vec{\beta}$ is the velocity of the Other Frame relative to the Home Frame.

Limitations: These equations assume that $t = t'$, which is not true unless both $\vec{v} \ll c$ and $\vec{\beta} \ll c$, for reasons we are about to discover.

Note: Equation R1.1 and the equation $t = t'$ together comprise the **galilean transformation equations;** equation R1.3 represents the **galilean velocity transformation equations.**

Equation R1.3 implies that Newton's second law is frame-independent: the vector sum of the physical forces acting on an object will be equal to its mass times its acceleration in all inertial frames. This is an example of the laws of physics being the same in two frames (the problems explore some others).

R1.1   Introduction to the Principle

If you have ever traveled on a jet airplane, you know that while the plane may be flying through the air at 550 mi/h, things inside the plane cabin behave pretty much as they would if the plane were sitting at the loading dock. A cup dropped from rest in the cabin, for example, will fall straight to the floor (even though the plane moves forward many hundreds of feet with respect to the earth in the time that it takes the cup to reach the floor). A ball thrown up in the air by a child in the seat in front of you falls straight back into the child's lap (instead of being swept back toward you at hundreds of miles per hour). Your watch, the attendants' microwave oven, and the plane's instruments behave just as they would if they were at rest on the ground.

The laws of physics in the cabin of a rapidly moving jet are the same as those on the ground.

Indeed, imagine that you were confined to a small, windowless, and soundproofed room in the plane during a stretch of exceptionally smooth flying. Is there any physical experiment that you could perform entirely within the room (i.e., that would not depend on any information coming from beyond the walls of the room) that would indicate whether or how fast the plane were moving?

The answer to this question appears to be no. No one has ever found a convincing physical experiment that yields a different result in a laboratory moving at a constant velocity than it does when the laboratory is at rest. The designers of the plane's electronic instruments do not have to use different laws of electromagnetism to predict the behavior of those instruments when the plane is in flight than they do when the plane is at rest. Scientists working to enhance the performance of the *Voyager 2* space probe tested out various techniques on an identical model of the probe at rest on earth, confident that if the techniques worked for the earth-based model, they would work for the actual probe, even though the actual probe was moving relative to the earth at nearly 72,000 km/h. Astrophysicists are able to explain and understand the behavior of distant galaxies and quasars by using physical laws developed in earth-based laboratories, even though such galaxies and quasars move with respect to the earth at substantial fractions of the speed of light.

In short, all available evidence suggests that we can make the following general statement about the way that the universe is constructed:

> The laws of physics are the same inside a laboratory moving at a constant velocity as they are in a laboratory at rest.

An informal statement of the principle of relativity

This is an unpolished statement of what we will call the **principle of relativity.** This simple idea, based on common, everyday experience, is the foundation of Einstein's **special theory of relativity.** All of that theory's exciting and mind-bending predictions about space and time follow as *logical consequences* of this simple principle! The remainder of this book is little more than a step-by-step unfolding of the rich implications of this statement.

The principle of relativity is both a very new and a very old idea. It was not first stated by Einstein (as one might expect) but by Galileo Galilei in his book *Dialog Concerning the Two Chief World Systems* (1632). (Galileo's vivid and entertaining description of the principle of relativity is a wonderful example of a style of discourse that has, unfortunately, become archaic.) In the nearly three centuries that passed between Galileo's statement and Einstein's first paper on special relativity in 1905, the principle of relativity as it applied to the laws of *mechanics* was widely understood and used (in fact, it was generally considered to be a consequence of the particular nature of Newton's laws).

Historical notes

What Einstein did was to assert the applicability of the principle of relativity to *all* the laws of physics, and most particularly to the laws of electromagnetism (which had just been developed and thus were completely unknown to Galileo). Thus Einstein did not *invent* the principle of relativity; rather, his main contribution was to reinterpret it as being *fundamental* (more fundamental than Newton's laws or even than the ideas about time that up to that point had been considered obvious and inescapable) and to explore insightfully its implications regarding the nature of light, time, and space.

Our task in this text is to work out the rich and unexpected consequences of this principle. Figure R1.1 illustrates how we will proceed to do this in the remaining chapters of this unit.

This unit is divided into four subsections, as shown in figure R1.1. The first (chapters R1 and R2) deals with the foundations of relativity theory: we will discuss the principle of relativity and both Newton's and Einstein's approaches to defining time. The second subsection explores the implications of the principle of relativity regarding the nature of time, with special emphasis on the *metric equation*, an equation that links space and time into a geometric unity that we call *spacetime*. The third examines how the metric equation and the principle of relativity together determine how observers in different reference frames will interpret the same sequence of physical events; how *disagreements* between such observers lead to the phenomenon of length contraction; and how making sure that all observers *agree* about certain crucial things implies that nothing can go faster than light. The final subunit explores how we have to redefine energy and momentum somewhat to make conservation of energy and conservation of momentum consistent with the principle of relativity.

Overview of the unit

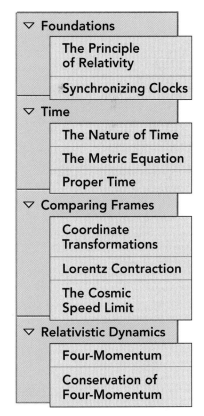

Our focus in this particular chapter is on the principle of relativity itself, and on developing an understanding of its meaning in the context of newtonian physics, before we proceed to explore the changes that Einstein proposes. It is important before we proceed , however, to understand two important things about the principle of relativity: (1) it is a *postulate*, and (2) it needs to be more precisely stated before we can extract any of its logical implications.

Figure R1.1
A chart illustrating the four subsections of unit R.

The principle of relativity
is a *postulate*

The principle of relativity is one of those core physical assumptions (like Newton's second law or the law of conservation of energy) that have to be accepted on faith: it cannot be *proved* experimentally or logically derived from more basic ideas (e.g., it is not possible even in principle to test *every* physical law in every laboratory moving at a constant velocity). The value of such a postulate rests entirely on its ability to provide the foundation for a model of physics that successfully explains and illuminates experimental results.

The principle of relativity has weathered nearly a century of intense critical examination. No contradiction of the principle or its consequences has ever been conclusively demonstrated. Moreover, the principle of relativity has a variety of unusual and unexpected implications that have been verified (to an extraordinary degree of accuracy) to occur exactly as predicted. Therefore, while it cannot be *proved*, it has not yet been *disproved*, and physicists find it to be something that can be confidently believed. The principle of relativity, simple as it is, is a very rich and powerful idea, and one that the physics community has found to be not only helpful but *crucial* in the understanding of much of modern physics.

Our informal statement of the
principle needs clarification

Turning to the other problem, we see that the principle of relativity as we have just stated it suffers from certain problems of both abstraction and ambiguity. For example, what do we *mean* by "the laws of physics are the same"? What exactly do we mean by "a laboratory at rest"? How can we tell if a laboratory is "at rest" or not? If we intend to explore the logical consequences of any idea, it is essential to state the idea in such a way that its meaning is clear and unambiguous.

Our task in the remaining sections of this chapter is to solve these problems. We will first replace the ambiguous phrases "laboratory," "at rest," and "constant velocity" with a single phrase involving more clearly defined terms. In the final sections, we will explore what we really mean by "the laws of physics are the same" in such laboratories. In so doing, we will provide a firm foundation for exploring the implications of the principle of relativity.

R1.2   Events and Spacetime Coordinates

The principle of relativity, as we have stated it so far, asserts that the laws of physics are the same in a laboratory moving at a constant velocity as they are in a laboratory at rest. A *laboratory* in this context is presumably a place where one performs experiments that test the laws of physics. How can we more carefully define what we mean by this term?

Our task is to specify what we
really mean by "laboratory"

The most fundamental physical laws describe how particles interact with one another and how they move in response to such interactions. Thus, perhaps what a physicist seeking to specify and test the laws of physics needs most is a means of mathematically describing the *motion* of a particle in space.

As we develop the theory of relativity, we need to be *very* careful about describing exactly *how* we will measure the motion of particles (hidden assumptions about the measurement process have plagued thinkers both before and after Einstein). In what follows, I will describe how we can measure the motion of a particle in terms of simple and well-defined concepts that are based on a minimum of supporting assumptions.

Definition of *event*

The first of these concepts is described by the technical term **event.** An *event* is any physical occurrence that we can consider to happen at a definite place in space and at a definite instant in time. The explosion of a small firecracker at a particular location in space and at a definite instant in time is a vivid example of an event. The collision of two particles or the decay of a single particle at a certain place and time also defines an event. Even the simple passage of a particle through a given mathematical point in space can be treated as an event (simply imagine that the particle sets off a firecracker at that point as it passes by).

Because an event occurs at a specific point in space and at a specific instant of time, we can quantify when and where the event occurs by four numbers: three numbers that specify the location of the event in some three-dimensional spatial coordinate system and one number that specifies what time the event occurred. We call these four numbers the event's **spacetime coordinates.**

Note that the exact values of an event's spacetime coordinates depend on certain arbitrary choices, such as the origin and orientation of the spatial coordinate axes and what time is considered to be $t = 0$. Once these choices are made and consistently used, however, specifying the coordinates of physical events provides a useful method of mathematically describing motion.

Specifically, we can quantify the motion of a particle by treating it as a *series* of events. For example, imagine an airplane moving along the x axis of some coordinate system. The airplane carries a blinking light. Each blink of the light is an event in the sense that we are using the word here: it occurs at a definite place in space and at a definite instant of time. We can describe the plane's motion by plotting a graph of the position coordinate of each "blink event" versus the time coordinate of the same, as illustrated in figure R1.2. If we decrease the time between blink events, we get an even more detailed picture of the plane's motion. We can in fact describe the plane's motion to whatever accuracy we need by listing the spacetime coordinates of a sufficiently large number of blink events distributed along its path.

The preceding is a specific illustration of a general idea: the motion of *any* particle can be mathematically described to arbitrary accuracy by specifying the spacetime coordinates of a sufficiently large number of events suitably distributed along its path. Studying the motion of particles is the most basic way to discover and test the laws of physics. Therefore, *the most fundamental task of a "laboratory" (as a place in which the laws of physics are to be tested) is to provide a means of measuring the spacetime coordinates of events.*

We can describe motion in terms of events

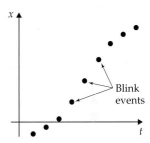

Figure R1.2
We can sketch out a graph of an object's motion (position versus time) by plotting the "blink events" that occur along the object's path.

R1.3   Reference Frames

We have already discussed in chapter C2 how we can quantify the *spatial* coordinates of an object (and thus presumably an event) using a cubical lattice of measuring sticks (or something equivalent). In our past discussions of reference frames, however, we did not really face the issue of how one might measure the *time* of an event: we simply *assumed* that this could be done in some simple and obvious way.

To proceed with our discussion of relativity, we now need to face this issue squarely. In what follows, we will extend the cubical lattice model of a reference frame to include a mechanism for measuring time in such a way that we can clearly distinguish the approaches to time implicit in newtonian mechanics and special relativity.

The trick is to take our cubical lattice and imagine that we attach a clock to every lattice intersection (see figure R1.3). We can then define the *time* coordinate of an event (such as a firecracker explosion) to be the time displayed on the lattice clock nearest the event (relative to some specified time $t = 0$) and the event's three *spatial* coordinates to be the lattice coordinates of that

The operational definition of spacetime coordinates

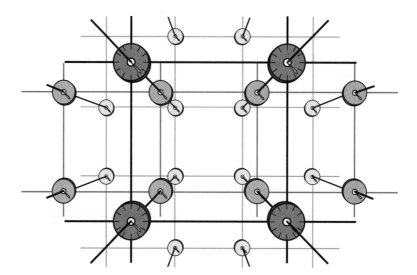

Figure R1.3

A reference frame visualized as a cubical lattice with a clock at every lattice intersection. This figure (and indeed the whole approach in this section) is adapted from E. F. Taylor and J. A. Wheeler, *Spacetime Physics*, San Francisco: Freeman, 1966, pp. 17–18.

nearest clock, specified in the usual way by stating the distances along the lattice directions that one has to travel (from some specified spatial origin) to the clock. We can determine these four numbers to whatever precision we want by sufficiently decreasing the lattice spacing and the time between clock ticks.

Why is it important to have a clock at *every* lattice intersection? The point is to make sure that there is a clock essentially *at* the location of any event to be measured. If we attempt to read the time of an event by using a clock located a substantial distance away, we need to make assumptions about how long it took the information that the event has occurred to *reach* that distant clock. For example, if we read the time when the *sound* from an event reaches the distant clock, we should correct that value by subtracting the time that it takes sound to travel from the event to the clock; but to do this, we have to know the speed of sound in our lattice. We can avoid this problem if we require that an event's time coordinate be measured by a clock essentially *present* at the event.

Note the clocks must all be *synchronized* in some meaningful and self-consistent manner if we are to get meaningful results. If these clocks are not synchronized, adjacent clocks might differ wildly, thus giving one totally incoherent picture of when a particle moving through the lattice passes various lattice points. What exactly we *mean* when we say that our "lattice clocks are synchronized" is precisely where Newton's and Einstein's models diverge. We will discuss this issue later: for now, it is sufficient to recognize that we have to synchronize the clocks *somehow*.

Once we have specified a synchronization method, the image of a clock lattice completely defines a *procedure* that we can use (in principle) to determine an event's spacetime coordinates. This amounts to an **operational definition** of spacetime coordinates: an *operational definition* of a physical quantity defines that quantity by describing how the quantity can be *measured*. Operational definitions provide a useful way of anchoring slippery human words to physical reality by linking the words to specific, repeatable procedures rather than to vague comparisons or analogies.

The procedure just described represents an admittedly idealized method for determining the spacetime coordinates of an event. The actual methods employed by physicists may well differ from this description, but these methods should be *equivalent* to what is described above: the clock lattice method defines a standard against which actual methods can be compared. It is such a simple and direct method that it is inconceivable that any actual technique could yield different results and still be considered correct and meaningful.

With this in mind, we define the following technical words to aid us in future discussions:

Technical terms involving reference frames

A **reference frame** is defined to be a rigid cubical lattice of appropriately synchronized clocks *or its functional equivalent.*

The **spacetime coordinates** of an event in a given reference frame are defined to be an ordered set of four numbers, the first specifying the *time* of the event as registered by the nearest clock in the lattice, followed by three that specify the spatial coordinates of that clock in the lattice. For example, in a frame oriented in the usual way on the earth's surface, a firecracker explosion whose spacetime coordinates are [3 s, −3 m, 6 m, −1 m] thus happened 3 m west, 6 m north, and 1 m below the frame's spatial origin, and 3 s after whatever event defines $t = 0$.

An **observer** is defined to be a (possibly hypothetical) person who interprets measurements made in a reference frame (e.g., the person who interprets the spacetime coordinates collected by a central computer receiving information from all the lattice clocks).

Note that the act of "observing" in the last definition is an act of *interpretation* of measurements generated by the frame apparatus, and that act may have little or nothing to do with what that observer sees with his or her own eyes. When we say that "an observer in such-and-such reference frame observes such-and-such," we are actually referring to *conclusions* that the observer draws from measurements performed using the reference frame lattice.

This image represents a computer-reconstructed "observation" of a particle decay process in one of the detectors at the Stanford Linear Accelerator (SLAC).

R1.4   Inertial Reference Frames

While exploring relativity theory, we will often speak of a reference frame in connection with some object. For example, one might refer to "the reference frame of the surface of the earth" or "the reference frame of the cabin of the plane" or "the reference frame of the particle." In these cases, we are being asked to imagine a clock lattice (or equivalent) fixed to the object in question. Sometimes the actual frame is referred to only obliquely, as in the phrase "an observer in the plane cabin finds . . . ." Since *observer* in this text refers to someone who is using a reference frame to determine event coordinates, this phrase presumes the existence of a reference frame attached to the cabin of the plane.

A reference frame may be moving or at rest, accelerating, or even rotating about some axis. The beauty of the definition of spacetime coordinates given above is that measurements of the coordinates of events (and thus measurements of the motion of objects) can be carried out in a reference frame no matter how it is moving (provided only that the clocks in the frame can be synchronized in some meaningful manner).

However, not all reference frames are equally useful for doing physics. We saw in chapter N9 that we can divide reference frames into two general classes: **inertial frames** and **noninertial frames.** An *inertial frame* is one in which an isolated object is always and everywhere observed to move at a constant velocity (as required by Newton's first law); in a *noninertial frame* such an object is observed to move with a nonconstant velocity in at least some situations.

We can operationally distinguish inertial from noninertial frames using a **first-law detector.** Figure R1.4 shows a simple first-law detector. Electrically actuated "fingers" hold an electrically uncharged and nonmagnetic ball at rest in the center of an evacuated spherical container. When the ball is released, it should remain at rest by Newton's first law; if it does not, the frame to which the detector container is attached is noninertial. (If we want to operate this detector in a gravitational field, we have to figure out a way to cancel the gravitational force on the ball without inhibiting its freedom to move; but in principle this can be done.) If we attach such a first-law detector to the clock at each lattice location in our reference frame and *none* of these detectors registers a violation of Newton's first law, then we can say with confidence that our frame is inertial.

This definition of an inertial frame is simple enough to apply to realistic examples. For example, while a gravity-compensated detector as shown in figure R1.4 would register no violation of the first law if it were attached to a plane at rest, we know without actually trying it that if the plane began to accelerate for takeoff, the detector's floating ball would be deflected toward the rear of the plane by the same (fictitious) force that pressed us back into our

Figure R1.4

(a) A cross-sectional view of a floating-ball first-law detector. Electrically actuated "fingers" hold the ball initially at rest in the spherical container. (b) After the fingers are retracted, the ball should continue to float at rest in the container, as long as the frame to which the container is attached is inertial.

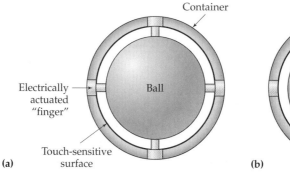

(a)   (b)

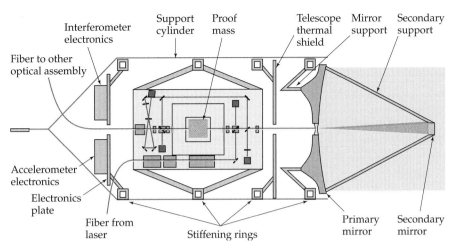

This is a schematic diagram of one of the spacecraft in the proposed Laser Interferometer Space Antenna (LISA), a space-based gravitational wave detector. The spacecraft's position is continually adjusted by very small thrusters so that the "proof mass" floats freely exactly in the spacecraft's center. This ensures that the spacecraft is an inertial reference frame.

chairs. Similarly, we might expect that detectors in a reference frame floating in deep space (far from any massive objects) would register no violation of Newton's first law; yet we know that detector balls in a similar frame that is rotating around its center will be deflected outward by the (fictitious) centrifugal force in that noninertial frame.

The following statement is an important and useful consequence of the definition of an inertial reference frame:

> *Any* inertial reference frame will be observed to move at a *constant velocity* relative to *any other* inertial reference frame. Conversely, a rigid, nonrotating reference frame that moves at a constant velocity with respect to any other inertial reference frame must *itself* be inertial.

Inertial frames move with constant velocities relative to each other

We first discussed this issue in chapter N9, but the *methods* that we used there unfortunately employ certain assumptions about the nature of time that turn out to be inconsistent with the principle of relativity (as we will see). We can, however, prove that the statement above follows *directly* from the definition of an inertial reference frame without having to make any assumptions about the nature of time. Here is an argument for the first part of the statement above; the proof of the converse statement is left as an exercise.

Consider two inertial reference frames (see figure R1.5), which we will call the **Home Frame** and **Other Frame,** respectively. (*Home Frame* and *Other Frame* are phrases that I will use in this text as *names* of inertial reference frames, which I will emphasize by the capitalization.) Since these are *inertial* reference frames, observers will measure an isolated object to move with a constant velocity in either frame *by definition.* Imagine a specific isolated object that happens to be at *rest* relative to the Other Frame. Since such an isolated object must move at a constant velocity in the Other Frame if the frame is inertial, if the object is initially at rest, it will have to *remain* at rest in that frame. Now let us observe the same object from the Home Frame. Since the object is isolated and the Home Frame is also inertial, the object must move at a constant velocity relative to that frame as well. But since that object is at rest with respect to the Other Frame, this means that the whole Other Frame must be observed to move relative to the Home Frame at the same constant

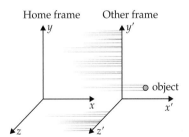

Figure R1.5
An isolated object at rest in the Other Frame must move at a constant velocity with respect to the Home Frame, so the whole Other Frame must move at the same constant velocity relative to the Home Frame.

velocity as the object! Therefore, the Other Frame will be observed to move at a constant velocity relative to the Home Frame, consistent with the statement above.

Exercise R1X.1

Using a similar approach, prove the converse part of the statement above (i.e., a rigid reference frame that moves at a constant velocity with respect to any other inertial reference frame must *itself* be inertial).

R1.5   The Final Principle of Relativity

Rest has no physical meaning in relativity

Our first informal statement of the principle of relativity stated that "the laws of physics are the same in a laboratory moving at a constant velocity as they are in a laboratory at rest." We have subsequently developed the idea of a *reference frame* to express the essence of what we mean by a *laboratory*. However, how can we physically distinguish a reference frame "moving at a constant velocity" from one "at rest"?

The short answer is that we cannot! The principle of relativity specifically states that a reference frame moving at a constant velocity is *physically equivalent* to a frame at rest. Therefore, there can be no physical basis for distinguishing a laboratory at rest from another frame moving at a constant velocity. Imagine that you and I are in spaceships coasting at a constant velocity in deep space. You will consider yourself to be at rest, while I am moving by you at a constant velocity. I, on the other hand, will consider myself to be at rest, while *you* are moving by me at a constant velocity. According to the principle of relativity, there is no physical experiment that can resolve our argument about who is "really" at rest. We could, of course, agree to *choose* one or the other of us to be at rest, but this choice is completely arbitrary. Therefore, if the principle of relativity is true, there is no basis for assigning an absolute velocity to any reference frame: only the *relative* velocity between two frames is a physically meaningful concept.

On the other hand, it is plausible that what we *really* mean by a reference frame "at rest" is an *inertial frame*. Moreover, we have just seen that a reference frame moving at a constant velocity relative to it must *also* be an inertial frame. Therefore, we can remove both the vague word *laboratory* and the ambiguity of the concepts *at rest* and *moving at a constant velocity* in our original statement of the principle of relativity by restating it as follows.

Our final statement of the principle of relativity

The Principle of Relativity
The laws of physics are the same in all inertial reference frames.

This is our final polished statement of the principle of relativity. It replaces the fuzzy and ambiguous ideas in our original statement with the sharply and operationally defined idea of an inertial reference frame. What this principle essentially claims is that if Newton's *first* law (which describes what happens to an isolated object) is the same in two given reference frames, then *all* the laws of physics are the same in both frames. (Note that the unit's "great idea" that appears on the front cover is a compressed version of this statement.)

But what *exactly* do we mean when we say that "the laws of physics are the same" in two frames? In section R1.6, we discuss the newtonian assumption about how we can synchronize clocks in an inertial reference frame. We then use this as a framework to explore what the principle of relativity means in newtonian mechanics.

R1.6   Newtonian Relativity

Imagine that we have an inertial frame floating in deep space, ready to use. We would like to use this frame to measure the coordinates of events happening in it so as to test the laws of physics. But an important problem remains to be solved: how do we synchronize its clocks?

"The solution is easy," says a newtonian physicist. "Everyone knows, as Newton himself asserted, that 'time is absolute and flows equably without regard to anything external.' *Any* good clock will therefore measure the flow of this absolute time. Therefore, we can simply designate one clock to be a master clock, carry it around to each of the lattice clocks, and synchronize each lattice clock to the master. Since the master clock and the lattice clocks all measure the flow of immutable absolute time, the motion of the master clock as it is carried from place to place in the lattice is irrelevant. Once a lattice clock is set to agree with the master clock, it will certainly remain in agreement with it, since both clocks measure the flow of absolute time. Indeed, if the master clocks in two different reference frames are in agreement at any given event, then all the clocks in the two frames will always agree. It doesn't matter whether the frames are in motion with respect to each other; it doesn't even matter if they are inertial or not. This follows from the self-evident absolute nature of time."

The newtonian approach to clock synchronization

This picture of the nature of time is straightforward and believable. It reflects the intuitive picture of time that most of us already hold. But what are its consequences?

Again consider two inertial frames that we will call the Home Frame and the Other Frame. We will often (but not always!) imagine ourselves to be in the Home Frame (so that this frame appears to *us* to be at rest). The Other Frame will move at a *constant* velocity $\vec{\beta}$ with respect to the Home Frame according to the proof given in section R1.5. (I will consistently use the Greek letter β to represent the relative velocity of frames, standing for the "boost" in velocity that one needs to go from being at rest in one frame to being at rest in the other.)

These frames might in principle have any relative orientation, but it is conventional in special relativity to use our freedom to choose the orientations to put the two frames in **standard orientation,** where the Home Frame's x, y, and z axes point in the same directions as the corresponding axes in the Other Frame. We conventionally distinguish the Home Frame and Other Frame axes by referring to the Home Frame axes as x, y, and z and the Other Frame axes as x', y', and z' (the mark is called a *prime*). It also is conventional to define the origin event (the event that defines $t = 0$ in both frames) to be the instant at which the spatial origin of one frame passes the origin of the other. Finally, we conventionally choose the common x axis so that the Other Frame moves in the $+x$ direction with respect to the Home Frame (implying that the Home Frame moves in the $-x$ direction with respect to the Other Frame). Signs in many equations in this text depend on this conventional choice, so it is wise to consistently follow it. Figure R1.6 illustrates two frames in standard orientation.

Standard orientation for inertial reference frames

Now consider an object moving in space that periodically emits blinks of light. Let the spatial position of a certain blink event as measured in the Home Frame be represented by the vector $\vec{r}(t)$ and the same measured in the Other Frame by $\vec{r}'(t)$ (we conventionally write symbols for quantities observed in the Other Frame with an attached prime). Since according to our assumption time is universal and absolute, observers in both frames should agree at what time this blink event occurs: $t = t'$. The position of the spatial origin of the Other Frame in the Home Frame at that time is simply $\vec{\beta}t$, since

Consequences of the newtonian view of time

Figure R1.6
A schematic drawing of two reference frames in standard orientation. The spatial origins of the frames coincided at $t = t' = 0$ just a little while ago. (You should imagine the frame lattices intermeshing so that events can be recorded in both frames.)

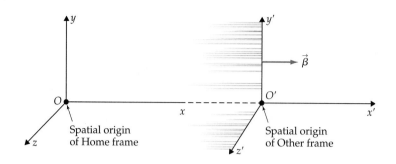

Figure R1.7
The relationship between $\vec{r}$ and $\vec{r}'$ for two inertial reference frames (assuming that time is universal and absolute).

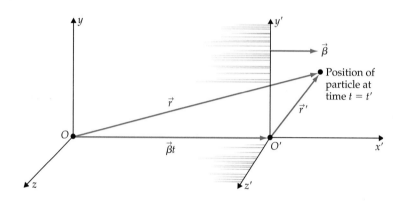

the Other Frame moves at a constant velocity $\vec{\beta}$ with respect to the Home Frame, and we conventionally take both frames' origins to coincide at $t = 0$. The relationship between the object's position vectors in the two frames at the time of the blink is (as shown in figure R1.7) given by $\vec{r}(t) = \vec{r}'(t') + \vec{\beta}t$, or

$$\vec{r}'(t') = \vec{r}(t) - \vec{\beta}t \qquad (R1.1)$$

For frames in standard orientation, $\vec{\beta}$ points in the $+x$ direction, meaning that we can write equation R1.1 in component form as follows:

The galilean transformation equations

$$t' = t \qquad \text{(reminding us that time is absolute)} \qquad (R1.2a)$$
$$x' = x - \beta t \qquad (R1.2b)$$
$$y' = y \qquad (R1.2c)$$
$$z' = z \qquad (R1.2d)$$

Physicists call these four equations the **galilean transformation equations**. These equations allow us to find the position of the object at a given time t' in the Other Frame if we know its position at time $t = t'$ in the Home Frame (*assuming*, of course, that time is universal and absolute).

If we take the time derivative of both sides of each of the last three equations, we get

The galilean *velocity* transformation equations

$$v'_x = v_x - \beta \qquad (R1.3a)$$
$$v'_y = v_y \qquad (R1.3b)$$
$$v'_z = v_z \qquad (R1.3c)$$

(Note that since $t' = t$, it really doesn't matter that we are taking a derivative with respect to t' on the left side and a derivative with respect to t on the right.) These equations tell us how to find the velocity of an object in the Other Frame, given its velocity in the Home Frame: we call these equations the **galilean velocity transformation equations**.

If we take the time derivative again, we get

$$a'_x = a_x \qquad (R1.4a)$$
$$a'_y = a_y \qquad (R1.4b)$$
$$a'_z = a_z \qquad (R1.4c)$$

An object's acceleration is the same in all inertial reference frames

Equations 1.4 tell us that *observers in both inertial frames agree about an object's acceleration at a given time,* even though they may well disagree about the object's position and velocity components at that time!

An illustration of how the laws of physics can be the same in different frames

Now we are in a position to discuss more fully what we might mean by "the laws of physics are the same" in every inertial reference frame in the newtonian context at least. Consider the following example. A child on an airplane throws a ball vertically into the air and catches it again. As measured in the plane cabin (which we will take to be the Other Frame), the ball appears to travel vertically along the vertical z axis. Now imagine that you watch this process from a nearby mountaintop as the plane passes by. Instead of observing the ball travel vertically up and down, *you* will instead observe the ball to follow a shallow parabolic trajectory, because in your frame (which we will take to be the Home Frame), the ball, plane, and child all have a considerable horizontal velocity (see figure R1.8).

The motion of the ball in these two reference frames looks very different: it looks entirely vertical in the plane's cabin but parabolic (almost horizontal!) in your frame. Even so, you and an observer on the plane would *agree* that (1) the ball has a certain mass m (which you and the observer could each measure with your own balances) and thus should experience a gravitational force of magnitude mg acting on it, (2) this force must be the net force on the object while it is in flight (since nothing else is in contact with the ball, ignoring air friction), and (3) the ball has the same acceleration in your

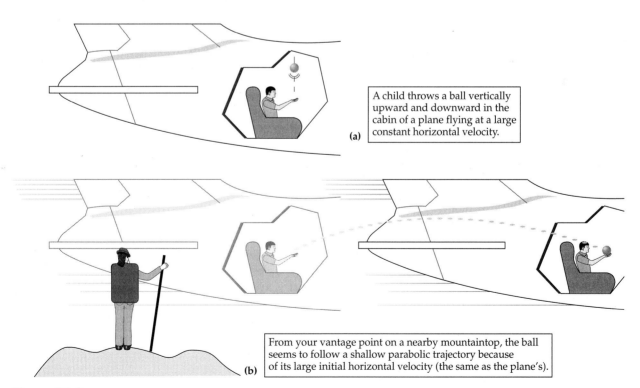

(a) A child throws a ball vertically upward and downward in the cabin of a plane flying at a large constant horizontal velocity.

(b) From your vantage point on a nearby mountaintop, the ball seems to follow a shallow parabolic trajectory because of its large initial horizontal velocity (the same as the plane's).

Figure R1.8
An example illustrating the application of the principle of relativity in newtonian physics. Newton's second law describes the motion of the ball in both frames.

respective reference frames (see equations R1.4). Since you agree on the value of m, the magnitude and direction of the net force on the ball, and the acceleration that the ball experiences, you will *both* agree that Newton's second law

$$\vec{F}_{net} = m\vec{a} \qquad \text{(with } \vec{F}_{net} = m\vec{g} \text{ here)} \tag{R1.5}$$

accurately predicts the ball's motion, even if you disagree about the ball's initial velocity and thus the exact character of its subsequent motion (i.e., whether it is vertical or parabolic).

In a similar fashion, you might imagine observing a game of pool in the plane cabin. You on the mountaintop and your friend on the plane will totally disagree about the initial and final velocities of the balls in any given collision (since you will observe them to have a large horizontal component of velocity that your friend does not observe). Even so, you both will find that the total momentum of the balls just before a given collision is equal to their total momentum just afterward, consistent with the law of conservation of momentum (see problems R1S.9 and R1S.10).

This is what it means to say that the "laws of physics are the same" in different inertial frames. Observers in different inertial frames may disagree about the values of various quantities (particularly positions and velocities), but each observer will agree that if one takes the mathematical equation describing a physical law (such as Newton's second law) and plugs in the values measured in that observer's frame, one will always find that the equation is satisfied. In other words, the same basic equations will be found to describe the laws of physics in all inertial reference frames.

TWO-MINUTE PROBLEMS

R1T.1 Which of the following are (at least nearly) inertial reference frames and which are not? (Respond T if the frame is inertial, F if it is noninertial. *Both* answers might be right in some cases: this could be an opportunity to discuss the issues involved.)
a. A nonrotating frame floating in deep space.
b. A rotating frame floating in deep space.
c. A nonrotating frame attached to the sun.
d. A frame attached to the surface of the earth.
e. A frame attached to a car moving at a constant velocity.
f. A frame attached to a roller-coaster car.

R1T.2 Which of the following physical occurrences fit the physical definition of an *event?*
A. The collision of two point particles.
B. A point particle passing a given point in space.
C. A firecracker explosion.
D. A party at your dorm.
E. A hurricane.
F. A, B, and C.
T. *Any* of the above could be an event, depending on the scale and precision of our reference frame.

R1T.3 Since the laws of physics are the same in every inertial reference frame, there is no meaningful physical distinction between an inertial frame at rest and one moving at a constant velocity, true (T) or false (F)?

Is this an event? (See problem R1T.2.)

R1T.4 Since the laws of physics are the same in every reference frame, an object must have the same kinetic energy in all inertial reference frames, T or F?

R1T.5 Since the laws of physics are the same in every inertial reference frame, an interaction between objects must be observed to conserve energy in every inertial reference frame, T or F?

R1T.6 Since the laws of physics are the same in every inertial reference frame, if you perform identical experiments in two different inertial frames, you should get *exactly* the same results, T or F?

R1T.7 Imagine two boats. One travels 3.4 m/s eastward relative to the earth and the other 5.0 m/s eastward relative to the earth. We set up a reference frame on each boat with the x axis pointing eastward. Which boat should we select to be the Home Frame, according to the convention established in this chapter?

A. The faster boat is the Home Frame.
B. The slower boat is the Home Frame.
C. We are free to choose either boat to be the Home Frame.

R1T.8 Imagine you are in a train traveling at one-half of the speed of light relative to the earth. Assuming that photons emitted by the train's headlight travel at the speed of light relative to you, they would (according to the galilean velocity transformation) travel at 1.5 times the speed of light relative to the earth, T or F?

HOMEWORK PROBLEMS

Basic Skills

R1B.1 A train moving with a speed of 55 m/s passes through a railway station at time $t = t' = 0$. Fifteen seconds later a firecracker explodes on the track 1.0 km away from the train station in the direction that the train is moving. Find the coordinates of this event in both the station frame (consider this to be the Home Frame) and the train frame. Assume that the train's direction of motion relative to the station defines the x direction in both frames.

R1B.2 Imagine that we define the rear end of a train 120 m long to define the origin $x' = 0$ in the train frame, and we define a certain track signal light to define the origin $x = 0$ in the track frame. Imagine that the rear end of the train passes this sign at $t = t' = 0$ as the train moves in the $+x$ direction at a constant speed of 25 m/s. Twelve seconds later, the engineer turns on the train's headlight.
(a) Where does this event occur in the train frame?
(b) Where does this event occur in the track frame?
(Please explain your response in both cases.)

R1B.3 Imagine that boat A is moving relative to the water with a velocity of 6 m/s due east and boat B is moving with a velocity of 12 m/s due west. Assume that observers on both boats use reference frames in which the x direction points east. What is the velocity of boat A relative to boat B? (*Hint:* Draw a picture! What is the Home Frame attached to here? What is the Other Frame attached to?)

R1B.4 Imagine that in an effort to attract more passengers, Amtrak trains now offer free bowling in a specially constructed "bowling alley" car. Imagine that such a train is traveling at a constant speed of 35 m/s relative to the ground. A bowling ball is hurled by a passenger on the train in the same direction as the train is traveling. The ball is measured in the ground frame to have a speed of 42 m/s. What is its speed in the frame of the train according to the galilean velocity transformation? (*Hints:* Draw a picture. How should you define the $+x$ direction? Which frame is the Home Frame? Which is the Other Frame?)

R1B.5 Consider the situation described in problem R1B.4, except assume that we know that the ball's velocity is 8 m/s in the train frame. What will be its speed relative to the ground?

Synthetic

R1S.1 Read Galileo's 1632 presentation of the principle of relativity. (If your professor does not make a copy available, you can find the core of Galileo's argument quoted in E. F. Taylor and J. A. Wheeler, *Spacetime Physics*, 2d ed., San Francisco: Freeman, 1992, pp. 53–55.) In a short paragraph, compare and contrast his presentation of the principle with the presentation in section R1.2. What for Galileo corresponds to a "laboratory moving at a constant velocity"? What for Galileo corresponds to the phrase "the laws of physics are the same" in such laboratories?

R1S.2 Imagine that Frames R Us, Inc., is constructing an economy reference frame whose price will be below every other frame on the market. Placing a clock at every point in the frame lattice is too expensive, so the company decides to place *one* clock at the origin. At all other positions, the company simply places a flag that springs up when an object goes by. The flag has the lattice location printed on it, so an observer sitting at the origin can assign spacetime coordinates to every event by noting *when* he or she sees the flag spring up (according to the clock at the origin) and noting the spatial coordinates indicated by the flag. Why *doesn't* this method yield the same spacetime coordinates as having a clock at every location would? Pinpoint the *assumption* that the engineers at Frames R Us are making that is incorrect. (See problem R1R.2 for a further exploration of the problems with this reference frame.)

R1S.3 Consider a floating-ball first-law detector like the one shown in figure R1.4. If the ball is 10 cm in diameter and is placed at $t = 0$ in the center of a spherical shell whose inside diameter is 12 cm, about how long will it take the ball to hit the shell if the shell accelerates at 0.1 m/s^2?

R1S.4 A totally symmetric way to orient a pair of reference frames is so that their $+x$ directions point in the direction that the other frame is moving. How is this different from the "standard" orientation (draw a picture)? How would the galilean position and velocity transformation equations be different if we were to use this convention?

R1S.5 Two firecrackers explode simultaneously 125 m apart along a railroad track, which we can take to define the x axis of an inertial reference frame (the Home Frame). A train (which defines the Other Frame) moves at a constant 25 m/s in the $+x$ direction relative to the track frame.
(a) According to the galilean transformation equations, do the firecrackers explode at the same time?
(b) How far apart are the explosions as measured in the train frame? (*Hint:* If $x_2 - x_1 = 125$ m, what is $x_2' - x_1'$?)
(c) Assume that instead of the explosions being simultaneous, the firecracker farther ahead in the $+x$ direction explodes 3.0 s before the other. Now how far apart would the explosions be as measured in the train frame?

R1S.6 Imagine that construction crews have set up a pair of blinking lights 60 m apart to mark a construction zone along a road. A car moves along the road with a speed of 30 m/s toward the lights. Let us take the road to be the Home Frame and the car to be the Other Frame.
(a) In the Home Frame, the light farther from the car (light 2) blinks 0.66 s before the other (light 1) blinks as the car approaches both lights. What is the time interval $t_2' - t_1'$ between the blink events in the car's frame? (*Hint:* Will this interval be positive or negative?)
(b) What is the x displacement $x_2' - x_1'$ between the events in the car frame?

R1S.7 In a certain particle accelerator experiment, two subatomic particles A and B are observed to fly away in opposite directions from a particle decay. Particle A is observed to travel with a speed of $0.6c$ relative to the laboratory, and particle B is observed to travel with a speed of $0.9c$, where c is the speed of light (3.0×10^8 m/s). For simplicity's sake, let's agree to take the direction of motion of particle B to define the $+x$ direction in both the laboratory frame and particle A's frame. We would like to calculate the relative velocity of particles A and B.
(a) What object should we choose to be the Home Frame in this problem, according to the conven-

tions established in this chapter? What object defines the Other Frame? What object is the object whose velocity we want to determine in both frames? Please explain.
(b) Use the galilean velocity transformation equations to calculate the relative velocity of the two particles as a fraction or multiple of c. Please explain your work.

R1S.8 A highway patrol officer on the ground uses a radar gun to measure the speed of a suspect's car to be 50 m/s. A patrol car is traveling at 40 m/s relative to the ground, but is moving in the opposite direction as it approaches the car. Assume that the direction in which the suspect is traveling defines the $+x$ direction for everyone. We would like to calculate the relative velocity of the suspect and the patrol car.
(a) What object should we choose to be the Home Frame in this problem, according to the conventions established in this chapter? What object defines the Other Frame? What object is the object whose velocity we want to determine in both frames? Please explain.
(b) Use the galilean velocity transformation equations to calculate the relative velocity of the suspect's car and the patrol car. Please explain your work.

R1S.9 Some people are playing a game of shuffleboard on an ocean cruiser moving down the Hudson River at a constant speed of 17 m/s in the $+x$ direction relative to the shore. During one shot, a puck (which has a mass $m = 750$ g and is traveling at 10 m/s in the $-x'$ direction in the boat frame) hits a puck having the same mass at rest. After the collision, the first puck comes to rest, and the other puck travels at 10 m/s in the $-x'$ direction in the boat frame. (Assume that the ground frame's x axis points in the same direction as the boat frame's x' axis.)
(a) Show that the total x-momentum of the two-puck system is conserved in the boat frame. Explain your calculation carefully.
(b) Imagine that someone sitting on a bridge under which the boat is passing takes a video of this important game. What x-velocity will each puck be measured to have relative to the shore? Explain carefully.
(c) Show that in spite of the fact that the puck's x-velocities have signs and magnitudes that are *different* from those measured on the boat, the total momentum of the two-puck system is still conserved in the shore frame. Explain your work.

R1S.10 Imagine two inertial reference frames in standard orientation, where the Other Frame moves with a speed β in the $+x$ direction relative to the Home Frame. Imagine that an observer in the Home Frame observes the following collision: an object with mass m_1 and velocity $\vec{v}_1$ hits an object with mass m_2 traveling with velocity $\vec{v}_2$. After the collision, the objects move off with velocities $\vec{v}_3$ and $\vec{v}_4$, respectively. Do *not* assume that all or even *any* of these velocities are in the x direction. Assume, though, that total momentum is measured to be conserved in the Home Frame, that is, that

$$m_1\vec{v}_1 + m_2\vec{v}_2 = m_1\vec{v}_3 + m_2\vec{v}_4 \qquad \text{(assume this!)}$$
(R1.6)

Using this equation and the galilean transformation equations, show that if the newtonian view of time is correct, then the total momentum of the two objects will also be conserved in the Other Frame

$$m_1\vec{v}_1' + m_2\vec{v}_2' = m_1\vec{v}_3' + m_2\vec{v}_4' \qquad \text{(prove this!)} \quad \text{(R1.7)}$$

even though the velocities measured in the two frames are very different. Please show your work in detail.

R1S.11 The engines on a 4000-kg jet plane accelerating for takeoff exert a constant thrust force of magnitude $F_{th} = 20{,}000$ N on the jet as it accelerates from rest a distance $D = 1000$ m down the runway before taking off. This takeoff is observed by someone riding in a train with a constant speed of 30 m/s alongside the runway. Assume that the train and the jet move in the same direction relative to the ground (which we will take to be the $+x$ direction), and assume that both the passenger and the plane are at $x = 0$ in the ground frame when the plane starts its run at $t = 0$.
  (a) In chapter C8, we saw that an object's change in kinetic energy during a given interval of time should be equal to the total k-work the object receives during that time. Use this to show that the jet's speed relative to the ground at takeoff is 100 m/s. Explain your reasoning.
  (b) Use the idea that the jet's acceleration is constant to show that it takes it 20 s to reach this takeoff speed. Explain your work.
  (c) What are the plane's initial and final x-velocities in the train frame? Explain.
  (d) Assume that the passenger's position defines the origin $x' = 0$ in the train's frame. What is the jet's initial x-position in the train frame? What is its x-position at takeoff? Explain.
  (e) Show that the jet's change in kinetic energy in the train frame is equal to the k-work it receives

in the train frame. (This shows that this law of physics is the same in both frames, even though the values of the kinetic energies and k-work are not the same in both frames.)

Rich-Context

R1R.1 Design a first-law detector that does *not* use a floating ball as the basic active element. Your detector should primarily test Newton's first law and not some other law of physics (although it is fine if other laws of physics are involved in addition to the first law). Preferably, your detector should be reasonably practical and (if at all possible) usable in a gravitational field. (*Note:* There are *many* possible solutions to this problem. Be creative!)

R1R.2 Consider the economy reference frame described in problem R1S.2. Prove that an object that actually moves at a constant velocity close to that of light will be observed in the economy frame to move faster as it approaches the origin and slower as it departs. Describe what happens if the object moves *faster* than light.

Advanced

R1A.1 A person in an elevator drops a ball of mass m from rest from a height h above the elevator floor. The elevator is moving at a constant speed β downward with respect to its enclosing building.
  (a) How far will the ball fall in the building frame before it hits the floor? (*Hint:* $> h$!)
  (b) What is the ball's initial vertical velocity in the building frame? (*Hint:* not 0!)
  (c) Use the law of conservation of energy in the building frame to compute the ball's final speed (as measured in that frame) just before it hits the elevator floor.
  (d) Use the galilean velocity transformation equations and the result of part (c) to find the ball's final speed in the elevator frame.
  (e) Use the result of part (d) to show that the law of conservation of energy apparently holds in the elevator frame (assuming that it holds in the building frame).

ANSWERS TO EXERCISES

R1X.1 Imagine that the Home Frame is an inertial frame. Consider a set of isolated objects arrayed around the Home Frame that happen to be initially at rest in that frame. Since their velocity with respect to the inertial Home Frame has to be constant, these objects will *remain* at rest relative to the Home Frame. Now, if the Home Frame moves at a constant velocity relative to the Other Frame (and the latter is

rigid and nonrotating so that all parts of the Other Frame move with a constant velocity relative to the Home Frame, and vice versa), then our set of isolated objects must also move with a constant velocity relative to the Other Frame. Since these isolated objects are observed to move with a constant velocity everywhere in the Other Frame, it must be an inertial frame as well.

R2

Synchronizing Clocks

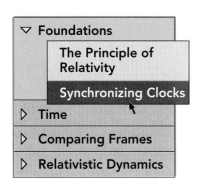
Chapter Overview

Introduction

In this chapter, we will see that (contrary to our assumption in chapter R1) time *cannot* be universal and absolute if the laws of electromagnetism are consistent with the principle of relativity, and we will take the first steps toward developing a conception of time that solves this problem. We will also develop some tools we will find useful later.

Section R2.1: The Problem of Electromagnetic Waves

We saw in chapter R1 that the newtonian assumption that time is absolute implies the galilean velocity transformation equation $\vec{v}' = \vec{v} - \vec{\beta}$. On the other hand, **Maxwell's equations** of electromagnetism predict that light (which consists of electromagnetic waves) must move with a certain speed c whose value is specified by the equations. In the 19th century, people assumed that electromagnetic waves moved through a hypothetical medium called the **ether** and that Maxwell's equations only really applied in the ether's rest frame. In *other* frames, then, the speed of light would be more or less than c according to the galilean velocity transformation equation. However, 19th-century experiments seeking to verify this assumption showed that the speed of light seemed to have the same numerical value in *all* reference frames and failed to find *any* evidence supporting the ether hypothesis.

    Einstein asserted that the simplest way to explain the evidence was to reject the ether hypothesis and instead assume that Maxwell's equations satisfy the principle of relativity. This means that the speed c of light in a vacuum must be the same in all inertial reference frames.

Section R2.2: Relativistic Clock Synchronization

This implies, as Einstein noted, that *any* method for **synchronizing** an inertial frame's clocks, if the method is to be consistent with the principle of relativity, should be equivalent to using light flashes to synchronize the clocks *assuming* that the speed of those flashes is c.

Section R2.3: SR Units

The speed of light in this approach is a frame-independent universal constant that connects time and space units. (Indeed, the meter is currently *defined* to be the distance that light moves in exactly 1/299,792,458 s.)

    In *Système Relativistique* **(SR) units,** we avoid the conversion factor entirely by measuring distances in **seconds,** where 1 s of distance is the distance that light moves in 1 s of time. The speed of light in this unit system is 1 s/1 s = 1; all other speeds are similarly unitless; and energy, momentum, and mass all have units of kilograms. To convert a quantity from SI to SR units, you multiply it by whatever power of the unit operator $1 = 3 \times 10^8$ m/1 s makes the units of meters go away; to convert from SR to SI units, you multiply by whatever power gives you the appropriate power of meters.

Section R2.4:   Spacetime Diagrams

A **spacetime diagram** is an important visual aid for displaying the relationship between events. Such a diagram is simply a graph with a vertical time axis and horizontal spatial axes. *Events* in spacetime correspond to *points* on such a diagram. The connected sequence of events that describe a particle's motion in spacetime is represented by a curve on the diagram called a **worldline,** which displays how the particle's position varies with time. If the axes on the diagram have the same scale and the object moves only along the x axis, then the slope $\Delta t/\Delta x$ of its diagram worldline at any instant of time is $1/v_x$, where v_x is the particle's x-velocity at that instant. Note that the slope of any light-flash worldline will be ± 1.

Section R2.5:   Spacetime Diagrams as Movies

If you view a spacetime diagram through a horizontal slit moving upward at a constant rate, you see a one-dimensional movie that displays the events and moving particles depicted on the diagram (see figure R2.7).

Section R2.6:   The Radar Method

Section R2.2 describes how we can correctly synchronize clocks in a reference frame. However, there is an equivalent way to determine an event's spacetime coordinate that requires only *one* clock at the frame's origin. Imagine that a light flash sent from the master clock at time t_A is reflected by an event E and returns to the master clock at time t_B. Since light moves at speed 1 in all inertial frames, event E's coordinates must be

$$t_E = \tfrac{1}{2}(t_B + t_A) \qquad x_E = \tfrac{1}{2}(t_B - t_A) \tag{R2.6}$$

Purpose: This equation expresses an event E's spacetime coordinates t_E and x_E in terms of the time t_A that a light flash leaves a given frame's origin and the time t_B when it returns after being reflected by the event. (Both times are measured by a master clock at the frame's origin.)

Limitations: These equations assume that the event is located on the x axis, that the master clock is in an inertial reference frame, and that we are using SR units.

This approach to determining coordinates is called the **radar method,** because radar tracking of airplane positions is essentially done in this way.

R2.1   The Problem of Electromagnetic Waves

Review of newtonian
approach to relativity

To Newton and his followers, time was self-evidently universal and absolute, flowing "equably without relation to anything external," as he put it. This makes time a *frame-independent* quantity: every observer in every reference frame (inertial or not!) should be able to agree on what time it is. This assumption about time was considered self-evident because it is consistent with our immediate experience about the way that time works.

We saw in chapter R1 that this assumption implies that if an object is measured to have a velocity in one inertial reference frame (the Home Frame), then its velocity measured in another inertial frame (the Other Frame) is given by the vector equation

$$\vec{v}' = \vec{v} - \vec{\beta} \qquad (R2.1)$$

where $\vec{\beta}$ is the velocity of the Other Frame with respect to the Home Frame (this is equation R1.3). This equation is a direct consequence of the definition of an inertial reference frame and the assumption that time is frame-independent.

Maxwell's equations and light
waves

In 1873, James Clerk Maxwell published a set of equations (now called **Maxwell's equations**) that summarized the laws of electromagnetism in compact and elegant form. These equations (which we will study in depth in the latter part of unit E) were the culmination of decades of intensive and ingenious research by a number of physicists into the connection between the phenomena of electricity and magnetism, and the equations represent one of the greatest achievements of 19th-century physics.

Among the many fascinating consequences of these equations was the prediction that traveling waves could be set up in an electromagnetic field, much as ripples could be produced on the surface of a lake. The speed at which such waves would travel is *completely determined* by various universal constants appearing in the equations, constants whose values were fixed by experiments involving electrical and magnetic phenomena and were fairly well known in 1873. The predicted speed of such electromagnetic waves turns out to be about 3.00×10^8 m/s. Light was already known to have wavelike properties (as the result of research work done in the early 19th century by Thomas Young and Augustin-Jean Fresnel) and to travel at roughly this speed (as measured by Ole Rømer in 1675 and Jean-Bernard Leon Foucault in 1846). On the basis of this information, Maxwell concluded that light consisted of such electromagnetic waves. This assertion was supported by later experiments confirming that the value of the speed of light was indeed indistinguishable from the value predicted on the basis of the constants in Maxwell's equations, and particularly by the work of Heinrich Hertz, who was able to directly generate low-frequency electromagnetic waves (i.e., radio waves) and demonstrate that they had the properties predicted by Maxwell's equations.

The ether hypothesis

In short, Maxwell's equations of electromagnetism predicted that light waves must travel at a specific speed $c = 3.00 \times 10^8$ m/s. The question is, relative to what? The consensus in the physics community at the time (one that Maxwell shared) was that electromagnetic waves were oscillations of a hypothetical medium called the **ether,** just as sound waves are oscillations in air and water waves are oscillations in the surface of a body of water. Physicists therefore generally assumed that light waves would travel at the predicted speed c relative to this ether, and thus have this speed in a frame in which the ether is at rest.

In all other inertial reference frames, however, light waves must be observed to travel at a speed different from c. To see this, imagine a spaceship

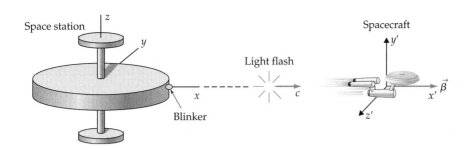

Figure R2.1
A light flash chasing a departing spaceship.

flying away from a space station at a speed β. A blinker on the space station emits a pulse of light waves toward the departing spacecraft (see figure R2.1). Let's imagine that the space station is at rest with respect to the ether and treat this as the Home Frame. Observers on the space station will then measure the emitted pulse of light to move away from the blinker at a speed of c. How rapidly would the pulse of light waves from the blinker be observed to travel in a frame fixed to the spaceship?

By construction, both the flash and the spacecraft move in the $+x$ direction. The x component of the galilean velocity transformation equation (equation R2.1) in this case implies that

$$v'_{\text{light},x} = v_{\text{light},x} - \beta = c - \beta \qquad (R2.2)$$

This means that the speed of the light flash, as measured in the frame of the spacecraft (i.e., the Other Frame), is $c - \beta$. This makes sense: if the spacecraft happened to travel at the speed of light (so that $\beta = c$), the flash should *intuitively* appear to be motionless in the frame of the spacecraft and thus never catch up with it, in agreement with equation R2.2. In any frame moving with respect to the ether, then, light waves would be measured to have a speed $\neq c$.

But this means that Maxwell's equations strictly apply only in a *certain* inertial reference frame (the frame at rest with respect to the ether), since they do specify that light waves move with a *specific* speed c. Presumably some small modifications would have to be made to these equations to make them work in frames that are *not* at rest with respect to the ether.

Now, the ether concept proved to be problematic from the beginning. This ether had to fill all space and permeate all objects, and yet be virtually undetectable. It had to have virtually zero density and viscosity, because it was not observed to significantly impede any object's motion. On the other hand, it had to be extraordinarily "stiff" with regard to oscillations because generally the speed of waves in a medium increases with the stiffness of the medium, and c is very large (for comparison, mechanical waves traveling through *solid rock* have speeds of only 6000 m/s).

In 1887, U.S. physicists Albert Michelson and Edward Morley performed a sensitive experiment designed to prove the existence of this seemingly indiscernible ether. If this ether filled all space, the earth must (as a result of its orbital motion around the sun) be moving through the ether at a speed comparable to its orbital speed of about 30 km/s. This "ether wind" would have the effect of making the speed of light depend on the direction of its travel: a light wave moving against the ether wind would move more slowly than a wave moving across the wind. Michelson and Morley therefore constructed a very sensitive experiment that compared the speed of two beams of light sent in perpendicular directions in a very clever way. The presence of an ether wind would manifest itself in a slight difference between the speeds of light in these two directions.

To the surprise of everyone involved, it turned out that there was no discernible difference in the speeds of the two light waves. The experiment was

Physicists fail to detect the ether

repeated with different orientations of the apparatus, at different times of the year (just in case the earth happened to be at rest with respect to the ether at the time of the first experiment), and by other physicists. In all cases, the result was that the speed of light was apparently a fixed value independent of the earth's motion.

In a modern version of the Michelson-Morley experiment performed in 1978, A. Brillet and J. L. Hall (see *Physical Review Letters*, vol. 42, p. 549, 1979) set up a laser that fed light into a Faby-Perot cavity, which is essentially a region of space bounded by two mirrors. The laser's frequency was continually adjusted so that a specific integer number of waves fit between the mirrors. Any variation in the speed of light would cause the number of waves of a given frequency that fit between the mirrors to increase or decrease, requiring the electronics to adjust the laser's frequency to keep the number of waves fixed. The laser and cavity were mounted on a granite turntable so that its orientation relative to any ether wind could be varied, and the frequency of the laser light was monitored by sending some light out along the axis of rotation to be compared with a stabilized stationary reference laser. (See figure R2.2 for a schematic diagram of the experiment.) While the table was rotated, Brillet and Hall observed that the fractional change in the laser frequency was $(1.5 \pm 2.5) \times 10^{-15}$. If an ether wind comparable in speed to the earth's orbital speed v existed, it should cause a variation in the fractional frequency on the order of magnitude of v^2/c^2, or about 10^{-8}, more than 10 million times any possible variation consistent with experiment!

The Michelson-Morley result (and other corroborating results) caused a ruckus in the physics community, as physicists strove to explain away these results while saving the basic ether concept. Many explanations were offered (some involving bizarre effects of the ether wind on measurement devices), but none provided a satisfactory explanation of all known experimental data.

Einstein's proposed solution to the problem

In 1905, Albert Einstein published a short paper on the subject in the European journal *Annalen der Physik* that changed the direction of physics. In that paper, Einstein proposed that since it seemed to be impossible to demonstrate the existence of the ether, the concept of the ether should be rejected and that we should simply accept the idea that light is able to move in a vacuum. But since the vacuum of empty space provides no anchor defining a special frame where the speed of light is c (what would such a frame be attached to?),

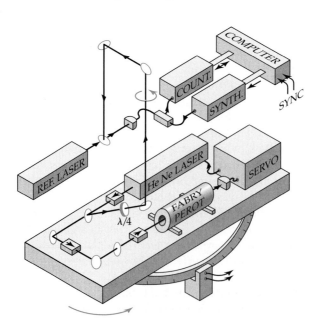

Figure R2.2

A schematic diagram of the Brillet-Hall experiment. Adapted from A. Brillet and J.L. Hall, *Phys. Rev. Left.*, Vol. 42, p. 550.

if we accept that the speed of light in a vacuum is c, then we must accept that this speed is c in *every* inertial reference frame, in direct contradiction to equation R2.2! The assumption that *the speed of light is a frame-independent quantity* is necessary, Einstein argued, to make Maxwell's equations consistent with the principle of relativity, since Maxwell's equations are laws of physics that predict a specific and well-defined value for the speed of light.

The frame independence of the speed of light goes deeply against our intuition. How can a pulse of light waves be measured to move at a speed c in the frame of the space station of our previous example and *also* in the frame of the spacecraft, when the two frames are not at rest with respect to each other? Einstein's bold idea, while neatly sidestepping the experimental difficulties associated with the ether idea, seemed impossible to most of his contemporaries.

However, there are really only three possibilities: the principle of relativity is wrong, Maxwell's equations are wrong, or the galilean velocity transformation equations are wrong. By accepting the ether concept, physicists before Einstein had opted to accept the idea that Maxwell's equations would have to be modified in frames moving with respect to the ether, thus keeping the galilean velocity transformation and implicitly rejecting the full principle of relativity. Even as evidence against the ether hypothesis became firm and incontrovertible, rejection of the galilean velocity transformation, so solidly based on simple and obvious ideas, seemed absurd.

On the other hand, what Einstein suggested did have a certain simplicity and elegance. Throw away the ether, he said. It is an unhelpful, ad hoc hypothesis with no experimental backing. Throw away the awkward and bizarre theories that arose to explain our inability to detect the ether. Embrace instead the beautiful simplicity of the principle of relativity and Maxwell's equations. The speed of light, then, is the same in all inertial reference frames as a matter of course, and the null results of experiments like the Michelson-Morely experiment are trivially explained.

The cost? *The galilean transformation equations must be wrong.* But what could possibly be wrong with their derivation? It is the idea of universal and absolute time that is wrong, argued Einstein. Let's look at what the principle of relativity requires of time. . . .

R2.2   Relativistic Clock Synchronization

Our basic problem is that the concept of universal and absolute time seems to be at odds with having the principle of relativity apply to Maxwell's equations, yet experimental results support the idea that the principle of relativity *does* apply to Maxwell's equations. If time is not "universal and absolute," how can we even define what time means?

The solution, as Einstein was the first to see, is that we must define what we mean by "time" *operationally* within each inertial frame by specifying a concrete and specific procedure for synchronizing that frame's clocks that is consistent with both the principle of relativity and the laws of electromagnetism.

But how can we synchronize clocks in such a manner? Here is one simple method. Maxwell's equations imply that light moves through a vacuum at a certain fixed speed c. The principle of relativity requires that this speed be the same in every inertial reference frame, as we discussed in chapter R1. Therefore, *any* synchronization method consistent with the principle of relativity will lead to light being found to have speed c in any inertial reference frame.

Since the speed of light has to be c in every inertial frame anyway, let us in fact synchronize the clocks in our inertial reference frame by *assuming* that

Einstein's method of clock synchronization

light always has the same speed of c! How do we do this? Imagine that we have a master clock at the spatial origin of our reference frame. At exactly $t = 0$, we send a light flash from that clock that ripples out to the other clocks in the frame. Since light is assumed to travel at a speed of $c = 299{,}792{,}458$ m/s, this flash will reach a lattice clock exactly 1.0 m from the master clock at exactly $t = (1.0 \text{ m})/(299{,}792{,}458 \text{ m/s}) = 3.33564095 \times 10^{-9}$ s $= 3.33564095$ ns. Therefore, if we set this clock to read 3.33564095 ns exactly as the light flash passes, we know that it is synchronized with the master clock. The process is similar for all the other clocks in the lattice.

So here is a first draft of a description of Einstein's approach to synchronize the clocks in an inertial reference frame:

> A light flash is emitted by clock A in an inertial frame at time t_A (as read on clock A) and received by clock B in the same frame at time t_B (as read on clock B). These clocks are defined to be **synchronized** if $c(t_B - t_A)$ is equal to the distance between the clocks. That is, the clocks are synchronized if they measure the speed of a light signal traveling between them to be c.

Exercise R2X.1

Imagine that we have a clock on the earth and a clock on the moon. How can we tell if these clocks are synchronized according to this definition? Imagine that we send a flash of light from the earth's clock toward the moon at exactly noon, as registered by the earth's clock. What time will the clock on the moon read when it receives the flash if the two clocks are synchronized? (The distance between the earth and the moon is 384,000,000 m.)

Note that we are assuming that Maxwell's equations obey the principle of relativity

"Now, wait a minute!" I hear you cry. "Isn't all this a bit circular? You claim that Maxwell's equations predict that light will be measured to have the same speed c in every inertial reference frame. But then you go, setting up the clocks so that this result is *ensured*. Is this fair?"

This *is* fair. We are not trying here to *prove* that Maxwell's equations obey the principle of relativity—we are *assuming* that they do, so that we can determine the consequences of this assumption. To make this clear in his original paper, Einstein actually stated the frame independence of the speed of light as a separate postulate, *emphasizing* that it is an assumption. However, this is not really a *separate* assumption: it is a consequence of the assumption that the principle of relativity applies to all laws of physics, *including* Maxwell's equations. The point is that if the principle of relativity is true, the speed of light will be measured to have speed c no matter *what* valid synchronization method we use, so why not use a method based on that fact?

Moreover, there *are* other valid ways of synchronizing the clocks in a given inertial reference frame, ways that make no assumptions whatsoever about the frame independence of the speed of light.[†] If such a method were

---

[†]One of the simplest is described by Alan Macdonald (*Am. J. Phys.*, vol. 51, no. 9, 1983). Macdonald's method is as follows. Assume that clocks A and B emit flashes of light toward each other at $t = 0$ (as read on each clock's *own* face). If the readings of the two clocks also agree when they receive the light signal from the other clock, they are synchronized. This method only assumes that the light flashes take the same time to travel between the clocks in each direction (i.e., there is no preferred direction for light travel): it does *not* assume light has any frame-independent speed. Achin Sen (*Am. J. Phys.*, vol. 62, no. 2, 1994) presents a particularly nice example of an approach that side-steps the synchronization issue, showing mathematically that the results of relativity follow directly from the principle of relativity. Sen's article also contains an excellent list of references.

used to synchronize clocks in an inertial frame, such a frame could be used to verify *independently* that the speed of light is indeed frame-independent. Theorists have shown that these alternative methods yield the same consequences as one gets assuming the frame-independence of the speed of light. These methods are, however, also more complex and abstract: the definition of synchronization in terms of light is much more vivid and easy to use in practice.

R2.3   SR Units

In ordinary SI units, the speed of light c is equal to 299,792,458 m/s, a somewhat ungainly quantity. The definition of clock synchronization given in section R2.2 means that we will often need to calculate how long it would take light to cover a certain distance or how far light would travel in a certain time. You can perhaps see how messy such calculations will be.

For this reason (and many others) it will be convenient when we study relativity theory to measure distance not in the conventional unit of meters but in a new unit, called a *light-second* or just *second* for short. A **light-second** or **second** is defined to be the distance that light travels in 1 s of time. Since 1983, the meter has in fact been officially *defined* by international agreement as the distance that light travels in 1/299,792,458 s. Therefore there is exactly 299,792,458 m in 1 light-second *by definition*.

Definition of the "second" of distance

We can, of course, measure distance in any units that we please: there is nothing magically significant about the meter. Choosing to measure distance in light-seconds has some important advantages. First, light travels exactly 1 s of distance in 1 s of time by definition. This allows us to talk about clock synchronization much more easily. For example, if clock A and clock B are 7.3 light-seconds apart in an inertial reference frame and a light flash leaves clock A when it reads $t_A = 4.3$ s, the flash should arrive at clock B at a time $t_B = (4.3 + 7.3)$ s $= 11.6$ s if the two clocks are correctly synchronized, since light travels exactly 1 light-second in 1 s of time by definition. You can see that we need to perform no ungainly unit conversions if we measure distance in this way.

Advantages of using this unit of distance

Indeed, agreeing to measure distance in seconds allows us to state the definition of clock synchronization in an inertial frame in a particularly nice and concise manner (this will be our *final* draft of this description):

> *Two clocks in an inertial reference frame are defined to be synchronized if the time interval (in seconds) registered by the clocks for a light flash to travel between them is equal to their separation (in light-seconds).*

In spite of the tangible advantages that measuring distance in seconds yields when it comes to talking about synchronization, this is not the most important reason for choosing to do so. In the course of working with relativity theory, we will uncover a deep relationship between time intervals and distance intervals, akin to the relationship between distances measured north and distances measured east on a plane. One would not think of measuring northward distances in feet, say, and eastward distances in meters: that would obscure the fundamental similarity and relationship of these measurements. Similarly, measuring time intervals in seconds while measuring distance intervals in meters obscures the fundamental similarity in these measurements that will be illuminated by relativity theory. Choosing to measure time in the same units as distance will make this beautiful symmetry of nature more apparent.

The standard unit system used by scientists studying ordinary phenomena comprises the *Système International,* or SI, units. In this system, the units for mechanical quantities (such as velocity, momentum, force, energy, angular momentum, and pressure) are based on three fundamental units: the *meter*, the *second*, and the *kilogram*. In this text, however, we will use a slightly modified version of SI units (let's call it *Système Relativistique,* or **SR, units**) where distance is measured in *seconds* (i.e., light-seconds) instead of in meters (with the other basic units being the same).

As discussed in chapter C1, we can use *unit operators* to convert a quantity from one kind of unit to another. We first write down an equation stating the basic relationship between the units in question: 1 mi = 1.609 km, for example. We then rewrite this in the form of a ratio equal to 1: either $1 = 1\ \text{mi}/1.609\ \text{km}$ or $1 = 1.609\ \text{km}/1\ \text{mi}$. Since multiplying by 1 doesn't change a quantity, you can multiply the original quantity by whichever ratio leads you to the correct final units upon cancellation of any units that appear in both the numerator and denominator. For example, to determine how many miles there are in 25 km, you simply multiply 25 km by the first of the two ratios described above, as follows:

$$25\ \text{km} = 25\ \text{km} \cdot 1 = 25\ \text{km} \left(\frac{1\ \text{mi}}{1.609\ \text{km}} \right) = \frac{25}{1.609}\ \text{mi} \approx 16\ \text{mi} \quad (R2.3)$$

The factors used to convert from SI distance units to SR distance units are based on the fundamental definition of the light-second: $299{,}792{,}458\ \text{m} \equiv 1\ \text{s}$. Thus the basic conversion factors that we need are $1 = (1\ \text{s}/299{,}792{,}458\ \text{m})$, or $1 = (299{,}792{,}458\ \text{m}/1\ \text{s})$. For example, a distance of 25 km can be converted to a distance in light-seconds as follows:

$$25\ \text{km} = 25\ \text{km} \cdot 1 \cdot 1 = 25\ \text{km} \left(\frac{1000\ \text{m}}{1\ \text{km}} \right) \left(\frac{1\ \text{s}}{2.998 \times 10^8\ \text{m}} \right)$$

$$= 8.3 \times 10^{-5}\ \text{s} = 83\ \mu\text{s} \quad (R2.4)$$

meaning that 25 km is equivalent to the distance that light travels in 83 millionths of a second.

The light-second is a rather large unit of distance (the moon is only about 1.3 light-seconds away from the earth!). A more appropriate unit on the human scale is the light-nanosecond, where 1 light-nanosecond $\equiv 10^{-9}$ light-second ($= 0.2998\ \text{m} \approx 1\ \text{ft}$). On the astronomical scale, the light-year (where 1 light-year $\approx 3.16 \times 10^7\ \text{s} \approx 0.95 \times 10^{16}\ \text{m}$) is an appropriate (and commonly used) distance unit. The dimensions of the solar system are conveniently measured in light-hours (it is about 10 light-hours in diameter). All these units represent extensions of the basic unit of the light-second.

In SR units, the light-second is considered to be *equivalent* to the second of time, and both units are simply referred to as *seconds*. This means that these units can be canceled if one appears in the numerator of an expression and the other in the denominator. For example, in SI units, velocity has units of meters per second; but in SR units it has units of seconds per second = unitless(!). Thus an object that travels 0.5 light-second in 1.0 s has a speed in SR units of 0.5 s/1.0 s = 0.5 (no units!). The bare number representing a speed in this case actually represents a comparison of the particle's speed to the speed of light, since light covers 1.0 s of distance in 1.0 s of time by definition. Thus a particle traveling at a speed of 0.5 (in SR units) is traveling at one-half the speed of light.

In a similar manner, one can find the natural units for any physical quantity in SR units. For example, the kinetic energy of a particle has the same units as mass times speed squared. In SI, these units, abbreviated, would be $kg \cdot m^2/s^2$. Thus the natural unit of energy in SI units is the joule, defined to be 1 $kg \cdot m^2/s^2$. In SR units, mass times speed squared has units of $kg \cdot s^2/s^2 = kg$. Thus the natural SR unit for energy is the kilogram (the same as the unit of mass!). How much energy is represented by an SR kilogram of energy? We can determine this by using the standard conversion factor to convert the SR distance unit of seconds to the SI distance unit of meters:

$$1 \text{ kg (energy)} = 1 \text{ kg} \frac{s^2}{s^2} \left(\frac{2.998 \times 10^8 \text{ m}}{1 \text{ s}} \right)^2 \left(\frac{1 \text{ J}}{1 \text{ kg} \cdot m^2/s^2} \right) = 8.988 \times 10^{16} \text{ J}$$

$$(R2.5)$$

This unit is roughly equal to the yearly energy output of 10 full-size electric power plants!

Exercise R2X.2

What is the natural unit for momentum in SR?

In general, what we need to do to convert from SR units to SI units is to multiply the SR quantity by whatever power of the conversion factor $1 = (299,792,458 \text{ m}/1 \text{ s})$ yields the correct power of meters in the units of the final result. Similarly, to convert from SI units to SR units, we simply multiply the SI quantity by whatever power of this factor causes the units of meters to disappear. (See appendix RA for a complete discussion of how to convert units and equations from one system to the other.)

R2.4   Spacetime Diagrams

The clock synchronization method described at the beginning of this chapter completes the description of how to build and operate an inertial reference frame. We now know how to assign spacetime coordinates to any event occurring within that inertial frame in a manner consistent with the principle of relativity.

Problems in relativity theory often involve studying how physical events relate to one another. We can conveniently depict the coordinates of events by using a special kind of graph called a **spacetime diagram**. Throughout the unit, we will find spacetime diagrams to be indispensable in helping us to express and visualize the relationships between events.

How to draw events on a spacetime diagram

Consider an event A whose spacetime coordinates are measured in some inertial frame to be t_A, x_A, y_A, z_A. To simplify our discussion somewhat, assume that $y_A = z_A = 0$ (that is, the event occurs somewhere along the x axis of the frame). Now imagine drawing a pair of coordinate axes on a sheet of paper. Label the vertical axis with a t and the horizontal axis with an x: it is conventional to take the t axis to be vertical in spacetime diagrams. Choose an appropriate scale for each of these axes. Now you can represent when and where event A occurs by plotting the event as a point on the diagram (in the usual manner that you would use in plotting a point on a graph), as shown in figure R2.3a. Any event that occurs along the x axis in space can be plotted as a point on such a diagram in a similar manner.

Figure R2.3

(a) How to plot an event *A* on a spacetime diagram. In this particular case, event *A* has a time coordinate $t_A \approx 5.4$ s and an *x* coordinate $x_A \approx 4.5$ s. (b) Plotting an event *A* that has nonzero *x* and *y* coordinates (but zero *z* coordinate).

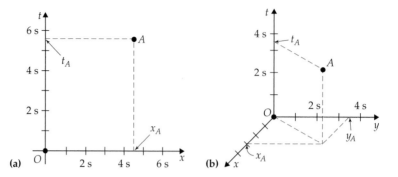

Note that the point marked *O* in figure R2.3a also represents an event. This event occurs at time $t = 0$ and at position $x = 0$. We call this event the **origin event** of the diagram.

If we need to draw a spacetime diagram of an event *A* that occurs in space somewhere in the *xy* plane (i.e., which has $z_A = 0$ but nonzero t_A, x_A, y_A), we must add another axis to the spacetime diagram (as shown in figure R2.3b). The resulting diagram is somewhat less satisfactory and more difficult to draw because we are trying to represent a three-dimensional graph on a two-dimensional sheet of paper. A spacetime diagram showing events with *three* nonzero spatial components is impossible to draw, as that would involve trying to represent a four-dimensional graph on a two-dimensional sheet of paper. A four-dimensional graph is hard to visualize at all, much less draw! Fortunately, diagrams with one or two spatial dimensions will be sufficient for most purposes.

Worldlines

In chapter R1 we visualized describing the motion of an object by imagining the object to carry a strobe light that blinks at regular intervals. If we can specify when and where each of these blink events occurs (by specifying its spacetime coordinates), we can get a pretty good idea of how the object is moving. If the time interval between these blink events is reduced, we get an even clearer picture of the object's motion. In the limit that the interval between blink events goes to zero, the object's motion can be described in unlimited detail by a list of such events. Thus the motion of any object can be described in terms of a connected *sequence* of events. We call the set of all events occurring along the path of a particle the particle's **worldline.**

On a spacetime diagram, an event is represented by a point. Therefore a worldline is represented on a spacetime diagram by an infinite set of infinitesimally separated points, which is a *curve.* This curve is nothing more than a graph of the position of the particle versus time (except that the time axis is conventionally taken to be vertical on a spacetime diagram). Figure R2.4 illustrates worldlines for several examples of objects moving in the *x* direction.

A worldline's slope is the *inverse* of its particle's *x*-velocity

Note that because the time axis is taken to be vertical, the slope of the curve on a spacetime diagram representing the worldline of an object traveling at a constant velocity in the *x* direction is not its *x*-velocity v_x (as one might expect) but rise/run $= \Delta t / \Delta x = 1/v_x$! Thus the slope of the curve representing the worldline of an object at rest is infinity and *decreases* as *vx* increases. The worldline of an object traveling at a constant *x*-velocity has a constant slope.

Occasionally we need to draw a spacetime diagram of the worldline of an object moving in two spatial dimensions. The spacetime diagram in such a case is necessarily three-dimensional, which is hard to draw on two-dimensional paper. Figure R2.5 shows an example of such a spacetime diagram.

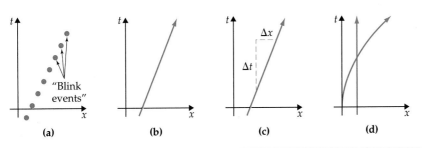

(a) | **(b)** | **(c)** | **(d)**

| A sequence of events along an object's path marks out its motion. | As the time between the events goes to zero, the set of events becomes a curve. | The slope of this *worldline* is equal to the object's inverse x-velocity ($1/v_x$). | The vertical line is the worldline of a particle at rest. The other is the worldline of a particle accelerating from rest. |

Figure R2.4

A sequence of spacetime diagrams illustrating various important things to know about worldlines.

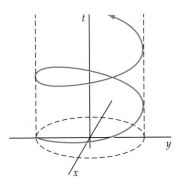

Figure R2.5

The worldline of a particle traveling in a circular path in the xy plane.

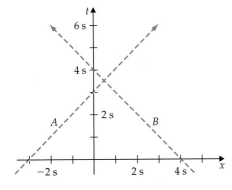

Figure R2.6

Light-flash worldlines on a spacetime diagram. Worldline *A* represents a light flash moving in the $+x$ direction. Worldline *B* represents a flash traveling in the opposite direction.

In drawing spacetime diagrams, it is also convenient and conventional to use the same-size scale on both axes. If this is done, the worldline of a flash (i.e., a very short pulse) of light always has a slope of either 1 (if the flash is moving in the $+x$ direction) or -1 (if the flash is moving in the $-x$ direction), since light travels 1.0 s of distance in 1.0 s of time by definition in every inertial reference frame. It is also conventional to draw the worldline of a flash of light with a dashed line instead of a solid line (see figure R2.6).

Light-flash worldlines have slopes of ± 1

Exercise R2X.3

On figure R2.6, draw the worldline of a particle moving in the $-x$ direction through the origin event with a speed of 0.2 (1 light-second per 5 s).

Exercise R2X.4

Assume that the worldline of the center of the earth coincides with the vertical t axis in figure R2.6. If at $t = 0$ the moon lies in the $+x$ direction relative to the earth, draw its worldline as well (the moon is 384,000 km from earth). The moon actually orbits the earth once a month. Are you going to be able to see the curvature of the moon's worldline on this diagram? Explain.

R2.5   Spacetime Diagrams as Movies

It is easy to get confused about what a spacetime diagram really represents. For example, in the spacetime diagram shown in figure R2.6, it is easy to forget that the light flashes shown are moving in only one dimension (along the x axis), not in two. The velocity vectors of the light flashes shown above point opposite to each other, not perpendicular to each other.

There is a technique that you can use to make the meaning of any spacetime diagram clear and vivid: turn it into a movie! Here's how. Take a 3×5 index card and cut a slit about $\frac{1}{16}$ in. wide and about 4 in. long, using a knife or a razor blade. This slit represents the spatial x axis at a given instant of time. Now place the slit over the x axis of the spacetime diagram. What you see through the slit is what is happening along the spatial x axis at time $t = 0$. Now slowly move the slit up the diagram, keeping it horizontal: you will see through the slit what is happening along the spatial x axis at successively later times. You can watch the objects whose worldlines are shown on the diagram move to the left or right as the slit exposes different parts of their worldlines. Events drawn as dots on the diagram will show up as flashes as you move the slit past them. What you see through the slit as you move it up the diagram is essentially a movie of what happens along the x axis as time passes. Figure R2.7 illustrates the process.

If you employ this technique, you cannot fail to interpret a spacetime diagram correctly. After a bit of practice with the card, you will be able to convert diagrams to movies in your head.

Making a movie viewer for a spacetime diagram

R2.6   The Radar Method

If we are willing to confine our attention to events occurring only along the x axis (and thus to objects moving only along that axis), it is possible to determine the spacetime coordinates of an event with a single master clock and some light flashes: we don't need to construct a lattice at all! The method is analogous to locating an airplane by using radar.

Imagine that at the spatial origin of our reference frame (i.e., at $x = 0$), we have a master clock that periodically sends flashes of light in the $\pm x$ directions. Imagine that a certain flash emitted by the master clock at t_A happens to illuminate an event E of interest that occurs somewhere down the x axis. The reflected light from the event travels back along the x axis to the master clock, which registers the reception of the reflected flash at time t_B (see the spacetime diagram of figure R2.8).

The values of the emission and reception times t_A and t_B are sufficient to determine both the location and the time that event E occurred! Consider first how we can determine the location. The light flash's round-trip time

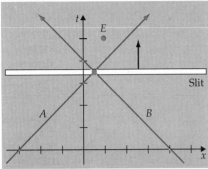

READ THIS FIGURE FROM THE
BOTTOM UP!

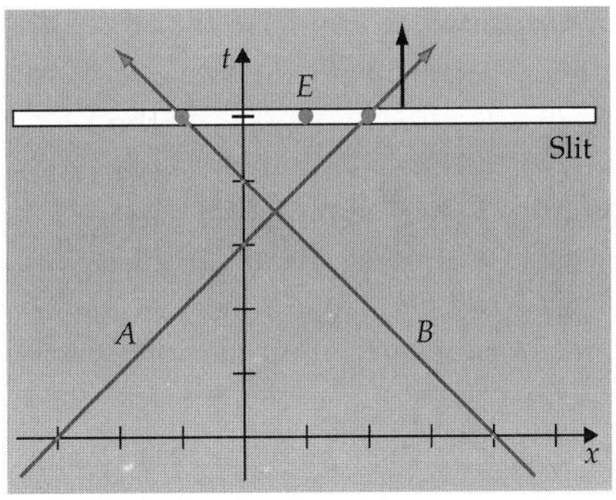

Figure R2.7d
Now the light flashes have passed
each other and are moving away
from each other. You can also see
through the slit the momentary flash
representing the firecracker
explosion (event E).

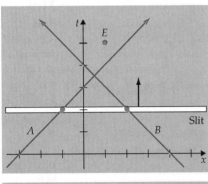

Figure R2.7c
At this instant the light flashes pass
through each other at a position of
about x = +1 m.

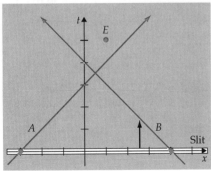

Figure R2.7b
As time passes (and you move the
slit up the diagram), the dots
representing the light flashes
approach each other.

Figure R2.7a
The spacetime diagram is basically
the same as figure R2.6 with a
firecracker explosion (event E)
thrown in to make things more
interesting. At time t = 0, the light
flashes are represented by black
dots that you can see through the
slit at x = −3 m and x = +4 m.

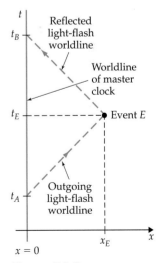

Figure R2.8
At time t_A, the master clock
at rest at x = 0 in the Home
Frame sends out a flash of
light, which reflects from
something at event E and
returns to the master clock
at time t_B.

is $t_B - t_A$. Since in this time the light covered the distance from x = 0 to event
E and back, and since the light flash travels 1 s of distance in 1 s of time by
definition, the distance to event E (in seconds) must be one-half of the
round-trip time (in seconds). The x coordinate of event E is therefore
$x_E = \pm\frac{1}{2}(t_B - t_A)$. We can determine the sign of x_E by noting whether the
reflected flash comes from the −x or +x direction. (In this case, the reflected
flash comes from the +x direction, so we select the plus sign.)

We can determine the event's time coordinate t_E as follows. Since the light flash traveled the same distance to the event as back from the event, and since the speed of light is a constant, the event must have occurred exactly halfway between times t_A and t_B. The midpoint in time between times t_A and t_B can be found by computing the average, so $t_E = \frac{1}{2}(t_B + t_A)$.

In summary, therefore, the spacetime coordinates of event E are

$$t_E = \tfrac{1}{2}(t_B + t_A) \qquad x_E = \tfrac{1}{2}(t_B - t_A) \tag{R2.6}$$

Purpose: This equation expresses an event E's spacetime coordinates t_E and x_E in terms of the time t_A at which a light flash leaves a given frame's origin and the time t_B when it returns after being reflected by the event. (Both times are measured by a master clock at the frame's origin.)

Limitations: This equation assumes that the event is located on the x axis, that the master clock is in an inertial reference frame, and that we are using SR units.

This method yields the same result as the clock lattice method

Equations R2.6 represent a method of determining the spacetime coordinates of an event that does *not* require the use of a complete lattice of synchronized clocks. But you should be able to convince yourself that *this method produces exactly the same coordinate values as you would get from a clock lattice.* For example, imagine that you actually had a lattice clock at x_E (the location of event E). The distance between that clock and the master clock at the origin must be equal to one-half the time that it would take a flash of light to go from one clock to the other and back, since light travels 1 s of distance in 1 s of time by definition. The lattice clock at x_E at the time of event E must read t_A + (the light travel time between the two clocks) $= t_A + \frac{1}{2}(t_B - t_A) = \frac{1}{2}(t_B + t_A)$, since two clocks are only defined to be synchronized if they measure a light flash to travel between them at the speed of light (1 s of distance in 1 s of time).

Using this method to determine spacetime coordinates is therefore equivalent to using a lattice of synchronized clocks. In some cases in this text, we will find it clearer or more convenient to use one method, and in some cases the other. The important thing to realize is that *either* the radar method or the clock lattice method provides specific, well-defined, and equivalent ways of assigning time coordinates to events, and both methods are based on the assumption that the speed of light is a frame-independent constant. These methods essentially define what time means in special relativity, and thus they will provide the foundation for most of the arguments in the remainder of the text. *Make sure that you thoroughly understand both these methods.*

Adapting the method for three spatial directions

We call this method the **radar method** because the method that radar installations actually use to track the trajectories of aircraft is a three-dimensional generalization of this approach (the impracticality of using a clock lattice to do the same is obvious!). We can precisely locate an object in three spatial dimensions if we record not only the time that the outgoing pulse was sent and the time that the reflected pulse was received but also the *direction* from which the reflected pulse was received. The analysis is a bit more complicated (see problems R2S.9 and R2S.10), but the basic method is the same.

TWO-MINUTE PROBLEMS

R2T.1 Imagine that in the distant future there is a space station on the planet Pluto. While you (on earth) are watching a video transmission from Pluto, which at the time is known to be 5.0 light-hours from earth, you notice that a clock on the wall behind the person speaking in the video reads 12:10 p.m. You note that your watch reads exactly the same time. Is the station clock synchronized with your watch?

    A.   Yes it is.
    B.   No it isn't.
    C.   The problem doesn't give enough information to tell.

R2T.2 Imagine that you receive a message from a starbase that is 13 light-years from earth. The message is dated July 15, 2127. What year does your calendar indicate at the time of reception if your calendar and the station's calendar are correctly synchronized?

    A.   2127
    B.   2114
    C.   2140
    D.   Other (specify)

R2T.3 The speed of a typical car on the freeway expressed in SR units is most nearly

    A. 10^{-7}
    B. 10^{-10}
    C. 10^{-8}
    D. 10^{-6}
    E. 10^{-4}
    F.   Other (specify)
    T.   None of these answers is right: we have to state units!

R2T.4 The spacetime diagram in figure R2.9 shows the worldlines of various objects. Which object has the largest speed at time $t = 1$ s?

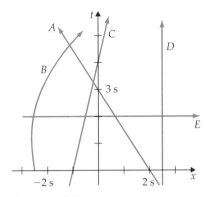

Figure R2.9
See problems R2T.4 through R2T.7.

R2T.5 The spacetime diagram in figure R2.9 shows the worldlines of various objects. Which object has the largest speed at time $t = 4$ s?

R2T.6 The spacetime diagram in figure R2.9 shows the worldlines of various objects. Which worldline cannot possibly be correct? (Explain why.)

R2T.7 In figure R2.9, the object whose worldline is labeled B is moving along the x axis, true (T) or false (F)?

R2T.8 A light flash leaves a master clock at $x = 0$, at time $t = -12$ s is reflected from an object a certain distance in the $-x$ direction from the origin, and then returns to the origin at $t = +8$ s. From this information, we can infer that the spacetime coordinates of the reflection event are $[t, x] =$

    A.   [4 s, 20 s]
    B.   [−4 s, −20 s]
    C.   [10 s, −2 s]
    D.   [2 s, −10 s]
    E.   [−2 s, −10 s]
    F.   Other (specify)

HOMEWORK PROBLEMS

Basic Skills

R2B.1 (Practice with SR units.)
    (a) What is the diameter of the earth in seconds?
    (b) A sign on the highway reads "Speed Limit 6×10^{-8}," meaning speed in SR units. Translate this to miles per hour.
    (c) Argue that the SR unit of acceleration is s^{-1}. What is 1 s^{-1} expressed as a multiple of g?

R2B.2 (Practice with SR units.)
    (a) A sign on a hiking trail reads "Viewpoint: 5.5 μs." About how long would it take you to walk to this viewpoint at a typical walking speed of 1 m/s?
    (b) Section R1.2 mentions that the *Voyager 2* spacecraft achieved speeds in excess of 72,000 km/h. What is this speed in SR units?

(more)

(c) In SI units, power is measured in watts (where $1\text{ W} = 1\text{ J/s} = 1\text{ kg·m}^2/\text{s}^3$). Argue that the natural SR units of power are kilograms per second. Let's define 1 SR-watt = 1 SRW ≡ 1 kg/s. A large electric power plant produces energy at a rate of about 10^9 W. What is this in SRW?

R2B.3   (a) Argue that in SR units, mass, momentum, and kinetic energy all are measured in kilograms.
        (b) Imagine a truck with a mass of 25 metric tons (t) (25,000 kg) barreling down a highway at a speed of 59 mi/h. What is the truck's momentum in kilograms? What is its kinetic energy in kilograms?

R2B.4   For each of the worldlines shown in figure R2.10, describe in words what the particles are doing, giving numerical values for velocities when possible. For example, you might say, "Particle A is traveling in the $+x$ direction with a constant speed of $\frac{1}{3}$."

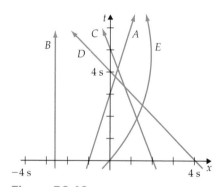

Figure R2.10
See problem R2B.4.

R2B.5   Imagine that you send out a light flash at $t = 3.0$ s as registered by your clock, and you receive a return reflection showing your kid brother making a silly face at $t = 11.0$ s as registered by your clock.
        (a) At what time did your brother actually make this silly face?
        (b) How far is your brother away from you (express in seconds and kilometers)?
        (c) Is this far enough away that he can't really be a nuisance?

R2B.6   Say that you send out a radar pulse at $t = -22$ h as registered by your clock and receive a return reflection from an alien spacecraft at $t = +12$ h, as registered by your clock. Is the spaceship inside or outside the solar system? When did the spaceship reflect the radar pulse?

R2B.7   Draw a spacetime diagram that shows the following worldlines.
        (a) Draw the worldline of a particle traveling at a constant speed of $\frac{3}{4}$ in the $+x$ direction that passes the point $x = 0$ at time $t = 2$ s.

(b) Draw the worldline of a particle that at time $t = 0$ is at position $x = +4$ s and is traveling at a speed of $\frac{1}{2}$ in the $-x$ direction, but which slows down as time passes, eventually coming to rest at $x = 0$ at time $t = 8$ s.
        (c) Draw the worldline of a light flash that passes the position $x = 0$ at time $t = -2$ s as it travels in the $+x$ direction.

R2B.8   Draw a spacetime diagram that shows the following worldlines.
        (a) Draw the worldline of a particle traveling at a constant speed of $\frac{3}{5}$ in the $-x$ direction that passes the point $x = 0$ at time $t = -2$ s.
        (b) Draw the worldline of a particle that at time $t = 0$ is at rest at position $x = 0$ and but accelerates in the $+x$ direction asymptotically toward the speed of light as time passes.
        (c) Draw the worldline of a light flash that passes the position $x = 0$ at time $t = +3$ s as it travels in the $-x$ direction.

Synthetic

R2S.1   A firecracker explodes 30 km away from an observer sitting next to a certain clock A. The light from the firecracker explosion reaches the observer at exactly $t = 0$, according to clock A. Imagine that the flash of the firecracker explosion illuminates the face of another clock B which is sitting next to the firecracker. What time will clock B register at the moment of illumination if it is correctly synchronized with clock A? Express your answer in milliseconds.

R2S.2   Imagine that you are in an inertial frame in empty space with a clock, a telescope, and a powerful strobe light. A friend is sitting in the same frame at a very large (unknown) distance from your clock. At precisely 12:00:00 noon according to your clock, you set off the strobe lamp. Precisely 30.0 s later, you see in your telescope the face of your friend's clock illuminated by your strobe flash. How far away is your friend from you (in seconds)? What should you see on the face of your friend's clock if that clock is synchronized with yours? Describe your reasoning in a few short sentences.

R2S.3   The spacetime diagram in figure R2.11 shows the worldline of a rocket as it leaves the earth, travels for a certain amount of time, comes to rest, and then explodes.
        (a) The rocket leaves the earth; the rocket comes to rest in deep space; the rocket explodes. What are the coordinates of each of these three events?
        (b) What is the constant speed of the rocket relative to the earth before it comes to rest?
        (c) A light signal from the earth reaches the rocket just as it explodes. Indicate on the diagram exactly where and when this light signal was emitted.

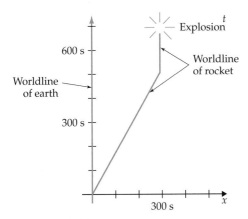

Figure R2.11
See problem R2S.3.

R2S.4 Imagine that a spaceship leaves the earth at event A and travels in the $+x$ direction, accelerating at a constant rate from rest at time $t = 0$ to a final speed of $\frac{4}{5}$ at time $t = 1$ hour. It remains at that speed thereafter. Just as the spaceship reaches that speed, it emits a laser signal back toward the earth (call this event B), which reaches the earth at event C. Draw a *quantitatively* accurate spacetime diagram of this situation (as observed in a frame attached to the earth, with the earth at $x = 0$) that shows the worldlines of the earth, the spaceship, the returning laser signal, and events A, B, and C.

R2S.5 An alien spaceship traveling at a constant velocity of $\frac{1}{3}$ in the $+x$ direction passes the earth (call this event A) at time $t = 0$. Just as the spaceship passes, people on the earth launch a probe, which accelerates from rest toward the spaceship at such a rate that it catches up to and passes the alien spaceship (call this event B) when both are 10 minutes of distance from the earth. As it passes the alien spaceship, the probe takes a photo and sends it back toward the earth as an encoded radio message that travels at the speed of light. The message reaches the earth at event C. Draw a *quantitatively* accurate spacetime diagram of this situation (as observed in a frame attached to the earth, with the earth at $x = 0$) that clearly shows the worldlines of the earth, the alien spaceship, the probe, the returning radio message, and events A, B, and C. In particular, clearly indicate when the people on earth receive the photo.

R2S.6 A rocket is launched from the moon and travels away from it at a speed of $\frac{2}{5}$. Call the event of the rocket's launching event A. After 125 s as measured in the reference frame of the moon, the rocket explodes: call this event B. The light from the explosion travels back to the moon: call its reception event C. Let the moon be located at $x = 0$ in its own reference frame, and let event A define $t = 0$. Assume that the rocket moves along the $+x$ direction.

(a) Draw a spacetime diagram of the situation, drawing and labeling the worldlines of the moon, the rocket, and the light flash emitted by the explosion and received on the moon.

(b) Draw and label events A, B, and C as points at the appropriate places on the diagram. Write down the t and x coordinates of these events.

R2S.7 Imagine that a spaceship in deep space is approaching a space station at a constant speed of $v = \frac{3}{4}$. Let the space station define the position $x = 0$ in its own reference frame. At time $t = 0$, the spaceship is 16.0 light-hours away from the station. At that time and place (call this event A), the spaceship sends a laser pulse of light toward the station, signaling its intention to dock. The station receives the signal at its position of $x = 0$ (call this event B), and after a pause of 0.5 h, emits another laser pulse signaling permission to dock (call this event C). The spaceship receives this pulse (call this event D) and immediately begins to decelerate at a constant rate. It arrives at rest at the space station (call this event E) 6.0 h after event D, according to clocks in the station.

(a) Carefully construct a spacetime diagram showing the processes described above. On your diagram, you should show the worldlines of the space station, the approaching spaceship, and the two light pulses and indicate the time and place (in the station's frame) of events A through E by labeling the corresponding points on the spacetime diagram. Scale your axes, using the hour as the basic time and distance unit, and give your answers in hours.

(b) In particular, exactly when and where does event D occur? Event E? Write down the coordinates of these events in the station frame, and explain how you *calculated* them (reading them from the diagram is a useful check, but is not enough).

(c) Compute the magnitude of the average acceleration of the spaceship between events D and E in natural SR units (s^{-1}) and in g's (1 $g \equiv$ 9.8 m/s^2). Note that a shockproof watch can typically tolerate an acceleration of about $50g$.

R2S.8 Imagine that a spaceship is docked at a space station floating in deep space. Assume that the space station defines the origin in its own frame. At $t = 0$ (call this event A) the spaceship starts accelerating in the $+x$ direction away from the space station at a constant rate (as measured in the station frame). The spaceship reaches a cruising speed of 0.5 after 8 h as measured in the station frame (call this event B). [*Hint:* You will find it helpful to construct a spacetime diagram as you go through the parts of this problem: such a diagram is required in part (e) anyway.]

(a) *Where* does event B occur? Explain your reasoning.   *(more)*

(b) Compute the magnitude of the ship's acceleration in SR units (s^{-1}) and in g's ($1\,g \equiv 9.8\ \text{m/s}^2$). Note that a shockproof watch can tolerate an acceleration of about $50g$.

(c) At event B, the spaceship sends a radio signal to the station, reporting that it has reached cruising speed. This signal reaches the station at event C. When and where does this event occur? (*Hint:* Radio signals are electromagnetic waves and so travel at the speed of light.)

(d) The technician responds to this message 0.5 h later after returning from a coffee break: call this event D. When and where does the ship receive this acknowledgment (event E)?

(e) Draw a careful spacetime diagram of this situation, showing the worldlines of the space station, the spaceship, the radio signals, and events A through E. Be sure to label all these items appropriately.

R2S.9   An air traffic control radar installation (see figure R2.12) receives a radar pulse reflected from a certain jet plane 280 μs after the pulse was sent. The signal comes from a direction that is 35° north of west and 5.5° up from horizontal. If the sending of the outgoing pulse defines $t = 0$, in a frame fixed to the earth's surface and oriented in the usual way with the installation at the spatial origin of the frame, what are the spacetime coordinates $[t, x, y, z]$ of the plane at the instant it reflects the pulse? (*Hint:* Radar pulses are electromagnetic waves, so they travel at the speed of light.)

R2S.10   Imagine that you are in a spaceship prospecting for asteroids. Your radar installation receives a pulse reflected from a nice large asteroid 1.24 s after it was sent, and the returning signal comes from a direction 25° to the right and 18° up from the direction your ship is facing. Assuming that the direction your ship is facing defines $+x$ direction, and

Figure R2.12
An air traffic control installation showing the radar display consoles (see problem R2S.9).

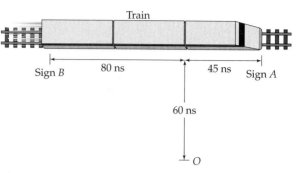

Figure R2.13
See problem R2S.11.

the up direction is the $+z$ direction, and that we define $t = 0$ to be when the pulse is sent, what are the spacetime coordinates of the asteroid at the time the pulse is reflected? (*Hint:* Radar pulses are electromagnetic waves, so they travel at the speed of light.)

R2S.11   (*Seeing* is not the same as *observing!*) Imagine that a bullet-train running at a *very* high speed passes two track-side signs (A and B), as shown in the aerial view in figure R2.13. Let event A be the passing of the front end of the train by sign A and event B be the passing of the rear end of the train by sign B. An observer is located at the cross marked by an O in the diagram.

(a) This observer *sees* (i.e., receives light with her eyes) event A to occur at time $t = 0$ and sees event B to occur at time $t = 25$ ns. When does she *observe* these events to occur? That is, what would a clock present at sign A read at event A, and what would a clock present at sign B read at event B if these clocks were correctly synchronized with the clock at O?

(b) In what way is the diagram above misleading about the implied time relationship between events A and B?

(*Hint:* Remember that the clocks at signs A and B must be synchronized with the clock at O in such a way that they would read the speed of a light signal traveling between them and O to be 1 s/s. Given this, the distance between O and A, and the time that light from event A reached O, can you *infer* when event A must have happened?)

Rich-Context

R2R.1   Imagine that an advanced alien race, bent on keeping humans from escaping the solar system into the galaxy, places an opaque spherical force field around the solar system. The force field is 6 light-hours in diameter, is centered on the sun, and is formed in a single instant of time as measured by synchronized clocks in an inertial frame attached to the sun. That instant corresponds to 9 p.m. on a certain night in your time zone. When does the opaque sphere appear to start blocking light from the stars from your vantage point on earth (8 light-minutes from the sun)? Does the opaque sphere appear all at once? If not, how long does it take for the sphere to appear, and what does it look like as it appears? Describe things as completely as you can. (This is inspired by the novel *Quarantine* by Greg Egan.)

R2R.2 Two radar pulses sent from the earth at 6:00 a.m. and 8:00 a.m. one day bounce off an alien spaceship and are detected on earth at 3:00 p.m. and 4:00 p.m. (but you aren't sure which reflected pulse corresponds to which emitted pulse). Is the spaceship moving toward earth or away? If its speed is constant (but less than *c*), when will it (or did it) pass by the earth? (*Hint:* Draw a spacetime diagram.)

R2R.3 (*Seeing* is not the same as *observing*!) Imagine that you are sitting immediately adjacent to a set of train tracks. A certain bullet-train running at a speed of 0.5 on these tracks has three blinking lights, one at

each end, and one in the middle. The end lights are 200 ns of distance from the middle light according to measurements in your ground frame. As the train's center passes you (a negligible distance away), you *see* all three lights blink simultaneously. Do you *observe* them to blink simultaneously? Explain. If you do *not* observe them to blink simultaneously, which light blinks first, and how much in time advance of the center light does it blink? Explain carefully.

Advanced

R2A.1 A meter stick moves at a speed of 0.5 (SR units) along a line parallel to its length that passes within 1 m of camera. The camera shutter opens for an incredibly brief instant just as the meter stick's center passes closest to the camera. Explain why the marks on the meter stick do not look equally spaced in the resulting picture, and describe what they look like. (Ignore length contraction.)

ANSWERS TO EXERCISES

R2X.1 The distance between the earth and the moon in seconds is

$$384,000 \, \text{km} \left(\frac{1000 \, \text{m}}{1 \, \text{km}} \right) \left(\frac{1 \, \text{s}}{3.0 \times 10^8 \, \text{m}} \right) = 1.28 \, \text{s} \tag{R2.7}$$

So the clock on the moon should read 1.28 s after noon if it is synchronized with the earth clock.

R2X.2 In SI units, the abbreviated units of momentum are kg·m/s, so in SR units, the units of momentum are kg·s/s, or simply kg, just as for energy.

R2X.3 and R2X.4 With the added worldlines, figure R2.6 looks like figure R2.14. Note that the particle worldline discussed in exercise R2X.3 has a slope of −5 on the diagram, since it moves 1 s of distance in the −*x* direction per 5 s of time. Since the moon is 1.28 s of distance from the earth, we should draw its

worldline 1.28 s from the *t* axis. In 6 s of time, the moon is not going to move appreciably on this diagram, so its worldline will be essentially straight and vertical.

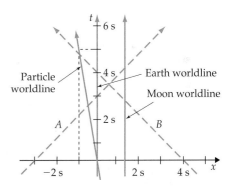

Figure R2.14
See the answers to exercises R2X.3 and R2X.4.

R3

The Nature of Time

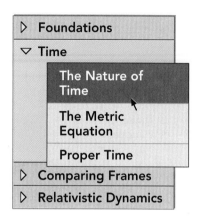
Chapter Overview

Introduction

In chapter R2, we developed a method of synchronizing clocks consistent with the principle of relativity and learned to use spacetime diagrams to display the coordinates of events and the motion of particles in a vivid way. We now have the tools to unwrap the package that the principle of relativity represents and to examine the surprises within.

What *is* the nature of time (now that we have unhooked ourselves from the notion of a universal and absolute time)? We have defined *procedures* for measuring the spacetime coordinates of an event. In the three-chapter subsection of the unit that this chapter opens, we will explore what these procedures *mean* about the concept of time.

Section R3.1:   Three Kinds of Time

There are three different ways that we can measure the time interval between two events, and it turns out that in special relativity, these three methods generally yield different results.

Section R3.2:   Coordinate Time Is Frame-Dependent

The **coordinate time** Δt between two events is the time measured in the context of an inertial reference frame. Usually, the two events will occur at different places in a given frame, and in such a case the coordinate time is the difference between the event times as recorded by two synchronized clocks in the frame, one present at each event. If the events happen to occur at the same place in the frame, then a *single* clock present at that position suffices to measure Δt.

The coordinate time between a given pair of events is a *frame-dependent* number, because although the method of clock synchronization described in chapter R2 is self-consistent *within* a given inertial frame, an observer in another inertial reference frame will *not* observe these clocks to be synchronized when comparing them to synchronized clocks in her or his own frame! This can cause, for example, an observer in one frame to think that two events are simultaneous, while an observer in a different frame concludes that they are not.

Section R3.3:   The Same Result from the Radar Method

This is not due to problems with the process of synchronizing multiple clocks. One can show that the radar method of defining coordinates leads to exactly the same results. The disagreement is a simple consequence of defining time so that the speed of light is the same in all inertial reference frames.

Section R3.4:   The Geometric Analogy

We have no trouble with analogous frame-dependent quantities in a related but more familiar guise. Imagine superimposing two xy cartesian coordinate systems on a town, one rotated with respect to the other. We know that surveyors using properly established but different coordinate systems will disagree about both the north-south displacement Δy and the east-west displacement Δx between two given locations in the town. The way that different inertial reference frames define spacetime coordinates for events is directly analogous to the way that different coordinate axes define x,y coordinates for points on the two-dimensional plane; so we should *expect* observers in different reference frames to disagree about the value of the coordinate Δt between events (as well as about their spatial separation Δd).

Note that we can use a tape measure to *directly* quantify the separation of two points on a plane, either by measuring the *pathlength* along some arbitrary path between the points or by measuring the *distance* between them along the unique straight line connecting those points. Since we can measure these quantities directly without setting up a coordinate system, any person who *calculates* these quantities using coordinate-based quantities must get the same result as any other: The pathlength along a path and the distance between two points are *coordinate-independent* quantities.

Section R3.5:   Proper Time and the Spacetime Interval

The analogy to a tape measure in spacetime is a single clock that moves in such a way as to be present at both events. Any clock that is present at both of two events measures a frame-independent **proper time** $\Delta\tau$ (short for *proprietary time*) along its particular worldline. If the worldline happens to be the unique constant-velocity worldline between the events (so that the clock is inertial), the clock also measures the frame-independent **spacetime interval** Δs between the events. *Proper time* along a worldline is analogous to the pathlength along a path on a plane, and the *spacetime interval* between events to the *distance* between points on a plane. This analogy is summarized in table R3.2

Note that the spacetime interval Δs between two events is *always* also a coordinate time Δt between those events, as measured in the particular inertial frame where the events happen at the same place as well as a particular proper time $\Delta\tau$ measured between the events along a worldline that happens to be straight.

If the analogy is true, it should not be any more surprising that Δt, $\Delta\tau$, and Δs between two given events have different values than it is that Δy, the pathlength, and the distance Δd have different values between two given points. Experimentally, these times *are* in fact different. We will see *why* in chapters R4 and R5.

Table R3.2   The geometric analogy

Plane Geometry		Spacetime Geometry
Map	$\leftrightarrow$	Spacetime diagram
Points	$\leftrightarrow$	Events
Paths or curves	$\leftrightarrow$	Worldlines
Coordinate systems	$\leftrightarrow$	Inertial reference frames
Relative rotation of coordinate systems	$\leftrightarrow$	Relative velocity of inertial reference frames
Differences between spatial coordinate values	$\leftrightarrow$	Differences between spacetime coordinate values
Pathlength along a path	$\leftrightarrow$	Proper time along a worldline
Distance between two points	$\leftrightarrow$	Spacetime interval between two events

R3.1   Three Kinds of Time

A firecracker explodes. Some time later and somewhere else, another fire-cracker explodes. How much time passes between these two events? How can we measure that time interval?

There is only one kind
of newtonian time

In a newtonian universe, such questions would be easy to answer. We might measure the time between the events with a pair of synchronized clocks, one present at each event. We might instead measure the time between the events with a single clock that travels in such a way that it arrives at each event's location just as the event happens. Since all clocks in a new-tonian universe register the same universal, absolute time, these methods (and any of a number of other valid methods) will yield the same result. It is unimportant what method we actually use.

There are three kinds of time
in relativity theory

In chapter R2, however, we argued that in *our* universe, universal and absolute time does not exist, and measuring the time interval between two events is thus somewhat more problematic. In this chapter, we will discover that there are *three* fundamentally different ways to measure the time inter-val between two events in the theory of special relativity, and that these dif-ferent methods yield different results, even if applied to the same two events!

Because it is important in the real universe to distinguish between the various methods used to measure the time interval between two events, we will refer to the time interval determined by each method with a special technical name: we speak of the **coordinate time,** the **proper time,** and the **spacetime interval** between the events. Our purpose in this chapter is to de-scribe and understand these distinct ways of measuring time and to begin the process of uncovering the quantitative relationships between them.

R3.2   Coordinate Time Is Frame-Dependent

Once the clocks in an inertial reference frame have been satisfactorily syn-chronized, we can use them to measure the time coordinates of various events that occur in that frame. In particular, we can measure the time be-tween two events A and B in our reference frame by subtracting the time read by the clock nearest event A when it happened from the time read by the clock nearest event B when it happened: $\Delta t_{BA} \equiv t_B - t_A$. Note that this method of measuring the time difference between two events requires the use of two synchronized clocks in an established inertial reference frame. Such a measurement therefore cannot be performed in the absence of an inertial frame.

In short, we define the **coordinate time** between two events as follows:

The definition of *coordinate
time* between events

We call the time measured between two events either by a pair of synchronized clocks at rest in a given inertial reference frame (one clock present at each event) or by a single clock at rest in that inertial frame (if both events happen to occur at that clock in that frame) the *coordinate time* Δt between the events in that frame.

The coordinate time between
events depends on your choice
of frame

Now, imagine that the observer in some inertial reference frame (let's call this frame the Other Frame: we'll talk about a Home Frame in a bit) sets out to synchronize its clocks. In particular, let us focus on two clocks in that frame that lie on the x axis an equal distance to the left and right of the mas-ter clock at $x' = 0$. At $t' = 0$, the observer causes the center clock to emit two flashes of light, one traveling in the $+x'$ direction and the other in the $-x'$ direction. Let's call the emission of these flashes from $x' = 0$ at $t' = 0$ the ori-gin event O.

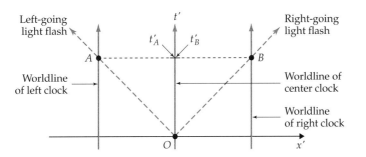

Figure R3.1
The synchronization of two clocks equally spaced from a center clock, as observed in the Other Frame. If the right and left clocks are set to agree at events A and B, they will be synchronized with each other.

Now, since both of the other clocks are the *same distance* from the center clock and since the speed of light is 1 (light-)second/second (s/s) in every inertial reference frame, the left-hand clock will receive the left-going light flash (call the event of reception event A) at the *same* time as the right-hand clock receives the right-going flash (event B). By the definition of synchronization, both clocks should thus be set to read the *same* time at events A and B (a time in seconds equal to their common distance from the center clock).

The spacetime diagram in figure R3.1 illustrates this process. Note that since all three clocks are at rest in this frame, their worldlines on the spacetime diagram are vertical. Moreover, since the speed of light is 1 s/s in this (and every other inertial) frame, the worldlines of the light flashes will have slopes of ± 1 on the spacetime diagram (i.e., they make a 45° angle with each axis) as long as the axes have the same scale. On this diagram it is clear that events A and B really do occur at the same time in the Other Frame.

Now consider a second inertial reference frame (the Home Frame) within which the Other Frame is observed to move in the $+x$ direction at a speed β. Let us look at the same events from the vantage point of the Home Frame. For convenience, let us take the event of the emission of the flashes to be the origin event in this frame also (so event O occurs at $t = x = 0$ in the Home Frame).

The observer in the Home Frame will agree that the right and left clocks in the Other Frame are always equidistant from the center clock in the Other Frame. Moreover, at $t = 0$, when the center clock passes the point $x = 0$ in the Home Frame as it emits its flashes, the right and left clocks are equidistant from the emission event. But as the light flashes are moving to the outer clocks, the observer in the Home Frame observes the left clock to move up the x axis *toward* the flash coming toward it, and the right clock to move up the x axis *away* from the flash coming toward it. As a result, the left-going light flash has less distance to travel to meet the left clock than the right-going flash does to meet the right clock. Since the speed of light is 1 in the Home Frame as well as in the Other Frame, this means that the left clock receives its flash first. *Therefore, event A is observed to occur **before** event B in the Home Frame.*

Figure R3.2 shows a spacetime diagram of the process as observed in the Home Frame. In drawing the worldlines of the clocks in question, it is important to note that the clocks are *not* at rest in the Home Frame. Their worldlines on a Home Frame spacetime diagram will be equally spaced lines with slopes of $1/\beta$, indicating that the clocks are moving to the right at a speed β. The light flashes have a speed of 1 s/s in the Home Frame (and every other frame), so their worldlines have to be drawn with a slope of ± 1 on the spacetime diagram.

In summary, the coordinate time between events A and B as measured in the Other Frame is $\Delta t' = 0$ (by construction in this case), but the coordinate time between these events as measured in the Home Frame is $\Delta t \neq 0$. We see

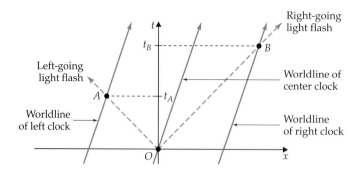

Figure R3.2
The same events as observed in the Home Frame. In this frame, event *B* is measured to occur *after* event *A*.

that the coordinate times between the *same two events* measured in different reference frames are *not* generally equal. Thus we say that the coordinate time differences are **relative** (i.e., they depend on one's choice of inertial reference frame).

Why? If each observer synchronizes the clocks in his or her own reference frame according to our definition, *each will conclude that the clocks in the other's frame are not synchronized.* Notice that the Other Frame observer has set the right and left clocks to read the same time at events *A* and *B*. Yet these events do *not* occur at the same time in the Home Frame. Therefore the observer in the Home Frame will claim that the clocks in the Other Frame are not synchronized. (Of course, the observer in the Other Frame will make the same claim about the clocks in the Home Frame.) The definition of synchronization that we are using makes perfect sense *within* any inertial reference frame, but it does not allow us to synchronize clocks in *different* inertial frames. In fact, the definition *requires* that observers in different inertial frames measure *different* time intervals between the same two events, as we have just seen.

In general, two observers in different frames will also disagree about the *spatial* coordinate separation between the events. Consider two events *C* and *D* that both occur at the center clock in the Other Frame, but at different times. Since the center clock defines the location $x' = 0$ in the Other Frame, the events have the same x' coordinate in that frame, so $\Delta x' = 0$. But in the Home Frame, the center clock is measured to move in the time between the events, and so the two events do *not* occur at the same place: $\Delta x \neq 0$ (see figure R3.3.)

Spatial coordinate differences are relative also

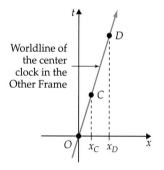

Figure R3.3
Events *C* and *D* both occur at the same place in the Other Frame ($\Delta x' = 0$) but not in the Home Frame ($\Delta x = 0$). (This diagram is drawn by an observer in the Home Frame.)

Exercise R3X.1

Note that the frame dependence of the *spatial* coordinate difference between two events has nothing to do with clock synchronization or relativity: this would be true even if time were universal and absolute. Show, using the galilean transformation equations, that if the separation between two events in the Other Frame is $\Delta x' = 0$ but $\Delta t' \neq 0$, then the separation between these events in the Home Frame is *not* zero ($\Delta x \neq 0$).

R3.3   The Same Result from the Radar Method

The reason why observers in different inertial frames disagree about whether the clocks in a given frame are synchronized is that *synchronization is defined so that light flashes are measured to have a speed of 1 in every inertial frame:* the frame dependence of coordinate time differences is a logical consequence of this assertion. This can be illustrated by considering the radar method of

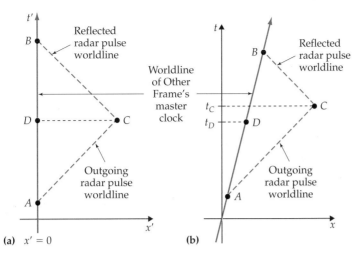

Figure R3.4

(a) In the Other Frame, events C and D are defined to be simultaneous if D occurs at the master clock at a time halfway between the emission event A and the reception event B. The coordinate time difference between events C and D in the Other Frame is thus $\Delta t' = 0$. (b) In the Home Frame, the Other Frame's master clock moves to the right as time passes, so its worldline is slanted. On the other hand, radar pulse worldlines still have slope ± 1, as shown. This means that an observer in the Home Frame will conclude that event C happens after event D, so the coordinate time difference between the events is $\Delta t \neq 0$.

assigning spacetime coordinates: although the radar method does not involve the use of synchronized clocks, it does depend on the assumption that the speed of light is the same in every inertial frame. Does the radar method also imply that the coordinate time difference between two events is frame-dependent?

Figure R3.4 shows that it does. Figure R3.4a shows the observer in the Other Frame using the radar method to determine the space and time coordinates of the event C. The observer in that frame will conclude that event C and event D occur at the same time, where D is the event of the master clock at $x' = 0$ registering $t'_D = \frac{1}{2}(t'_A + t'_B)$, that is, at the instant of time halfway between the emission of the radar pulse at t_A and its reception at t_B. According to the radar method, then, the coordinate time between events C and D is $\Delta t' = 0$. [Radar and visible light are both electromagnetic waves (with different frequencies), so both will move at a speed of 1 light-second/second = 1.]

When the same sequence of events is viewed from the Home Frame, though, a different conclusion emerges (see figure R3.4b). According to observers in the Home Frame, the Other Frame's master clock is moving along the x axis with some speed β; so in a spacetime diagram based on measurements taken in the Home Frame, the worldline of that clock will appear as a slanted line (with slope $1/\beta$) instead of being vertical. Radar pulse worldlines, on the other hand, still have slopes of ± 1, just as they did in the Other Frame spacetime diagram. The inevitable result (as you can see from the diagram) is that observers in the Home Frame are forced to conclude that event C occurs *after* event D does, and thus that the time difference between events C and D in the Home Frame is $\Delta t \neq 0$.

The point is that the relativity of coordinate time interval between events is a direct consequence of the fact that we are *defining* coordinate time by

Radar method yields the same conclusion

Frame dependence of coordinate time follows from principle of relativity assuming that the speed of light is 1 in every inertial reference frame. Remember, though, that we *must* make this assumption if the laws of electromagnetism are to be consistent with the principle of relativity!

R3.4   The Geometric Analogy

You may find it troubling that coordinate differences between events are not absolute but are instead frame-dependent. This is particularly true of the time coordinate separation: it is not easy to let go of the newtonian notion of absolute time! The fact is, though, that *we have no trouble at all with these ideas when they appear in a related but more familiar guise.*

An illustration of alternative coordinate systems in plane geometry

Consider Askew, a hypothetical town somewhere in the western United States (the residents wish its precise location to remain secret). Most towns in the rural United States have streets that run directly north/south or east/west. The surveyor who laid out the streets of Askew in 1882, however, *tried* to calibrate his compass against the North Star the night before, but in fact had forgotten exactly where it was (it was a long time since he had this stuff in high school after all) and ended up choosing a star that turned out to be 28° east of the true North Star. Therefore, all the streets of Askew are twisted 28° from the standard directions.

Now, if we would like to assign x and y coordinates to points of interest in this town (or any town), we have to set up a cartesian coordinate system. It is conventional to orient coordinate axes on a plot of land so that the y axis points north and the x axis points east (see figure R3.5a). This is usually convenient as well, since the streets will be parallel to the coordinate axes. There is no reason why this *has* to be done, though, and in Askew's case, it is actually more convenient to use a coordinate system oriented 28° to the east (figure R3.5b). Note that City Hall is the origin of both coordinate systems.

We can of course use any coordinate system we like to quantify the position of points of interest in the town: coordinate systems are arbitrary human artifacts that we impose for our convenience on the physical reality of the town. However, the coordinates we actually *get* for various points certainly *do* depend on the coordinate system used. For example, the coordinate differences between City Hall and the Statue of the Unknown Physicist in

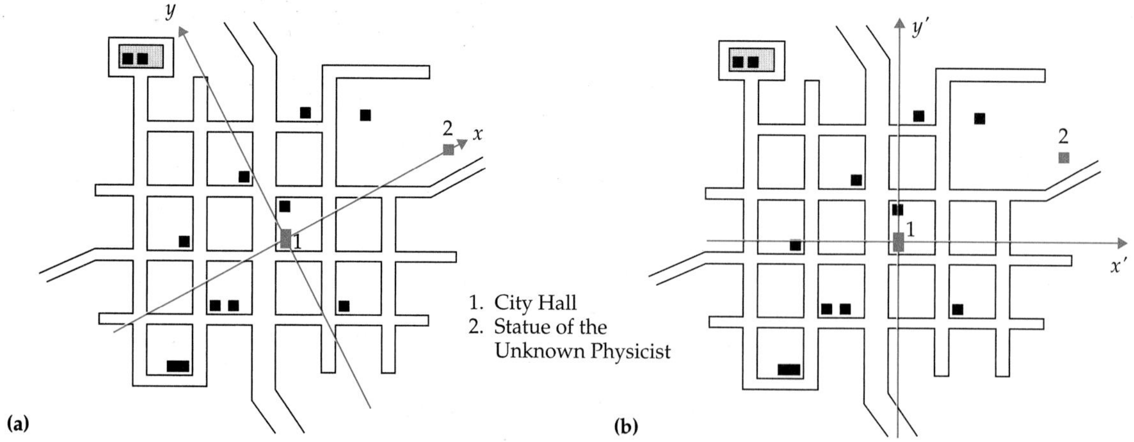

1. City Hall
2. Statue of the Unknown Physicist

(a)   (b)

Figure R3.5
(a) A standard northward-oriented cartesian coordinate system superimposed on the town of Askew.
(b) A more convenient coordinate system oriented 28° clockwise.

Memorial Park might be $\Delta y = 0$, $\Delta x = 852.0$ m in the standard coordinate system, but $\Delta y' = 399.9$ m, $\Delta x' = 752.3$ m in the conveniently oriented coordinate system.

Is it surprising that the results are different? Do the differences in the results cause us to suspect that one or the other coordinate system has been set up incorrectly? Hardly! We can accept the fact that both coordinate systems are perfectly correct and legal. We already *know and expect* that differently oriented coordinate systems on a plane will yield different coordinate measurements. This causes no discomfort: we understand that this is the way that things are.

In an entirely analogous way, we have carefully and unambiguously defined a procedure for setting up an inertial reference frame and synchronizing its clocks. This definition happens to imply that spacetime coordinate measurements in different frames yield different results. This should be no more troubling to us than the fact that Askew residents who use different sets of coordinate axes will assign different coordinates to various points in town. *Coordinates have meaning only in the context of the coordinate system or inertial frame that we use to observe them.*

The only reason that the relativity of time coordinate differences is a difficult idea is that we don't have *common experience* with inertial reference frames moving with high enough relative speeds to display the difference. The frames that we typically experience in daily life have relative speeds below 300 m/s, or about one-millionth of the speed of light. If for some reason we could only construct cartesian coordinate systems on the surface of the earth that differed in orientation by no more than one-millionth of a radian, then we might think of cartesian coordinate differences as being "universal and absolute" as well!

So, to summarize, the coordinate differences between points on a plane (or events in spacetime) are "relative" because coordinate systems (or inertial reference frames) are human artifacts that we *impose* on the land (or spacetime) to help us quantify that physical reality by assigning coordinate numbers to points on the plane (or events in spacetime). Because we are free to set up coordinate systems (or reference frames) in different ways, the coordinate differences between two points (or events) reflect not only something about their real physical separation, but also something about the artificial choice of coordinate system (or reference frame) that we have made.

So, is it true, then, that *everything* is relative? Is there nothing that we can measure about the physical separation of the points (or events) that is absolute (i.e., independent of reference frame)?

There *is* in fact a coordinate-independent quantity that describes the separations of two points on a plot of land: the *distance* between those two points. For example, the distance between City Hall and the Statue of the Unknown Physicist in Askew is $\Delta d = (\Delta x^2 + \Delta y^2)^{1/2} = [(852.0 \text{ m})^2 + 0]^{1/2} = 852.0$ m in the standard (north-oriented) coordinate system and $\Delta d' = [(\Delta x')^2 + (\Delta y')^2]^{1/2} = [(399.9 \text{ m})^2 + (752.3 \text{ m})^2]^{1/2} = 852.0$ m in the convenient coordinate system. It doesn't matter what coordinate system you use to calculate the distance: you always will get the same answer.

The distance between two points on a plot of ground thus reflects something that is deeply real about the nature of the plot of ground itself, without reference to the human coordinate systems imposed on it. This independence from coordinate systems arises because there is in fact a method of determining the distance between two points *without* using a coordinate system at all: simply lay a tape measure between the points! Since this method yields a certain definite result for the distance, calculations of this distance in *any* coordinate system should yield the same result if they are valid.

We are not surprised by frame dependence of coordinates in this case!

Distance and pathlength, on the other hand, are coordinate-*independent*

Of course, there are many ways that one could lay a tape measure between City Hall and the Statue of the Unknown Physicist. One could lay the tape measure along a straight path between the two points: this would measure the distance "as the crow flies," which is what is usually meant by the phrase "the distance between two points." But there are other possibilities. One might, for example, lay the tape measure two blocks down Elm Street from City Hall, then one block over along Grove Avenue, then up Maple Street, and so on. This would measure a different kind of distance between the two points that we might call a *pathlength*.

Both the straight-line distance and the more general pathlength between two points can be measured directly with a tape measure, and thus they are quantities independent of any coordinate system. But the distance and the pathlength between two points may not be the same. In general, the pathlength between two points will depend on the path chosen, and will always be greater than (or at best equal to) the straight-line distance.

The three kinds of spatial separation

To summarize, we have at our disposal three totally different ways to quantify the separation of two points on a plane. We can measure the *coordinate separations* between the points, using a coordinate system. (The results will depend on our choice of coordinate system.) We can measure the *pathlength* between the points with a tape measure laid along a specified path. (The result here will depend on the path chosen, but is independent of coordinate system.) Or we can measure the *distance* between the points with a tape measure laid along the unique path defined by the straight line between the points. Because in this last case the path of the tape is unique, the distance between two points is a unique number that quantifies in a very basic way the separation of the points in space.

R3.5   Proper Time and the Spacetime Interval

Consider two events. Label them A and B. Is there any way that we can measure the time between events A and B *without* using a reference frame, analogous to the way that we can measure the pathlength between two points on the plane without using a coordinate system?

We can avoid the use of a reference frame lattice if we measure the time interval between these events with a clock that is *present* at both events. In a manner analogous to laying a tape measure between two points so that it passes close by each point, we can send a clock between the events along just the right path so that it is very close to each event as it occurs (see figure R3.6).

A tape measure stretched between two points marks off the distance between those points and presents a scale that can be laid right next to the two points for easy and unambiguous reading. In an entirely similar manner, a clock that travels between two events marks off the time between those events and presents its face at each event for easy and unambiguous reading. Since the clock's face is right there at each event, *everyone* looking at the clock will agree as to its reading as each event happens. The quantity measured by this clock is therefore **frame-independent**: it is measuring something basic about the *absolute* physical relationship between the events.

We call a time interval measured this way a *proper time:*

Definition of the *proper time* between two events

The time between two events measured by any clock present at both events is a **proper time** $\Delta\tau$ between those events. The numerical value of a proper time measured by a given clock between two given events is an absolute, *frame-independent* quantity.

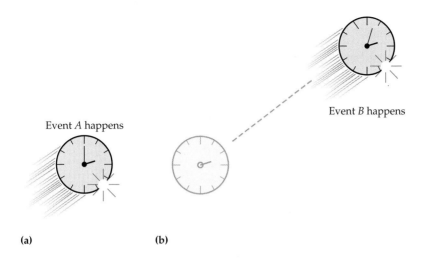

Figure R3.6

Measuring the *proper time* between two events with a clock present at both events. (Events *A* and *B* here are represented by firecracker explosions.)

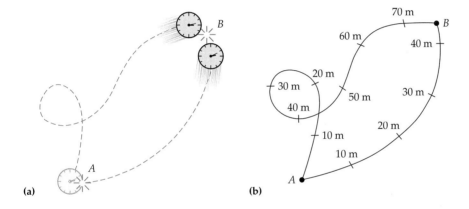

Figure R3.7

The proper time measured between two events may depend on exactly how the measuring clock travels between those events (a), just as the pathlength between two points depends on the path along which it is measured (b).

(Note that *proper* is a misleading adjective here. In English, this word has fairly recently come to mean almost exclusively "appropriate," or "correct in moral or manners." But the meaning intended here is more accurately "proprietary," that is, the time between the events measured specially by the *particular* clock in question. *Path time* might be a more appropriate term.)

There is, however, one thing that the proper time between two events *might* well depend on other than the events themselves. It might depend on the *worldline* that the clock follows in traveling from one event to the other, just as a pathlength measured by a tape measure depends on the path along which it is laid (see figure R3.7). We will see in chapter R4 that the path dependence of proper time is a straightforward consequence of the principle of relativity: for now it is enough to understand that this path dependence is a *possibility* suggested by the geometric analogy.

In an experiment performed in 1971 by J. C. Hafele and R. E. Keating (*Science*, vol. 117, p. 168, July 14, 1972), a pair of very accurate atomic clocks were synchronized, and then one was put on a jet plane and sent around the world, while the other remained in the laboratory (see figure R3.8). These clocks were both present at the same two events (the event of the departure of the plane clock from the laboratory and the event of its return): both thus measured proper times between these events. But the worldlines followed by each of these clocks were very different: one clock's worldline was simply a straight line at constant position (in the reference frame of the surface of the

Proper time depends on the clock worldline

Figure R3.8
Hafele and Keating carry one of their atomic clocks into an airplane in the process of performing their 1971 experiment concerning proper time.

earth), while the other's worldline went around the world. When these initially synchronized clocks were again brought together, it was found that they disagreed by several hundred nanoseconds (a difference that is in fact consistent with the quantitative prediction based on the theories of special and general relativity). The point is that it is not merely *possible* that the time that a clock measures depends on the nature of its worldline: it is an established experimental *fact*.

What is the analog of distance?

When measuring distances on a plane, we distinguish between the *pathlength* between two points measured along a certain path and the *distance* between the points: the distance is measured along the special path that is the unique straight line between the two points. Because the straight-line path is unique, the distance between two points along a straight line is a unique number reflecting something definite about the separation of those points in space.

Consider an **inertial clock** present at both events (a clock is *inertial* if an attached first-law detector registers no violation of Newton's first law). Such a clock follows a unique and well-defined worldline through spacetime between two events. Observed in an inertial frame, a clock would travel between the events in a straight line at a constant velocity. Such a worldline defines a unique path in space, and since there is only one value of a constant velocity that will be just right to get the clock from one event to the other, the clock's velocity along that worldline is also uniquely specified.

Definition of the *spacetime interval* between events

The **spacetime interval** between two events is defined to be the proper time measured by an *inertial* clock present at both events. This quantity is a unique, frame-independent number that depends on the separation of the events in space and time and nothing else. We conventionally use the symbol Δs to represent the spacetime interval between two events.

Spacetime interval is a proper time and also a coordinate time

It is important to note that the definitions of *coordinate time, proper time,* and the *spacetime interval* between two events overlap in certain special cases. The definition above makes it clear that the spacetime interval between two events is a special case of a proper time between two events, just as the distance between two points is a special case of the more general concept of pathlength between two points. An inertial clock present at both events also measures the *coordinate time* between those events in the clock's own reference frame, since the time interval measured between two events by a clock or clocks at rest in any inertial reference frame is a coordinate time

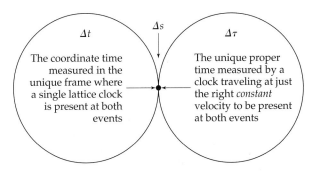

Figure R3.9
Let points in the left circle represent the set of all possible coordinate times that observers in inertial frames moving at various different velocities might measure between two given events. Let points in the right circle represent the set of all possible proper times measured between the same events by clocks present at both events but moving between the events along various different worldlines. The single point in common between these sets is the spacetime interval Δs between the events.

Table R3.1   Three kinds of time

	Coordinate Time	**Proper Time**	**Spacetime Interval**
Definition	The time between two events measured in an inertial reference frame by a *pair of synchronized* clocks, one present at each event. (If both events happen to occur at the same place, a single clock suffices.)[†]	The time between two events as measured by a single clock present at both events. (Its value depends on the worldline that the clock follows in getting from one event to the other.)	The time between two events as measured by an inertial clock present at both events. (Because an inertial clock follows a unique worldline between the events, the spacetime interval's value is unique for a given pair of events.)
Conventional symbol	Δt	$\Delta \tau$	Δs
Is value frame-independent?	No	Yes	Yes
Geometric analogy	Spatial coordinate differences	Pathlength	Distance

[†]*Note:* Alternatively, the coordinate time difference between two events might be inferred from measurements of the spacetime coordinates of these events using the radar method.

by definition. So the spacetime interval between two events is a special case of a proper time *and* a special case of a coordinate time (see figure R3.9).

Tables R3.1 and R3.2 summarize and organize the ideas presented in this chapter. In chapter R4, we will explore the relationship between the coordinate time and the spacetime interval between two events with the help of an equation called the *metric equation*. In chapter R5, we will use the metric equation to link coordinate time to proper time.

Table R3.2   The geometric analogy

Plane Geometry		Spacetime Geometry
Map	↔	Spacetime diagram
Points	↔	Events
Paths or curves	↔	Worldlines
Coordinate systems	↔	Inertial reference frames
Relative rotation of coordinate systems	↔	Relative velocity of inertial reference frames
Differences between spatial coordinate values	↔	Differences between spacetime coordinate values
Pathlength along a path	↔	Proper time along a worldline
Distance between two points	↔	Spacetime interval between two events

TWO-MINUTE PROBLEMS

R3T.1 Coordinate time would be frame-independent if the newtonian concept of time were true, true (T) or false (F)?

R3T.2 Observers in the Home Frame will conclude that the clocks in an Other Frame moving relative to the Home Frame in the $+x$ direction will be out of synchronization, even if the observers in the Other Frame have carefully synchronized clocks using the Einstein prescription, T or F? Specifically, Home Frame observers will see that for events farther and farther up the common $+x$ axis, the times registered by Other Frame clocks at the event
  A. Become further and further ahead.
  B. Become further and further behind.
  C. Remain the same.
  D. Have no clear relationship to the values that Home Frame clocks register for the same events.

R3T.3 A person riding a merry-go-round passes very close to a person standing on the ground once (event A) and then again (event B). Which person's watch measures *proper time* between these two events? (Assume that the ground is an inertial frame and that the merry-go-round rider moves at a constant speed.)

  A. The rider in the merry-go-round
  B. The person standing on the ground
  C. Both
  D. Neither

R3T.4 In the situation described in problem R3T.3, which person (if any) measures the *spacetime interval* between the events? (Select from the same answers.)

R3T.5 In the situation described in problem R3T.3, which person (if any) measures coordinate time between the two events in some inertial reference frame? (Select from the same answers.)

R3T.6 A spaceship departs from the solar system (event A) and travels at a constant velocity to a distant star. It then returns at a constant velocity, finally returning to the solar system (event B). A clock on the spaceship registers which of the following kinds of time between these events?
  A. Proper time
  B. Coordinate time
  C. Spacetime interval
  D. A and C
  E. B and C
  F. A, B, and C

HOMEWORK PROBLEMS

Basic Skills

R3B.1 Two firecrackers A and B are placed at $x' = 0$ and $x' = 100$ ns, respectively, on a train that is moving in the $+x$ direction relative to the ground frame. According to synchronized clocks on the train, both firecrackers explode simultaneously. Which firecracker explodes first according to synchronized clocks on the ground? Explain carefully. (*Hint:* Study figure R3.2.)

R3B.2 Figure R3.2 implies that an observer in the Home Frame concludes that clocks that have been carefully synchronized in the Other Frame are *not* synchronized in the Home Frame. Would an observer in the Other Frame conclude that clocks that have

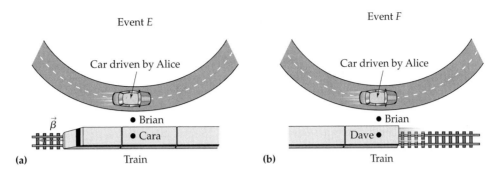

Figure R3.10

(a) Event *E* in the situation described in problem R3B.5. (b) Event *F* in the situation described in problem R3B.5.

been carefully synchronized in the Home Frame are not synchronized in the Other Frame? Justify your response.

R3B.3 Clock *P* is at rest alongside a racetrack. A jockey on horseback checks her watch against clock *P* as she passes it during the first lap (call this event *A*) and then checks her watch again as she passes clock *P* the second time (call this event *B*). Which clock (clock *P* or the watch) measures the spacetime interval between events *A* and *B*? Which measures proper time? (Be careful!) Do either of the clocks measure coordinate time between the events in the ground frame? Discuss.

R3B.4 Alyssa is a passenger on a train moving at a constant *velocity* relative to the ground. Alyssa synchronizes her watch with the station clock as she passes through the Banning town station, and then she compares her watch with the station clock as she passes through the Centerville town station farther down the line.
(a) Is the time that she measures between the events of passing through these towns a proper time? Is it a coordinate time? Is it the spacetime interval between the events?
(b) If one subtracts the Centerville station clock reading from the Banning station clock reading, what kind of time interval between the events does one obtain? Defend your answers carefully. (Treat the ground as an inertial frame, and assume that the Banning and Centerville clocks are synchronized.)

R3B.5 Alice is driving a race car around an essentially circular track. Brian, who is sitting at a fixed position at the edge of the track, measures the time that Alice takes to complete a lap by starting his watch when Alice passes by his position (call this event *E*) and stopping it when Alice passes his position again (call this event *F*). This situation is also observed by Cara and Dave, who are passengers in a train that passes very close to Brian. Cara happens to be passing Brian just as Alice passes Brian the

first time, and Dave happens to pass Brian just as Alice passes Brian the second time. Assume that the clocks used by Alice, Brian, and Cara are close enough together so that we can consider them all to be "present" at event *E*, and similarly that those used by Alice, Brian, and Dave are "present" at event *F*. Assume that the ground frame is an inertial reference frame.
(a) Who measures a proper time between events *E* and *F*?
(b) Who measures a coordinate time between events *E* and *F*?
(c) Who (if anyone) measures the spacetime interval between events *E* and *F*? Carefully explain your reasoning in each case. See figure R3.10.

Synthetic

R3S.1 Redraw figure R3.2, assuming that the newtonian concept of time is true. How does your redrawn diagram differ from the original, and how is this difference related to the behavior of light according to the newtonian and relativistic models?

R3S.2 After reading this chapter, your roommate exclaims, "Relativity cannot be right! This chapter claims that events that are simultaneous in one inertial reference frame are not simultaneous in another. Yet it is clear that two events are really simultaneous or really not simultaneous! This is not something that different observers could disagree about; or if they do, one has to be right and the other wrong!" Carefully and politely argue to your roommate that relativity *could* be right even so, and pinpoint the assumption that your roommate makes that could be debated. (*Hint:* You might be able to use the geometric analogy to good effect. Two different surveyors set up differently oriented coordinate systems on a plot of land. In one system, two rocks both lie along the *x* axis; in the other they do not. Is this a problem?)

R3S.3 Imagine that in the year 2065, you are watching a live broadcast from the space station at the planet

Neptune, which is 4.0 light-hours from earth at the time. (Assume that the TV signal from Neptune is sent to earth via a laser light communication system.) At exactly 6:17 p.m. (as registered by the clock on your desk), you see a technician on the TV screen suddenly exclaim, "Hey! We've just detected an alien spacecraft passing by here." Let this be event A. Exactly 1 h later, the alien spaceship is detected passing by earth: let this be event B. Assume that the earth and Neptune stations can be considered parts of the inertial reference frame of the solar system, and assume that the spaceship travels at a constant velocity.

  (a) During the broadcast, you can see on your TV screen the face of a clock sitting on the technician's desk. What time should you see on this clock face at 6:17 p.m. your time if that clock is synchronized with yours?

  (b) What is the coordinate time between events A and B in the solar system frame? Defend your response carefully.

  (c) What is the speed of the alien spaceship, as measured in the solar system frame?

  (d) What kind(s) of time would the spaceship's clock measure between events A and B?

R3S.4 Imagine two clocks, P and Q. Both clocks leave the spatial origin of the Home Frame at time $t = 0$: call this the origin event O. Both clocks move along the $+x$ axis, with clock P originally traveling at a speed of about $\frac{4}{5}$, while Q travels at a speed of about $\frac{1}{5}$. After a while, however, clock P decelerates, comes to rest, and then begins to move back toward the origin. A short time later, clock P collides with the slower clock Q, which has been moving with constant speed up the x axis during all this. Let the collision of the clocks be event A.

  (a) Draw a qualitatively accurate spacetime diagram of the situation described above, labeling the worldlines of clocks P and Q and the locations of events O and A.

  (b) Assume that clocks P and Q were both synchronized with the clock at the origin of the Home Frame when they left the origin. Will P and Q necessarily agree when they collide? Explain.

  (c) An observer in the Home Frame measures the time between events O and A with a pair of synchronized clocks (one at the spatial location of event O and one at the spatial location of event A). Clocks P and Q each also register a time between these events. Which clock(s) measure proper time between the events? The spacetime interval between the events? The coordinate time between the events?

R3S.5 A *particle accelerator* is a device that boosts subatomic particles to speeds close to that of light. Such an accelerator is typically shaped like a ring (which may be several kilometers in diameter): the particles are constrained by magnetic fields to travel

inside the ring. Imagine such an accelerator having a radius of 2.998 km. Assume that there are two synchronized clocks (P and Q) located on opposite sides of the ring. A certain particle in the ring is measured to travel from clock P to clock Q in 34.9 μs, as registered by those clocks. Let event A be the particle's departure from clock P and event B be the particle's arrival at clock Q. Assume that the particle contains an internal clock that measures the time between these events, and that the particle travels at a constant speed.

  (a) What is the speed of the particle in the laboratory frame?

  (b) Does the synchronized pair of laboratory clocks measure the proper time, the coordinate time, or the spacetime interval between events A and B?

  (c) Does the particle's internal clock measure the proper time, the coordinate time, or the spacetime interval between events A and B?

R3S.6 At $t = 0$, an alien spaceship passes by the earth: let this be event A. At $t = 13$ min (according to synchronized clocks on earth and Mars) the spaceship passes by Mars, which is 5 light-minutes from earth at the time: let this be event B. Radar tracking indicates that the spaceship moves at a constant velocity between earth and Mars. Just after the ship passes earth, people on earth launch a probe whose purpose is to catch up with and investigate the spaceship. This probe accelerates away from earth, moving slowly at first, but moves faster and faster as time passes, eventually catching up with and passing the alien ship just as it passes Mars. In all parts of this problem, you can ignore the effects of gravity and the relative motion of earth and Mars (which are small) and treat earth and Mars as if they were both at rest in the inertial reference frame of the solar system. Also assume that both the probe and the alien spacecraft carry clocks.

  (a) Draw a quantitatively accurate spacetime diagram of the situation, including labeled worldlines for the earth, Mars, the alien spacecraft, and the probe. Also label events A and B.

  (b) Whose clocks measure coordinate times between events A and B? Explain carefully.

  (c) Whose clocks measure proper times between these events? Explain.

  (d) Does any clock in this problem measure the spacetime interval between the events? If so, which one and why? If not, why not?

R3S.7 After reading this chapter, your roommate exclaims, "I know how to tell a moving frame from one at rest. This chapter clearly shows that after observers in a moving frame carefully try to synchronize clocks in that frame, they are still out of synchronization with clocks in a frame at rest. Therefore, the frame with the clocks that are out of synchronization is the moving frame, and a frame whose clocks are in synchronization must be at rest." Is your roommate right? Does this provide a way to distinguish *physically* a frame that is at rest from one that is not? Why or why not?

Rich-Context

R3R.1   A train is moving due east at a large constant speed on a straight track. Imagine that Harry is riding on the train exactly halfway between its ends. Sally is sitting by the tracks only a few feet from the train. Let the event of Harry passing Sally be the origin event O in both frames. At this same instant, both Harry and Sally receive the light from lightning flashes that have struck both ends of the train, leaving scorch marks on both the train and the track. Harry concludes that since he is in the middle of the train and he received the light from the strikes at the same time, the lightning strikes must have occurred at the same time in his reference frame. Is he right? If not, which strike (the one at the front of the train or the one at the rear) really happened first? Can Sally conclude from her seeing the flashes at the same time that the strikes happened at the same time in the ground frame? Why or why not? If not, which strike happened first in her frame? (This problem is adapted from one of Einstein's own illustrations of the implications of the constant speed of light.)

R3R.2   Imagine that you are looking through a telescope at a distant spaceship that is moving perpendicular to your line of sight at a constant speed that is a significant fraction of the speed of light. Imagine that the spaceship is several kilometers long in its direction of motion and has lights in a row along its length. Imagine that the spaceship's captain blinks

these lights every so often, and that all the lights blink at once in the ship's frame. Do the lights all blink at once as you see them? If not, explain why not, and which lights you see blinking first. (*Hint:* Since the spaceship is very distant, it takes essentially the same time for light to travel from all parts of the spaceship to your telescope. So if the lights actually do all blink simultaneously in your reference frame, they will also *look* to you as if they blink simultaneously from your vantage point. Differing light travel times from various parts of the ship are *not* the issue here.)

Advanced

R3A.1   Consider figure R3.2. We used that spacetime diagram to argue that clocks synchronized in the Other Frame will *not* be synchronized in the Home Frame. Imagine that the spatial separation between each side clock and the center clock is $L = 12$ ns as measured in the Home Frame, and that the speed of the clocks relative to the Home Frame is $\beta = 0.40$ in SR units. Find the time separation $t_B - t_A$ between events A and B as measured in the Home Frame. [*Hints:* This is a tricky problem, but it is not as impossible as it looks. Consider the left clock. In the time between $t = 0$ and $t = t_A$, this clock moves a distance βt_A toward the light flash coming toward it. Thus the total distance that the light flash has to cover in this time interval is $L - \beta t_A$. But since the light flash travels with unit speed in the Home Frame (and every frame), the *time* that it takes to travel to the clock is equal to the distance that it has to travel (in SR units). Write this last sentence as an equation, and then solve the equation for t_A. Repeat for the right-hand clock.]

ANSWERS TO EXERCISES

R3X.1   Consider events C and D with x coordinates x_C and x_D, respectively, in the Home Frame and x'_C and x'_D, respectively, in the Other Frame. According to the galilean transformation equations (equations R1.2), we have

$$x'_D - x'_C = (x_D - \beta t_D) - (x_C + \beta t_C)$$
$$= (x_D - x_C) - \beta(t_D - t_C) \qquad (R3.1)$$

If $\Delta x' = x'_D - x'_C = 0$, then

$$0 = (x_D - x_C) - \beta(t_D - t_C)$$
$$\Rightarrow \qquad x_D - x_C = \beta(t_D - t_C) \qquad (R3.2)$$

So if $\Delta t' = \Delta t = t_D - t_C \neq 0$, then we must have $\Delta x = x_D - x_C \neq 0$.

R4

The Metric Equation

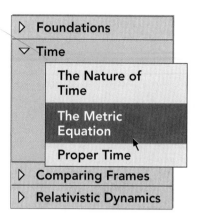
Chapter Overview

Introduction

Chapter R3 provided conceptual foundations for our discussion of time in the theory of relativity. In this chapter we begin building quantitative links between these methods of measuring time by introducing the *metric equation,* an equation that is essentially the pythagorean theorem for spacetime.

Section R4.1: Introduction to the Metric Equation

We saw in chapter R3 that there are three fundamentally distinct ways to measure the time interval between two events that can yield different results even for the same pair of events. However, given the coordinate-dependent coordinate differences Δx and Δy between two points on a plane, we can use the pythagorean theorm to compute the coordinate-independent *distance* $\Delta d = (\Delta x^2 + \Delta y^2)^{1/2}$ between those points. Our goal in this chapter is to find the analogous equation that links the frame-dependent coordinate differences Δt and Δd between two events in a given inertial reference frame with the frame-independent *spacetime interval* between those events.

Section R4.2: Deriving the Metric Equation

A **light clock** measures time by counting round trips of a flash of light bouncing between two mirrors a given distance L apart. In the clock's rest frame, a round trip takes a time $2L$. Now imagine a light clock that moves at a constant velocity in the Home Frame with its beam path perpendicular to its velocity. Let event A be the event of flash bouncing off one of the mirrors, and let event B be the event when the flash hits the same mirror again. The light clock itself is present at both events and is inertial, so the time $2L$ it measures between the events is the spacetime interval Δs between these events. In the Home Frame, however, the light flash follows a zigzag path of length $2[L^2 + (\frac{1}{2}\Delta d)^2]^{1/2}$, where Δd is the distance between the events in that frame. Since the speed of light is 1 in the Home Frame, this is equal to the coordinate time Δt between the events in that frame, so $\Delta t = 2[L^2 + (\frac{1}{2}\Delta d)^2]^{1/2} > \Delta s$. If we re-arrange terms in this equation, we arrive at the **metric equation** for spacetime

$$\Delta s^2 = \Delta t^2 - \Delta d^2 = \Delta t^2 - \Delta x^2 - \Delta y^2 - \Delta z^2 \qquad (R4.5)$$

Purpose: This equation specifies the frame-independent spacetime interval Δs between two events, given their coordinate separations Δt, Δx, Δy, and Δz in any given inertial frame.

Limitations: This equation applies only in an inertial reference frame.

Notes: This equation is to spacetime what the pythagorean theorem is to space.

We have proved this equation only for cases where $\Delta t > \Delta d$, but in fact it yields a frame-independent value of Δs^2 for *all* pairs of events, as we will see in chapter R8.

Section R4.3: About Perpendicular Displacements

The proof of the metric equation *assumes* that the distance between the light clock mirrors is L in both the clock frame and the Home Frame. An argument by contradiction presented in this section shows that the principle of relativity *requires* that the magnitude of any displacement measured in two different inertial reference frames have the same value in both frames as long as the displacement is perpendicular to the direction of the frames' relative motion.

Section R4.4: Evidence Supporting the Metric Equation

Muons are subatomic particles that decay with characteristic half-life of 1.52 μs as measured in their rest frame, as if they had an internal clock telling them when to decay. We can measure the time that the internal clocks of a batch of muons measure as they move between two detectors by observing the fraction of muons that decay. Since these muons are present at both detection events, their clocks measure the spacetime interval Δs between the events. Observers in the detector frame can measure the coordinate time Δt and distance Δd between the detection events. This section describes a muon experiment of this type that supports the predictions of the metric equation by a substantial margin.

Section R4.5: Spacetime Is *Not* Euclidean

The most important difference between the metric equation and the pythagorean equation is that the former has minus signs between terms, while latter has only plus signs. This implies, for example, that $\Delta s = 0$ if $\Delta t = \Delta d$ for a given pair of events, even though the events might look like two distinct points on a spacetime diagram. This is analogous to two points at 90° north latitude that look distinct on a flat map of the earth, but are in fact a single point (the north pole) on the real earth. Just as a flat map cannot accurately represent the noneuclidean geometry of the earth, so a flat spacetime diagram cannot accurately represent the noneuclidean geometry of spacetime: a pair of events whose separation on a spacetime diagram may look larger than that of another pair may in fact have a *smaller* spacetime interval between them.

Section R4.6: More About the Geometric Analogy

In two-dimensional plane geometry, the set of all points equidistant from a given point is a circle. Because of the minus signs in the metric equation, however, the set of all events that are the same spacetime interval from a given event forms hyperbolas instead.

Section R4.7: Some Examples

This section presents worked examples displaying different applications of the metric equation.

R4.1   Introduction to the Metric Equation

Review of the three kinds of time

In chapter R3, we discussed three different ways to measure the time interval between two events. An observer in an inertial reference frame can measure the *coordinate time* Δt between the events by comparing the reading of the clock present at one event with the reading of the synchronized clock present at the other event. Since observers in different reference frames will disagree about whether a given pair of clocks is synchronized, the coordinate time Δt measured between the events depends on the reference frame used.

One can avoid this problem by measuring the time between the events by a single clock that moves between the events so that it is present at both. Such a clock measures a *proper* (proprietary) time $\Delta \tau$ between the events. The magnitude of such a proper time, while frame-independent (in the sense that all observers will agree on what a given clock present at both events actually registers between those events), is known experimentally to depend on the worldline traveled by the clock as it moves from one event to the other.

There is, however, one and only one worldline that takes a clock from one event to the other at a *constant velocity*. The proper time between the events measured by an *inertial* clock is thus a unique, frame-independent number that depends on the spacetime separation of the two events and nothing else. This unique proper time is called the *spacetime interval* Δs.

Analogy between spacetime interval and distance

We also discussed in chapter R3 an analogy that compared these different ways of measuring the time between events in spacetime with different ways of measuring the separation of two points on a plane. The coordinate time corresponds to the north-south (or east-west) *coordinate displacement* between the points. Since surveyors using different coordinate systems will disagree about the exact direction of north, the value of the north-south displacement between two points will depend on one's choice of coordinate system. The proper time corresponds to the *pathlength* between the two points measured along a certain path with a tape measure. Since measuring the pathlength does not require determining where north is, its value is independent of one's choice of coordinate system, but does depend on the path chosen. The spacetime interval corresponds to the straight-line *distance* between the points. Since there is one and only one straight line between a given pair of points, the pathlength measured along this line is a unique, coordinate-independent number for characterizing the spatial separation of the points.

Analogy between metric equation and the pythagorean theorem

As discussed in section R3.5, we can actually *calculate* the distance Δd between two points on the plane, using the coordinate displacements Δx and Δy between the points (as measured in any given coordinate system) and the pythagorean theorem:

$$\Delta d^2 = \Delta x^2 + \Delta y^2 \qquad (R4.1)$$

The amazing thing about this formula is that while the values Δx and Δy between two points depend on one's choice of coordinate system, the distance Δd calculated from these does not.

It turns out that there is an analogous formula that links the coordinate time Δt and spatial coordinate displacements Δx, Δy, and Δz between two events measured in any given inertial reference frame with the frame-independent spacetime interval Δs between the events. This equation, which we call the *metric equation*, provides the crucial key needed to escape the "relativity" of inertial reference frames and quantify the separation of the events in *absolute* (frame-independent) terms.

Our purpose in this chapter is to derive the metric equation and describe some of its immediate consequences. In chapter R5, we show how the metric equation can be used to compute proper times along more general world-lines as well.

R4.2   Deriving the Metric Equation

The derivation that follows is the very core of the special theory of relativity. The metric equation is the key to understanding all the unusual and interesting consequences of the theory of relativity. You should make a special effort to understand this argument thoroughly.

What we want to do is to compare the time interval Δt between two events measured in some inertial frame (call it the Home Frame) with the time Δs measured by a clock moving at a constant velocity that is present at both events. To make this argument easier, I want to consider a special kind of clock we will call a **light clock**. An idealized light clock is shown in figure R4.1. It consists of two mirrors a fixed distance L apart and a flash of light that bounces back and forth between the mirrors. Each time the flash of light bounces off the bottom mirror, a detector in that mirror sends a signal to an electronic counter. The clock dial thus essentially registers the number of round trips that the light flash has completed. Since the speed of light is *defined* to be 1 s of distance per second of time in any inertial frame, we should calibrate this clock's face to register a time interval of $2L$ (where L is expressed in seconds) for each "tick" of the clock (i.e., each time the light flash bounces off the bottom mirror): the clock will then read correct time as long as it is inertial.

Now consider an arbitrarily chosen pair of events A and B. Let the coordinate time interval and spatial separation between these events (as measured in the Home Frame) be Δt and $\Delta d = (\Delta x^2 + \Delta y^2 + \Delta z^2)^{1/2}$, respectively. Also imagine that we have a light clock moving between these events (with its beam path oriented perpendicular to its direction of motion) at just the right constant velocity to be present at both events. To simplify our argument, let us also imagine that the length L between the light clock mirrors has just the right value so that events A and B happen to coincide with successive ticks of the light clock (in principle, we could always adjust L to make this true for the two given events).

In the inertial frame of the light clock, both events occur at the clock face, and the clock's light flash completes exactly one round trip. The time interval recorded by this clock between events A and B is thus exactly $2L$. Since this inertial clock is present at both events, it registers the spacetime interval between these events, so $\Delta s = 2L$.

In the Home Frame, the time of each event is registered by the clock nearest the event: since the events occur at different places, we will determine the coordinate time interval between the events by taking readings from a *pair* of clocks. In this frame, the light clock is observed to move the distance Δd in the time interval Δt. This means that the light flash will be observed in the Home Frame to follow the zigzag path shown in figure R4.2.

As you can see from figure R4.2, the total distance that the light flash travels in the Home Frame is (according to the pythagorean theorem)

$$2\sqrt{L^2 + \left(\frac{\Delta d}{2}\right)^2} = \sqrt{4L^2 + \Delta d^2} = \sqrt{(2L)^2 + \Delta d^2} \qquad \text{(R4.2)}$$

Since the synchronized clocks in the Home Frame must (by definition of *synchronization*) measure the speed of light to be 1, the coordinate time interval Δt registered on the pair of synchronized clocks in the Home Frame must be equal to the distance that the light flash traveled between the events:

$$\Delta t = \sqrt{(2L)^2 + \Delta d^2} \qquad \text{(R4.3)}$$

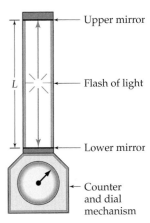

Figure R4.1
Schematic diagram of a light clock. Each "tick" of the light clock represents the passage of a time interval equal to $2L$ (in SR units).

Beginning of the derivation

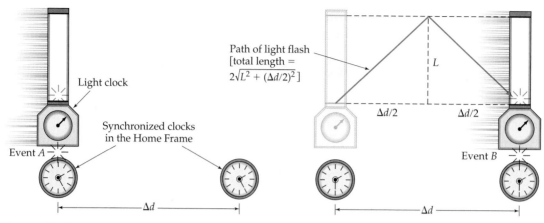

Figure R4.2
As the light clock moves from event *A* to event *B* in the Home Frame, its internal light flash will be observed to follow the zigzag path shown.

But we saw above that the light clock registers the spacetime interval between the two events to be $\Delta s = 2L$. Plugging this into equation R4.3 and squaring both sides, we get

$$\Delta t^2 = \Delta s^2 + \Delta d^2 \qquad \text{or} \qquad \Delta s^2 = \Delta t^2 - \Delta d^2 \qquad \text{(R4.4)}$$

Since $\Delta d^2 = \Delta x^2 + \Delta y^2 + \Delta z^2$ (where Δx, Δy, and Δz are the coordinate differences measured between the events in the Home Frame), we have finally

The metric equation

$$\Delta s^2 = \Delta t^2 - \Delta d^2 = \Delta t^2 - \Delta x^2 - \Delta y^2 - \Delta z^2 \qquad \text{(R4.5)}$$

Purpose: This equation specifies the frame-independent spacetime interval Δs between two events, given their coordinate separations Δt, Δx, Δy, and Δz in any given inertial frame.
Limitations: This equation applies only in an inertial reference frame.

This extremely important equation links the frame-*independent* spacetime interval Δs between any two events to the frame-*dependent* coordinate separations Δt, Δx, Δy, Δz measured between those events *in any arbitrary inertial reference frame!* Note that we have not sacrificed anything by using a light clock in this argument: since the speed of light is defined to be 1 in any inertial frame, any decent clock that we construct must agree with what the light clock says. The only real limitation to our argument is that Δt must be greater than Δd for the two events in question, so that it is possible for a light flash to travel between the events. (Note that if $\Delta t < \Delta d$, equation R4.5 yields an imaginary value for Δs, an absurd result indicating that the conditions of the proof have been violated.)

Since the spacetime interval Δs between two events in spacetime is analogous to the distance Δd between two points on a plane, the formula $\Delta s^2 = \Delta t^2 - \Delta x^2 - \Delta y^2 - \Delta z^2$ is directly analogous to the pythagorean theorem $\Delta d^2 = \Delta x^2 + \Delta y^2$. Note that the pythagorean theorem also relates a coordinate-independent quantity (the distance Δd between two points) with quantities whose values depend on the choice of coordinate system (the coordinate differences Δx and Δy). Indeed, the formula for the spacetime interval would be just like a four-dimensional version of the pythagorean theorem if it were not for the minus signs that appear. We will see that these minus signs have a variety of interesting and unusual consequences.

We call equation R4.5 the **metric equation.** It is the link between our human-constructed reference frames and the absolute physical reality of the separation between two events in space and time. It is difficult to overemphasize the importance of this equation: virtually all the rest of our study of the theory of relativity will be devoted to exploring the implications of this equation.

R4.3   About Perpendicular Displacements

The previous argument *assumes* that the vertical length L between the light clock mirrors is the same in both the light clock frame (where it was used to compute the spacetime interval) and the Home Frame (where it was used to compute the coordinate time). But how do we *know* that this is true? Since coordinate differences between events are generally frame-dependent, what gives us the right to assume that the displacement between the mirrors has the same value in both frames? This is not a trivial issue, because in chapter R6 we will see that observers in two different frames will disagree about the length of displacements measured parallel to the line of the frames' relative motion.

In this section, however, I will argue that if we have two reference frames in relative motion along a given line, any displacement measured *perpendicular* to that direction of motion *must* have the same value in both reference frames. I will demonstrate that this statement follows directly from the principle of relativity.

The proof presented here will be a proof by contradiction. We will assume that there *is* a contraction (or expansion) effect that applies to perpendicular lengths and then show that the existence of such an effect contradicts the principle of relativity. Turned around, this argument will then imply that if the principle of relativity is true, no such effect can exist.

Consider two inertial reference frames (a Home Frame and an Other Frame) in standard orientation, so that the line of relative motion is along the frames' common x and x' axes. In each frame we set up a measuring stick along the y or y' direction with spray-paint nozzles set 1.00 m apart (as shown in figure R4.3). Note in the diagram that the common x and x' axes (which lie along the line of relative motion of the frames) are perpendicular to the plane of the paper. This means that as the frames move relative to each other, one of the measuring sticks will move into the paper and the other out

What if L is not the same in both frames?

Proof that distances measured perpendicular to the line of relative motion of two inertial frames will be the *same* in both frames

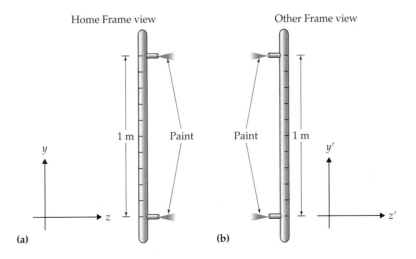

Figure R4.3
(a) The spray paint nozzles on the Home Frame measuring stick are 1.00 m apart in the y direction as measured in that frame. They point directly at their counterparts in the Other Frame so that stripes are painted on the Other Frame's measuring stick as it moves by. The x axis points directly into the plane of the paper here.
(b) Similarly, the paint nozzles on the Other Frame stick are 1.00 m apart in that frame and are pointed to paint stripes on the Home Frame measuring stick as it moves by.

of the paper. The paint nozzles in each frame are pointed in the direction of the secondary frame's measuring stick, so as the two measuring sticks pass each other, they will spray-paint stripes on each other.

Now imagine that there exists some kind of frame-dependent contraction effect so that an observer in the Home Frame measures the measuring stick at rest in the Other Frame (which is moving relative to the Home Frame) to be vertically contracted. This means that an observer in the Home Frame will measure the distance between the spray-paint nozzles on that stick to be *less* than 1.00 m apart. This in turn means that the stripes painted by these nozzles will be less than 1.00 m apart in the Home Frame: they will be painted *inside* the nozzles on the Home Frame's measuring stick. This also means that the nozzles on the measuring stick at rest in the Home Frame will paint stripes on the stick in the Other Frame which are *outside* the latter stick's nozzles (see figure R4.4a).

Now, the principle of relativity requires that the laws of physics be exactly the same in any inertial reference frame. More specifically, this means that if you perform exactly the same experiment in two inertial reference frames, you should get exactly the same result. There should be *no* way of experimentally distinguishing the two frames. How does this principle apply in this situation?

In the Other Frame, it is the Home Frame stick that is moving. Therefore, the principle of relativity requires that if a frame-dependent contraction effect exists, an observer in the Other Frame *must* measure the Home Frame stick to be contracted, just as the Home Frame observer measured the Other Frame stick to be contracted. This in turn means that the stripes painted by the Other Frame stick will be *outside* the Home Frame stick's nozzles, and the stripes painted by the Home Frame stick will be *inside* the Other Frame stick's nozzles, as shown in figure R4.4b.

Now, figure R4.4a and b describe a logical contradiction. In figure R4.4a, stripes are being painted on the Home Frame stick *inside* its nozzles. In figure R4.4b, stripes are being painted on the Home Frame stick *outside* its nozzles. These cannot be simultaneously true! The paint marks on the Home Frame stick are permanent and unambiguously visible to all observers in *every* reference frame. They cannot be "inside" the nozzles according to some observers and "outside" to others. So *either* figure R4.4a or figure R4.4b can be true, but *not both*. But the principle of relativity *requires* that both be true!

How can we resolve this conundrum? The only way is to reject the hypothesis that got us into this trouble in the first place, that is, the hypothesis

Figure R4.4

If the contraction effect is real, then the principle of relativity implies that observers in each frame must observe the stick in the Other Frame to be contracted. (a) So an observer in the Home Frame observes the Home Frame stick to paint stripes *outside* the Other Frame's nozzles as the latter moves by (into the plane of the drawing here). (b) Similarly, an observer in the Other Frame observes her or his stick to paint stripes *outside* the nozzles on the Home Frame's stick as the latter moves by (out of the plane of the paper here).

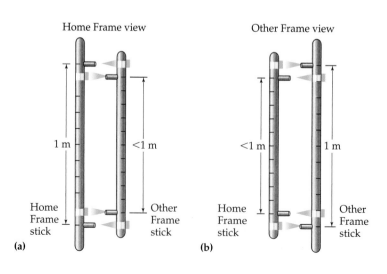

that distances measured perpendicular to the line of relative motion of the frames have different values in the two frames. If we assume that there is no contraction (or expansion) effect operating between the frames, then there is no problem with the principle of relativity. As shown in figure R4.5, both sticks will paint stripes across each other's nozzles. The situation is exactly the same in both frames, and the contradiction disappears.

This argument forces us to conclude:

Any displacement measured perpendicular to the line of relative motion of two inertial frames must have the same value in both frames.

This means that the distance L between the mirrors used in the derivation of the metric equation does in fact have the same value in the light clock frame as it does in the Home Frame, so our derivation should be correct.

Implication: L is the same in both frames in the proof of the metric equation

---

Exercise R4X.1

We will see in chapter R7 that a measuring stick *is* observed in a given reference frame to be contracted *parallel* to its direction of motion in that frame. Explain why the argument above *cannot* exclude contractions or expansions parallel to the line of motion. (*Hint:* What kind of stripes would the sticks paint on each other if they moved relative to each other in a direction parallel to their lengths?)

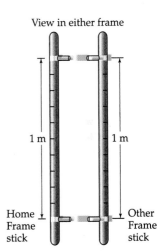

View in either frame

1 m   1 m

Home   Other
Frame   Frame
stick   stick

Figure R4.5

If we assume that there is no contraction, the contradiction disappears.

R4.4   Evidence Supporting the Metric Equation

Careful and compelling as the derivation of the metric equation in section R4.2 may be, we as physicists should not simply accept such an equation without some experimental confirmation. One of the classic experiments testing the validity of the metric equation involves **muons**. A muon is an elementary particle that is a more massive version of the electron (see chapter C1). Muons are continually generated in the upper atmosphere (at heights of approximately 60 km) by the interaction of cosmic rays with atmospheric gas molecules. Some of these muons stream downward toward the earth with speeds in excess of 0.99 (that is, 99% of c).

Now, muons are unstable, decaying after a short time into lighter particles. The half-life of muons at rest in a laboratory is about 1.52 μs, which means that if you have N muons at a certain time, after 1.52 μs you will have $N/2$ left; after another 1.52 μs you will have $N/4$ left, and so on. A batch of muons moving together in a bunch can thus serve as a clock: to determine how much time has passed in the muon frame, all that we need to do is to measure the number of undecayed muons in the bunch. You can imagine that each muon contains a built-in clock, and that each time the clock "ticks" the muon has a certain small probability of decaying, a probability such that after 1.52 μs worth of such ticks have passed, one-half of the muons in a bunch will have decayed.

It is possible to build a muon detector that will count the number of muons reaching it from a particular direction and traveling at a particular speed. Imagine building two detectors that register only muons traveling vertically downward at a speed of roughly 0.994 as measured in the earth's reference frame. We place one such detector at the top of a mountain and count the number of muons that it sees per unit time, and another at the foot of the mountain 1907 m ($\approx$ 6.36 μs of distance) lower and count the number of muons that it sees per unit of time.

Description of a muon experiment that tests the metric equation

Let's follow a single muon that happens to go through both detectors. Let event A be the event of this muon passing through the upper detector, and event B be this muon passing through the bottom detector. The distance Δd between these events in the earth's frame of reference is 6.36 μs in SR units. The coordinate time interval between these events measured in the earth's frame is simply the time required for a muon traveling at a speed of 0.994 to traverse this distance: $\Delta t = \Delta d/v = 6.36\,\mu\text{s}/0.994 = 6.40\,\mu\text{s}$.

Note that since the muon is present at each of these events by definition, and moves between them at a constant velocity of 0.994 downward, the clock inside this muon measures the spacetime interval Δs between the events. If the newtonian conception of time were true, all clocks would measure the same time interval between two events, implying that $\Delta s = \Delta t = 6.40\,\mu\text{s}$. This corresponds to 6.40 μs/1.52 μs ≈ 4.21 muon half-lives, so most of the hypothetical muon's co-moving siblings that make it through the top detector would decay before reaching the bottom detector. Specifically, if N muons make it through the top detector, we expect to see N times $(\frac{1}{2})^{4.21} \approx N/18.5$ make it to the bottom detector if the newtonian assumption about time is true.

Exercise R4X.2

Check that $(\frac{1}{2})^{4.21} \approx 1/18.5$.

But if the *metric* equation is true, the spacetime interval between the events would actually be $\Delta s^2 = (\Delta t^2 - \Delta s^2)^{1/2} \approx [(6.40\,\mu\text{s})^2 - (6.36\,\mu\text{s})^2]^{1/2} \approx 0.714\,\mu$s. This time, which is the time that our muon (and its moving siblings) measures between the events, is only 0.714 μs/1.52 μs ≈ 0.47 of a muon half-life, meaning that the internal clocks in most of the muons will *not* signal that it is time to decay before they reach the bottom detector. Specifically, if N muons pass through the upper detector, then you can show that about $N/1.38$ should make it to the bottom detector.

Exercise R4X.3

Verify this last statement.

In summary, the newtonian conception of time leads to the prediction that the ratio of the number of muons passing through the upper detector to the number passing through the lower detector should be 18.5, while the metric equation predicts that the same ratio should be 1.38. This is a substantial difference that can be easily measured.

This experiment was done in the early 1960s by D. H. Frisch and J. B. Smith (*Am. J. Phys.*, vol. 31, p. 342, 1963). They reported observing the ratio to be 1.38 (within experimental uncertainties), thus confirming the metric equation (and utterly refuting the newtonian conception of time by a substantial margin).

Actual results support the metric equation

R4.5   Spacetime Is *Not* Euclidean

We have found the analogy between ordinary euclidean plane geometry and spacetime geometry to be very illuminating, and this basic analogy will remain fruitful as we continue to develop the consequences of the theory of relativity. Nevertheless, it is important at this point to describe some of the important differences between euclidean geometry and spacetime geometry that are a result of the minus signs in the metric equation $\Delta s^2 = \Delta t^2 - \Delta x^2 - \Delta y^2 - \Delta z^2$ that do not appear in the corresponding pythagorean theorem equation $\Delta d^2 = \Delta x^2 + \Delta y^2$.

One important difference concerns the representation of distances on a map and spacetime intervals on a spacetime diagram. If one prepares a scale drawing (e.g., a map) of various points in a town, the distance between any two points on the map is *proportional* to the actual distance between those points in space. That is, distances on the drawing directly correspond to distances in the physical reality being represented. In figure R4.6a, for example, to determine the distance between City Hall and the Statue of the Unknown Physicist, one need merely measure the distance (in inches) between the two sites on the map shown above and multiply by the conversion factor (1000 m = 1 in.). It doesn't matter how the line between the two sites is oriented or where the sites are located on the drawing: the distance in the physical space being represented by the map is always proportional to the distance measured on that map.

However, it is *not* true that the displacement between two points on a spacetime diagram is proportional to the spacetime interval between the corresponding events. In fact, the spacetime interval between two events separated in space can even be zero (see figure R4.6b)!

A spacetime diagram thus may accurately display the spacetime *coordinates* of various events, but the distances between the points representing those events on the diagram are *not* proportional to the actual spacetime intervals between those events in spacetime.

This is very strange, and it may seem particularly strange that two events (such as *A* and *C* in figure R4.6b) can occur at different places and times and yet have zero spacetime interval between them. Nonetheless, there exists a useful analogy with something you may have seen before. Imagine a map of the world where the lines of longitude and latitude are drawn as equally spaced straight lines (see figure R4.7). Have you ever noticed how the shapes and sizes of the continents appear very warped near the north and south poles on such maps? For example, look at the continent of Antarctica on the map. It looks huge and seems to be shaped like a strip. But in fact it is not so large, and it has a nearly *circular* shape: its size and shape are quite distorted by the nature of the map. The shapes of Greenland and northern Canada are distorted as well. As a matter of fact, the two points marked *a* and *b* on the map are both at 90° north latitude, that is, at the *north pole*. Thus though these points are separated by a significant distance on the map, the physical distance between these points on the surface of the earth is zero!

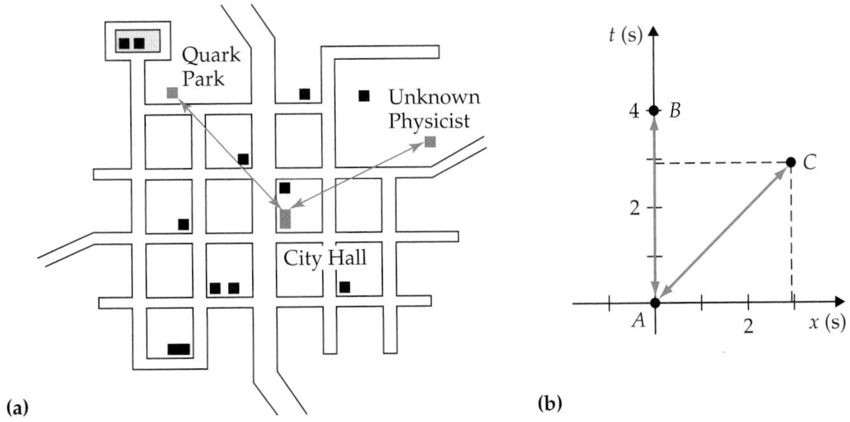

The distance between two points on a map of a flat space is proportional to actual distance in space

Distance between events on a spacetime diagram is *not* proportional to Δs

Analogy: distances on a flat map of curved earth

(a)

(b)

Figure R4.6
(a) A scale drawing (map) of Askew (1 in. = 1 km). The actual distance between City Hall and the Statue of the Unknown Physicist is 852 m. The distance between City Hall and the particle accelerator ride at Quark Park is also 852 m. The length of both double arrows on the map is thus 0.852 in. (b) Both events *B* and *C* are the same distance from event *A* on the spacetime diagram. The spacetime interval between *A* and *B* is 4 s, while the spacetime interval between *A* and *C* is *zero* (since $\Delta t = \Delta x$)!

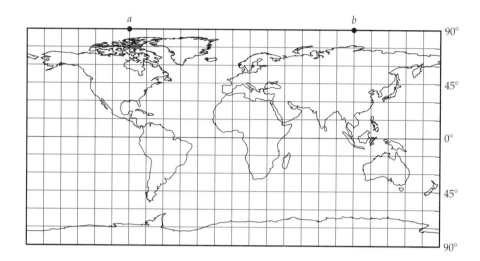

Figure R4.7
A Platte Carre projection map of the world, where the lines of longitude are represented as equally spaced straight lines.

The problem is that the geometry of the earth is not the same as that of map

Why does this map not accurately represent the distances between points on the earth's surface? The problem is that the surface of the earth as a whole is the surface of a *sphere*, which has a very different geometry from the euclidean geometry of a flat sheet of paper. For example, on a sheet of paper the interior angles of a triangle always add to 180°, and parallel lines never intersect. But on the surface of the earth, the interior angles of a triangle add to *more* than 180° (consider a triangle with one vertex at the north pole and two vertices at the equator), and initially parallel lines may converge or diverge (consider lines of longitude, which are parallel at the equator!). Because of these fundamental geometric differences between the surface of the earth and the sheet of paper, any flat map of the earth will *necessarily* be a distorted representation: one cannot make a map of the surface of the earth on a flat sheet of paper and have distances on the sheet correspond to actual distances on the earth.

Similarly, the geometries of spacetime and a spacetime diagram are different

Similarly, one cannot draw a spacetime diagram in such a way that distances between points on the drawing are proportional to the spacetime intervals between the corresponding events. Like the surface of the earth, spacetime has a different geometry from that of the flat sheet of paper on which a spacetime diagram is drawn. The minus signs in the metric equation where the corresponding pythagorean relation has plus signs is symptomatic of this difference.

The moral of the story is, Don't expect a spacetime diagram to give you *direct* information about the spacetime interval between events, any more than you would expect a flat map of the earth to give you accurate information about distances on the earth's surface. A spacetime diagram visually represents the *coordinates* of events and the worldlines of particles, nothing more. You can always compute the spacetime interval between two events from their coordinates, if necessary.

R4.6   More About the Geometric Analogy

In spite of the issue raised in section R4.5, we can further extend the analogy between the geometry of a plane and the geometry of spacetime by exploring the similarities (as well as differences) in how the metric equation describes the geometry of spacetime and how the pythagorean theorem describes the geometry of a plane.

When viewed in different coordinate systems, point B always lies somewhere on a *circle* around point A

The most important thing about both these equations is that they enable us to calculate an absolute quantity (Δs or Δd) in terms of frame-dependent

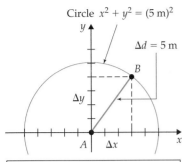

(a)

In this coordinate system, points A and B happen to have coordinate separations $\Delta y = +4$ m, $\Delta x = +3$ m.

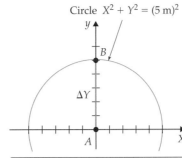

(b)

We can find a coordinate system in which A and B lie along the vertical axis (that is, $\Delta X = 0$). In this unique system, the coordinate separation ΔY is equal to the distance between the points.

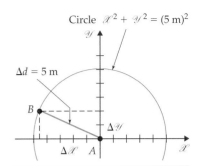

(c)

If we twist the axes further clockwise relative to the underlying space, we can find a coordinate system where $\Delta\mathcal{Y} = 2$ m and $\Delta\mathcal{X} = -4.6$ m. In *any* system, B will lie *somewhere* along the circle shown.

Figure R4.8

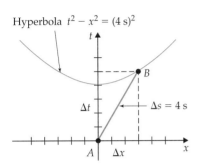

(a)

In this inertial frame, the events A and B happen to have coordinate separations $\Delta t = +5$ s, $\Delta x = +3$ s.

(b)

We can find an inertial frame in which A and B occur at the same place (that is, $\Delta X = 0$). In this unique system, the coordinate time ΔT is equal to the space-time interval Δs between the points.

(c)

If we check frames moving even faster to the right relative to the original frame, we can find an inertial reference frame where $\Delta\mathcal{T} = +6.4$ s and $\Delta\mathcal{X} = -5$ s. In any frame, B will lie *somewhere* along the hyperbola shown.

Figure R4.9

coordinate differences measured in an arbitrary inertial frame or coordinate system. This similarity is illustrated in figures R4.8 and R4.9. Figure R4.8 shows the same pair of points on the plane (A and B) that are 5 m apart, plotted in various coordinate systems having different orientations with respect to "north." Note that if we set up the coordinate systems so that point A is at the origin, then point B in each coordinate system lies somewhere on the circle defined by the equation $x^2 + y^2 = \text{constant} = \Delta d^2$, where Δd^2 is the squared distance between the points (since Δd is the distance between the points in all coordinate systems). Note that in these drawings, I have kept the axes of each coordinate system vertical and horizontal, and rotated the space containing the points A and B "underneath" these coordinate axes: please understand that the points A and B are meant to be the *same* physical points in all the diagrams.

Similarly, figure R4.9 shows a pair of events (A and B) separated by a spacetime interval of 4 s, plotted on spacetime diagrams drawn by observers in different inertial frames. If we choose the origin event in these frames so that event A occurs at $t = x = 0$, then event B lies somewhere on the curve defined by $t^2 - x^2 = \text{constant} = \Delta s^2$, where Δs^2 is the squared spacetime interval between the events (since Δs has a frame-independent value): such a

When viewed in different reference frames, event B lies somewhere on a *hyperbola* about event A

curve is a *hyperbola*, as shown. (Note that we are assuming that $\Delta y = \Delta z = 0$ for these two events.) Again, remember, that these spacetime diagrams are meant to show how different observers would plot the *same* physical events A and B on their diagrams.

The point is that the set of all points a given distance from the origin on the plane form a circle: the set of all events a given spacetime interval from the origin event in spacetime form a hyperbola. The reason that both curves aren't circles is the minus sign in the metric equation that doesn't appear in the pythagorean relation. But there is a nice one-to-one correspondence between circles in plane geometry and hyperbolas in spacetime geometry.

Comparing the magnitudes of the distance and spacetime interval with coordinate separations

Note that one consequence of the difference between the metric equation and the pythagorean relation is that in figure R4.8, we see that the north-south coordinate separation between a pair of points is always *less* than or equal to the distance between the points (the "hypotenuse" on the diagram): $\Delta y \leq \Delta d$. In figure R4.9, though, we see that the coordinate time Δt between a pair of events is always *greater* than the spacetime interval Δs between them: $\Delta t \geq \Delta s$, even though the "hypotenuse" that represents Δs on the diagram *looks* larger.

R4.7   Some Examples

Example R4.1

Problem   A firecracker explodes. A second firecracker explodes 25 ns away and 52 ns later, as measured in the Home Frame. In another inertial frame (the Other Frame), the two explosions are measured to occur 42 ns apart in space. How long a time passes between the explosions in the Other Frame?

Model   The key in this problem is to recognize that the spacetime interval between the two explosion events is frame-independent. That is, if we calculate it using the metric equation in the Home Frame, we must get the same answer that we would get if we calculated it in the Other Frame. That is,

$$\Delta t^2 - \Delta d^2 = \Delta s^2 = (\Delta t')^2 - (\Delta d')^2 \tag{R4.6}$$

Solution   Solving this equation for the unknown $\Delta t'$, we get

$$(\Delta t')^2 = \Delta t^2 - \Delta d^2 + (\Delta d')^2 = (52 \text{ ns})^2 - (25 \text{ ns})^2 + (42 \text{ ns})^2 = 3800 \text{ ns}^2$$
$$\Delta t' = \sqrt{3800 \text{ ns}^2} \approx 62 \text{ ns} \tag{R4.7}$$

Exercise R4X.4

Imagine that two events that are separated by 30 ns of distance in the Home Frame are also simultaneous in that frame. If in the Other Frame, the events are separated by 10 ns of time, what must their spatial separation in the Other Frame be?

Example R4.2

Problem   A certain physics professor fleeing the wrath of a set of irate students covers the length of the physics department hallway (a distance of about 120 ns) in a time of 150 ns as measured in the frame of the earth.

Assuming that the professor moves at a constant velocity, how much time does the professor's watch measure during the trip from one end of the hallway to the other?

Model   Part of the trick in many relativity physics problems is to rephrase a word problem in terms of *events*. In this case, let event A be the professor entering the hallway and event B be the professor's hasty departure from the other end. In the reference frame of the earth, these events occur a time $\Delta t = 150$ ns apart and a distance $\Delta d = 120$ ns apart. The professor's watch, however, is present at each of the events, so that watch registers the *spacetime interval* between these two events.

Solution   Therefore, by the metric equation, the professor's watch reads

$$\Delta s^2 = \Delta t^2 - \Delta d^2 = (150 \text{ ns})^2 - (120 \text{ ns})^2 = 8100 \text{ ns}^2$$
$$\Delta s = \sqrt{8100 \text{ ns}^2} = 90 \text{ ns} \tag{R4.8}$$

Example R4.3: A First Glance at the Twin Paradox

Problem   A spaceship departs from our solar system and travels at a constant speed to the star Alpha Centauri 4.3 light-years away, then instantaneously turns around (never mind about the impossible accelerations involved) and returns to the solar system at the same constant speed. Assume that the trip takes 13 years as measured by clocks here on earth. How long does the trip take as measured by clocks on the spaceship?

Model   Again, we need to translate the word problem into a problem about measuring the time between events. Let event A represent the ship's departure from the starbase, event B its arrival at Alpha Centauri, and event C its return to the solar system (see figure R4.10). A clock in the spaceship does *not* measure the spacetime interval between events A (departure from the solar system) and C (return to the solar system) even though the clock is present at both events. This is because the clock is accelerated when the spaceship turns around, and so the clock is not inertial. To find the total elapsed time registered on the ship clock, we can, however, consider each leg of the trip separately. The ship's clock *does* measure the spacetime interval between events A and B, and it also measures the spacetime interval between events B and C, as it is inertial during each individual leg of the trip and is present at the events in question. The total time registered by the ship's clock is thus the sum of the spacetime intervals between A and B and between B and C.

We can use the metric equation to compute these spacetime intervals from the coordinate differences for these events measured in the earth's frame. Events A and B occur $\Delta t = 6.5$ y apart in time and $\Delta d = 4.3$ y apart in space.

Solution   The spacetime interval between these events is

$$\Delta s_{AB} = \sqrt{(6.5 \text{ y})^2 - (4.3 \text{ y})^2} \approx 4.9 \text{ y} \tag{R4.9}$$

The spacetime interval between events B and C is the same. The total elapsed time for the trip as measured by a clock on the ship is thus $2(4.9 \text{ y}) = 9.8 \text{ y}$, which is somewhat shorter than the time of 13 y measured by clocks on earth.

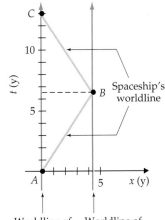

Figure R4.10
Spacetime diagram of a trip to Alpha Centauri and back.

Note that the line on the diagram connecting points A and B looks *longer* than 6.5 y, but the spacetime interval that this line represents is actually *shorter* than 6.5 y. This is an illustration of the issue discussed in section R4.6.

Exercise R4X.5

If spaceship clocks measure only 2 y for a round trip to Alpha Centauri, how long did it take according to clocks on earth?

TWO-MINUTE PROBLEMS

R4T.1   The spacetime interval Δs between two events can never be bigger than the coordinate time Δt between those events as measured in any inertial reference frame, true (T) or false (F)?

R4T.2   Two events occur 5.0 s apart in time and 3.0 s apart in space. A clock traveling at a speed of 0.60 can be present at both these events. What time interval will such a clock measure between the events?
  A.   8.0 s
  B.   5.8 s
  C.   5.0 s
  D.   4.0 s
  E.   2.0 s
  F.   Other (specify)

R4T.3   The spacetime coordinates of events A and B are shown in the spacetime diagram in figure R4.11. What is the spacetime interval between these events?
  A.   0 s
  B.   2 s
  C.   3 s
  D.   4 s
  E.   5 s
  F.   Other (specify)

R4T.4   The spacetime coordinates of events A and C are shown in this spacetime diagram in figure R4.11. What is the spacetime interval between these events?
  A.   0 s
  B.   2 s
  C.   3 s
  D.   4 s
  E.   5 s
  F.   Other (specify)

R4T.5   The spacetime coordinates of events A and D are shown in this spacetime diagram in figure R4.11. What is the spacetime interval between these events?
  A.   0 s
  B.   2 s
  C.   3 s
  D.   4 s
  E.   5 s
  F.   Other (specify)

R4T.6   Consider the spacetime diagram below. Let the spacetime interval between events O and A be Δs_{OA}, and let the spacetime interval between events O and B be Δs_{OB}. Which of these two spacetime intervals is larger? (Assume that the y and z coordinates of all these events are zero.)

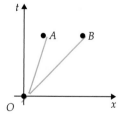

  A. $\Delta s_{OA} > \Delta s_{OB}$
  B. $\Delta s_{OA} < \Delta s_{OB}$
  C. $\Delta s_{OA} = \Delta s_{OB}$
  D.   There is no way to tell from this diagram.

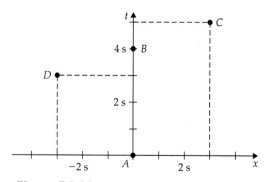

Figure R4.11
Diagram for problems R4T.3 through R4T.5.

R4T.7   An inertial clock present at two events always measures a *shorter* time than a pair of synchronized clocks in *any* inertial reference frame would register between the same two events (as long as the events don't occur at the same place in that frame), T or F?

HOMEWORK PROBLEMS

Basic Skills

R4B.1 In a certain inertial reference frame, two events are separated in time by $\Delta t = 25$ ns and by $\Delta d = 15$ ns in space. What is the spacetime interval between these events?

R4B.2 In the reference frame of the solar system, two events are separated by 5.0 h of time and 4.0 h of distance. What is the spacetime interval between these events?

R4B.3 In the Home Frame, two events are observed to occur with a spatial separation of 12 ns and a time coordinate separation of 24 ns.
(a) An inertial clock travels between these events in such a manner as to be present at both events. What time interval does this clock read between the events?
(b) What is the speed of this clock, as measured in the Home Frame?

R4B.4 An alien spaceship moving at a constant velocity goes from one end of the solar system to the other (a distance of 10.5 h) in 13.2 h as measured by clocks on earth. What time does a clock on the spaceship read for the passage? (*Hint:* Rephrase in terms of events.)

R4B.5 A space probe journeys at a constant velocity from earth to Tau Ceti (a distance of 11.3 y) in a time of 1.0 y as measured by the probe's internal clock. How long did the trip take according to clocks on earth?

R4B.6 In the Home Frame, two events are measured to occur 500 ns apart in time and 300 ns apart in space. In an Other Frame, these events are found to occur 400 ns apart in time. What is the spatial separation of these events in the Other Frame?

R4B.7 In the solar system frame, two events are measured to occur 3.0 h apart in time and 1.5 h apart in space. Observers in an alien spaceship measure the two events to be separated by only 0.5 h in space. What is the time separation between the events in the alien's frame?

Synthetic

R4S.1 Imagine that you and a friend are riding in trains that are moving relative to each other at relativistic speeds. As you pass each other, you both measure the time separation and spatial separation of two firecracker explosions that occur on the tracks between you. (You can measure the latter by measuring the distance between the scorch marks that the explosions leave on the side of your train.) You find the firecracker explosions to be separated by 1.0 μs of time and 0.40 μs of distance in your frame. By radio, your friend reports that the explosions were separated by only 0.60 μs of time in your friend's frame? Is this possible? If it is, find the spatial separation of the events in your friend's frame. If not, explain why not.

R4S.2 A muon is created by a cosmic-ray interaction at an altitude of 60 km. Imagine that after its creation, the muon hurtles downward at a speed of 0.998, as measured by a ground-based observer. After the muon's "internal clock" registers 2.0 μs (which is a bit longer than the average life of a muon), the muon decays.
(a) If the muon's internal clock were to measure the same time between its birth and death as clocks on the ground do (i.e., special relativity is not true and time is universal and absolute), about how far would this muon have traveled before it decayed?
(b) How far will this muon *really* travel before it decays?

R4S.3 A spaceship travels from one end of the Milky Way galaxy to the other (a distance about of 100,000 y) at a constant velocity of magnitude $v = 0.999$, as measured in the frame of the galaxy. How much time does a clock in the spaceship register for this trip? (*Hint:* Rephrase this problem in terms of events.)

R4S.4 In one inertial frame (the Home Frame), two events are observed to occur at the same place but $\Delta t = 32$ ns apart in time. In another inertial frame (the Other Frame), the same two events are observed to occur 45 ns apart in space.
(a) What is the coordinate time interval between the events in the Other Frame?
(b) Compute the speed of the Home Frame as measured by observers in the Other Frame. (*Hint:* The events occur at the same place in the Home Frame. So how far does the Home Frame appear to move in the time between the events as seen in the Other Frame? What is the time between the events in the Other Frame?)

R4S.5 The new earth-Pluto SuperShuttle line boasts that it can take you between the two planets (which are about 5.0 h apart at the time) in 2.5 h (according to your watch). Assume that acceleration and deceleration periods are very brief so that you spend essentially all the trip traveling at a constant velocity.
(a) What time interval must the synchronized space station clocks register between the shuttle's departure from earth and its arrival at Pluto if the advertisement is true?
(b) What is the shuttle's cruising speed?

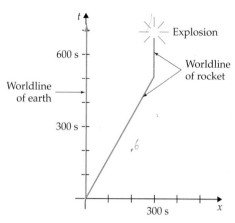

Figure R4.12

Spacetime diagram of the rocket discussed in problem R4S.6.

R4S.6   The spacetime diagram in figure R4.12 shows the worldline of a rocket as it leaves the earth, travels for a certain time, comes to a stop, and then explodes.
(a) What is the elapsed time between the rocket's departure and explosion, as measured by a clock on the rocket?
(b) What is the spacetime interval between these two events?

R4S.7   Imagine that a certain unstable subatomic particle decays with a half-life at rest of about 2.0 μs (that is, if at a certain time you have N such particles at rest, about 2.0 μs later you will have $N/2$ remaining). We can think of a batch of such particles as being a clock that registers the passage of time by the decreasing number of particles in the batch. Now imagine that with the help of a particle accelerator, we manage to produce in the laboratory a beam of these particles traveling at a speed $v = 0.866$ in SR units (as measured in the laboratory frame). This beam passes through a detector A, which counts the number of particles passing through it each second. The beam then travels a distance of about 2.08 km to detector B, which also counts the number of particles passing through it, as shown in figure R4.13.
(a) Let event A be the passing of a given particle through detector A and event B be the passing of the same particle through detector B. How much time will a laboratory observer measure between these events? (*Hint:* you don't need to

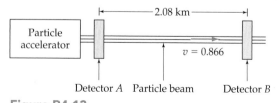

Figure R4.13

Laboratory setup for the particle decay experiment described in problem R4S.7.

know anything about relativity to answer this question!)
(b) How much time passes between these events as measured by the clock inside the particle, according to relativity theory? If relativity is true, about what fraction of the particles that pass through detector A survives to pass through detector B?
(c) According to the newtonian concept of time, the time measured by a particle clock between the events would be the same as the time measured by laboratory clocks. If this were so, what fraction of the particles passing through detector A survives to detector B?

Rich-Context

R4R.1   In 2095 a message arrives at earth from the growing colony at Tau Ceti (11.3 y from earth). The message asks for help in combating a virus that is making people seriously ill (the message includes a complete description of the viral genome). Using advanced technology available on earth, scientists are quickly able to construct a drug that prevents the virus from reproducing. You have to decide how much of the drug can be sent to Tau Ceti. The space probes available on short notice could either boost 200 g of drug (in a standard enclosure) to a speed of 0.95, 1 kg to a speed of 0.90, 5 kg to a speed of 0.80, or 20 kg to a speed of 0.60 relative to the earth. The only problem is that a sample of the drug in a standard enclosure at rest in the laboratory is observed to degrade due to internal chemical processes at a rate that will make it useless after 5.0 y. Is it possible to send the drug to Tau Ceti? If so, how much can you send?

R4R.2   Imagine that your boss is on the earth-Pluto shuttle, which travels at a constant velocity of 0.60 straight between earth and Pluto, a distance of 5.0 h in an inertial frame attached to the sun. An hour into the flight (according to your boss's watch) your boss sends a laser message to you on earth, asking you to send a wake-up call appropriately timed so that your boss can catch a 1-h nap (as measured on your boss's watch). You *immediately* reply with the wake-up call and an apology that the call is late, claiming in your defense that the laws of physics prevented a timely response.
(a) Explain why it is impossible to carry out your boss's request.
(b) How long has your boss slept (according to your boss's watch) by the time your message is received at the shuttle? Explain carefully. (*Hint:* Draw a spacetime diagram of the situation.)

Advanced

R4A.1   Just as we can describe the relationship between the hypotenuse of a triangle and the coordinate lengths

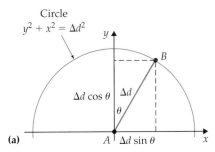

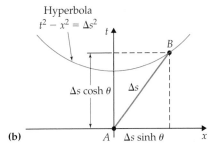

Figure R4.14

(a) Plane trigonometry. (b) Spacetime trigonometry. (Note that $\Delta s \cosh \theta > \Delta s$, even though the "hypotenuse" representing Δs *looks* bigger on the diagram.)

of its sides by using the sine and cosine functions, it turns out that we can describe the relationship between the spacetime interval Δs between two events in terms of the coordinate separations Δt and Δx between those events in terms of the hyperbolic sine and cosine functions. The *hyperbolic sine* and *hyperbolic cosine* functions of a quantity θ are defined as follows:

$$\sinh \theta = \tfrac{1}{2}(e^\theta - e^{-\theta}) \qquad \cosh \theta = \tfrac{1}{2}(e^\theta + e^{-\theta})$$

$$(R4.10)$$

(a) Prove that $\cosh^2 \theta - \sinh^2 \theta = 1$. This means that if the spacetime interval between two events occurring along the spatial x axis is Δs, then the coordinate separations Δt and Δx between these events can be written $\Delta t = \Delta s \cosh \theta$ and $\Delta x = \Delta s \sinh \theta$ for some appropriately chosen value of θ, just as in plane geometry $\Delta x = \Delta d \cos \theta$ and $\Delta y = \Delta d \sin \theta$ (see figure R4.14).

(b) Argue that θ in the hyperbolic case is *not* the angle that line AB makes with the t axis in the spacetime diagram of figure 4.13b. Argue in fact that as $\theta \to \infty$, the angle that AB makes with the t axis approaches $45°$.

(c) Argue that if v is the speed of an object that goes from event A to event B at a constant velocity, the "angle" θ is in fact $\tanh^{-1} v$.

(d) When $v = 0.80$, $\theta = 1.10$ (if your calculator can do inverse hyperbolic functions, verify this). What are the values of $\cosh \theta$ and $\sinh \theta$ for this value of θ? (Use the definitions of these functions given above if your calculator cannot evaluate hyperbolic functions). When $v = 0.99$, we have $\theta = 2.65$ (again, verify if you can). What are the values of $\cosh \theta$ and $\sinh \theta$ for this value of θ?

(e) Argue that as $\theta \to 0$, $\sinh \theta \to 0$ while $\cosh \theta \to 1$, just like the corresponding trigonometric functions. This also means that $\tanh \theta \to 0$ in this limit. Use this to argue that as $v \to 0$, $\Delta s \to \Delta t$ and $\Delta x \to 0$.

ANSWERS TO EXERCISES

R4X.1 Measuring sticks placed parallel to the line of relative motion will simply paint stripes down each other's length. There is no way to extract information about the length of a given stick from such stripes.

R4X.2 (Just type it into your calculator.)

R4X.3 If N muons travel through the top detector, the number that make it through the bottom detector after 0.47 times a muon half-life have passed will (by analogy to the calculation above exercise R4X.2) be $N(\tfrac{1}{2})^{0.47} = 0.722N = N/1.385$.

R4X.4 In this situation, we have $\Delta t = 0$ and $\Delta d = 30$ ns. Since we must have $(\Delta t')^2 - (\Delta d')^2 = \Delta t^2 - \Delta d^2$, and since we are given that $\Delta t' = 10$ ns, we can find the spatial separation $\Delta d'$ between the events in the Other Frame by solving the equation above for $\Delta d'$:

$$\Delta d' = \sqrt{(\Delta t')^2 - \Delta t^2 + \Delta d^2}$$

$$= \sqrt{(10 \text{ ns})^2 - 0 + (30 \text{ ns})^2} = 31.6 \text{ ns} \quad (R4.11)$$

R4X.5 If the round trip takes a total proper time of 2.0 y as measured on the ship, then each constant-velocity leg of the trip will take $\Delta \tau = \Delta s = 1.0$ y. Since we know that the distance between the trip's endpoints in the solar system frame is $\Delta d = 4.3$ y, we can find Δt for one leg of the trip by solving the metric equation $\Delta s^2 = \Delta t^2 - \Delta d^2$ for Δt:

$$\Delta t = \sqrt{\Delta s^2 + \Delta d^2} = \sqrt{(1.0 \text{ y})^2 + (4.3 \text{ y})^2} = 4.4 \text{ y}$$

$$(R4.12)$$

This means that the round trip takes about 8.8 y according to observers in the solar system frame, and the speed of the spaceship in that frame is $\Delta d/\Delta t \approx 4.3 \text{ y}/4.4 \text{ y} \approx 0.98$.

R5

Proper Time

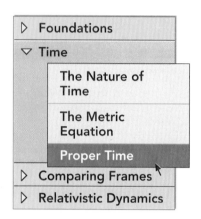
Chapter Overview

Introduction

In chapter R4, we derived the *metric equation*, which links the coordinate differences between two events measured in a given reference frame to the frame-independent spacetime interval between those events. In this chapter, we will learn how to use the metric equation to calculate the proper time along *any* worldline.

Section R5.1: A Curved Footpath

Imagine that we know the function $x(y)$ that describes a certain path on a two-dimensional plane. We can compute the pathlength ΔL_{AB} between any two points A and B along that path as follows. First, we divide the path into very small segments whose endpoints are coordinate displacements of dx and dy apart. Second, we compute the length of each segment, using the pythagorean theorem: $dL = (dx^2 + dy^2)^{1/2} = [(dx/dy)^2 + 1]^{1/2}\, dy$. Finally, we sum over all segments to find the total pathlength. In the limit that the segments become infinitesimally short, the sum becomes an integral.

Section R5.2: Curved Worldlines in Spacetime

We can do an exactly analogous calculation to find the proper time between two events A and B as measured by a clock following an arbitrary worldline. Imagine that we know the functions $x(t)$, $y(t)$, and $z(t)$ that describe the clock's position as a function of time as it moves along the worldline. (1) We divide the worldline into segments so short that the worldline is essentially straight between any segment's endpoints. (2) The proper time along each segment is then essentially equal to the spacetime interval along that segment: $d\tau \approx ds = (dt^2 - dx^2 - dy^2 - dz^2)^{1/2} = [1 - (dx/dt)^2 - (dy/dt)^2 - (dz/dt)^2]^{1/2}\, dt = (1 - v^2)^{1/2}\, dt$. (3) The total proper time along the worldline is the sum of these infinitesimal proper times. In the limit that the segments become infinitesimally short, the sum becomes an integral:

$$\Delta\tau_{AB} = \int_{t_A}^{t_B} (1 - v^2)^{1/2}\, dt \qquad (R5.6)$$

Purpose: This equation describes how we can use measurements performed in an inertial frame to compute the proper time $\Delta\tau_{AB}$ measured by a clock traveling between any two events A and B along an arbitrary worldline.

   Symbols: t_A is the time of event A; t_B is the time of event B; dt is the coordinate time differential; and $v = v(t)$ is the speed (as a function of time) of the clock, all measured in a given inertial reference frame.

   Limitations: This equation only works in an inertial reference frame.

   Note: *If* the clock's speed (not necessarily its velocity) is constant, then this equation reduces to

$$\Delta\tau_{AB} = (1 - v^2)^{1/2}\, \Delta t_{AB} \qquad \text{only if } v = \text{constant} \qquad (R5.7)$$

Section R5.3: The Binomial Approximation

The **binomial approximation** asserts that

$$(1+x)^a \approx 1 + ax \qquad \text{if } x \ll 1 \qquad \text{(R5.17)}$$

Purpose: This equation is a useful trick for simplifying calculations when we are trying to compute an arbitrary power a of a very small quantity x added to 1.

Limitations: You will not get very good results if $|ax|$ is much larger than 0.1.

Note: In this chapter, we will most often use this approximation to get

$$(1-v^2)^{1/2} = [1+(-v^2)]^{1/2} \approx 1 - \tfrac{1}{2}v^2 \qquad \text{(R5.18)}$$

This approximation usually works best when the problem is phrased (or the answer can be phrased) so that the 1 cancels out; but in the *worst* case, one can calculate $1 - ax$ by hand, something that is usually impossible for $(1+x)^a$.

Section R5.4: The Spacetime Interval Is the Longest Possible Proper Time

This section presents a proof of the following statement: The proper time measured by a clock traveling between two events A and B is *longest* if the clock follows a straight (constant-velocity) worldline.

Since the metric equation implies that $\Delta t^2 = \Delta s^2 + \Delta d^2 \geq \Delta s^2$ in any arbitrary reference frame, we have in general

$$\Delta t \geq \Delta s \geq \Delta \tau \qquad \text{(R5.21)}$$

Purpose: This equation describes the hierarchical relationship between the coordinate time Δt between two events (measured in any arbitrary inertial frame), the spacetime interval Δs between those events, and the proper time $\Delta \tau$ between those events (measured along any arbitrary worldline going between them).

Limitation: This equation applies only if $\Delta s^2 \geq 0$.

Note: The equality $\Delta t = \Delta s$ applies if Δt is measured where the events occur at the same place; the equality $\Delta s = \Delta \tau$ applies if $\Delta \tau$ is measured along a straight worldline.

Section R5.5: Experimental Evidence

This section discusses two relatively recent experiments that have tested the validity of equation R5.6.

Section R5.6: The Twin Paradox

Consider two twins. One leaves earth, travels to a distant star at a speed close to that of light, and then returns. A naive use of equation R5.6 might lead each twin to conclude that the other is younger (since each considers him- or herself to be at rest while the other is moving): this is the **twin paradox**. However, the twin's situations are not really symmetric: the traveling twin is *not* in an inertial reference frame but the twin on earth is (at least approximately). Therefore, the earth-based twin can legally use equation R5.6 (at least approximately), whereas the traveling twin cannot. The traveling twin really ends up being younger than the earth-based twin.

R5.1   A Curved Footpath

The metric equation $\Delta s^2 = \Delta t^2 - \Delta x^2 - \Delta y^2 - \Delta z^2$ connects the spacetime coordinate differences between two events measured in some inertial reference frame to the spacetime interval Δs between the same events measured by an inertial clock present at both events. In this section and section R5.2, we will use the metric equation to connect coordinate time in a given inertial frame to the proper time $\Delta \tau$ measured by *any* clock present at both events, inertial or not.

One might think that we would have to use the theory of *general* relativity to properly analyze the behavior of accelerating (noninertial) clocks. In fact, we can quite adequately analyze the behavior of such clocks by using only the metric equation if we remember the analogy between the proper time between events in spacetime and the pathlength between two points on a plane.

Consider a footpath around a small pond. Figure R5.1a shows a scale drawing of the path with a superimposed coordinate system. We could measure the length of this path from point A to point B with a long, flexible tape measure. But once we have set up a coordinate system, we can also *compute* the length of the path in the following manner. Imagine dividing up the path into a large number of infinitesimally small sections, as shown in figure R5.1b.[†] If we make these sections small enough, each will be approximately straight. In this limit, the pathlength dL of a given segment as measured by a flexible tape measure will be almost equal to the straight length computed (using the pythagorean theorem) from the coordinate differences of the segment's endpoints:

$$dL^2 \approx dx^2 + dy^2 \qquad \text{or} \qquad dL = \sqrt{dx^2 + dy^2} \qquad \text{(R5.1)}$$

The total length L_{AB} of the path from A to B is the sum of all the segment lengths, which in the limit where the segments are truly infinitesimal becomes the integral

$$\Delta L_{AB} = \int_{\text{path}} dL = \int_{\text{path}} \sqrt{dx^2 + dy^2} \qquad \text{(R5.2)}$$

Note that since the length dL of each segment is greater than its northward extension dy, the total pathlength between points A and B will be greater than the straight-line northward distance of 225 m between A and B: we can say quite generally, therefore, that $\Delta L_{AB} \geq \Delta d_{AB}$.

We can describe the path mathematically, using the function $x(y)$: this function specifies the path's x coordinate at each possible y coordinate. If we know this function, we can write the integral above as a single-variable integral over x by pulling a factor of dy out of the square root:

$$\Delta L_{AB} = \int_{y_A}^{y_B} \sqrt{1 + (dx/dy)^2}\, dy \qquad \text{(R5.3)}$$

[Note that we are considering y to be the independent variable and x to be the dependent variable in equation R5.3. This reversal of convention is necessary because $y(x)$ is not well defined for a path shown in figure R5.1.]

As we have discussed before, although this equation uses the coordinates x and y measured in a given coordinate system, the pathlength itself is an invariant quantity: we'll get the same answer (the answer that a flexible tape measure would give) no matter *what* coordinate system we use.

We compute the length of a curved path by breaking it up into tiny straight pieces

Formula for computing the length of a path $x(y)$

[†]The analogy with pathlength presented here follows E. F. Taylor and J. A. Wheeler, *Spacetime Physics*, San Francisco; Freeman, 1963, pp. 32–34.

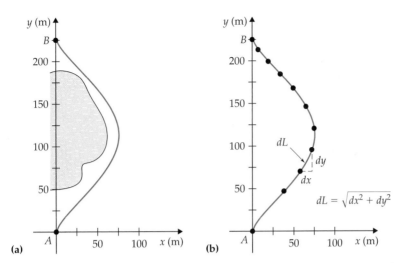

Figure R5.1

(a) Map of a footpath around a small pond, with a superimposed xy coordinate system. (b) We can find the length of the path by subdividing it into many infinitesimal, almost straight sections, finding the length dL of each section, and summing to find the total length L.

R5.2   Curved Worldlines in Spacetime

The analogy to events in spacetime is direct. Consider the worldline of a particle that travels out from the origin of some inertial reference frame a certain distance along the x axis and then returns. Such a worldline is shown on the spacetime diagram in figure R5.2a with the coordinate axes of that frame superimposed. Such a worldline describes an *accelerating* particle; we can see from the graph that the particle's x-velocity $v_x = dx/dt$ (which is the inverse slope of its worldline on the diagram) changes as time progresses.

A clock traveling with the particle measures the proper time $\Delta\tau_{AB}$ between events A and B along this worldline (by the definition of proper time). But once we have measured the worldline of the particle in an inertial reference frame (any inertial frame), we can calculate what this clock will read between events A and B by using the metric equation in a manner analogous to our determination of the pathlength between points A and B in section R5.1.

Imagine that we divide the particle's worldline up into many infinitesimal segments, each of which is nearly a straight line on the spacetime diagram (figure R5.2b). We choose each segment to be short enough that the particle's velocity is approximately constant as it traverses that segment. If this is true, then the proper time $d\tau$ that a clock would measure along each segment will be almost equal to the spacetime interval ds between the events

The situation in spacetime is directly analogous

We compute $\Delta\tau$ along a worldline by dividing the worldline into many tiny straight segments

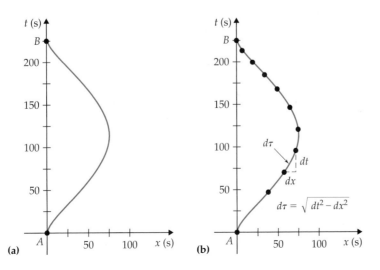

Figure R5.2

(a) A spacetime diagram of the motion of a particle's worldline based on measurements obtained in some inertial reference frame. (b) We can find the proper time along the worldline by subdividing it into many infinitesimal (almost straight) worldlines, finding the proper time dt along each segment, and then summing to find the total proper time. Note that because of the minus sign in the metric equation, $d\tau \leq dt$ here, even though it *looks* like $d\tau \geq dt$ on the diagram.

that mark the ends of the segment, since the clock is present at both these events and travels between them with an *almost* constant velocity. Therefore, by the metric equation

$$d\tau^2 \approx ds^2 = dt^2 - dx^2 - dy^2 - dz^2 \tag{R5.4}$$

If we take the square root and pull out a factor of the coordinate time dt, we find that

$$d\tau = \sqrt{1 - \left(\frac{dx}{dt}\right)^2 - \left(\frac{dy}{dt}\right)^2 - \left(\frac{dz}{dt}\right)^2}\, dt$$

$$= \sqrt{1 - v_x^2 - v_y^2 - v_z^2}\, dt = \sqrt{1 - v^2}\, dt \tag{R5.5}$$

This equation expresses the infinitesimal proper time $d\tau$ measured by a clock traveling between two *infinitesimally separated* events in terms of the coordinate time dt between those events measured in some inertial frame and the clock's instantaneous *speed* v measured in that frame. The clock may be moving along *any* smooth worldline (v does not *have* to be constant).

To find the *total* proper time measured between events A and B by a clock traveling along the worldline, we sum the proper times measured for each nearly straight segment of the worldline, which in the limit of truly infinitesimal segments amounts to integrating equation R5.5:

General formula for computing the proper time along a worldline

$$\Delta\tau_{AB} = \int_{t_A}^{t_B} (1 - v^2)^{1/2}\, dt \tag{R5.6}$$

Purpose: This equation tells us how we can use measurements performed in an inertial frame to compute the proper time $\Delta\tau_{AB}$ measured by a clock traveling between any two events A and B along an arbitrary worldline.

Symbols: t_A is the time of event A; t_B is the time of event B; dt is the coordinate time differential; and $v = v(t)$ is the speed (as a function of time) of the clock, all measured in a given inertial reference frame.

Limitations: This equation only works in an inertial reference frame.

If we know the speed v of the clock as a function of time $v(t)$, then the integral is simply an ordinary one-variable integral with respect to t, which we can evaluate (at least in principle). This equation links the total proper time $\Delta\tau_{AB}$ between two events measured by a clock traveling between events A and B to the events' coordinate times t_A and t_B and to the clock's speed $v(t)$ as a function of coordinate time, as measured in some given (but arbitrary) inertial reference frame. Although we are using an inertial frame to measure v, remember that the *result* of equation R5.6 is frame-independent, since the clock in question measures $\Delta\tau_{AB}$ *directly* without reference to any reference frame.

If the *speed* v of the clock happens to be constant, the integral in equation R5.6 can be done very easily:

Formula for the special case where the object's *speed* is constant

$$\Delta\tau_{AB} = \sqrt{1 - v^2} \int_{t_A}^{t_B} dt = \sqrt{1 - v^2}(t_B - t_A)$$

$$\equiv \sqrt{1 - v^2}\, \Delta t_{AB} \qquad \text{if } v = \text{constant} \tag{R5.7}$$

Please note that *constant speed* here does not necessarily imply *constant velocity*, as the *direction* of a particle's velocity may change without changing its speed. Thus equation R5.7 can be applied to clocks traveling along straight or curved worldlines, as long as the speed of the clock remains fixed. Equation R5.6 must be used whenever the speed changes.

Note that since $(1 - v^2)^{1/2} \leq 1$ always, any proper time measured between two events will be *smaller* than (or at most equal to) the coordinate time between those events measured in any inertial frame: $\Delta \tau_{AB} \leq \Delta t_{AB}$ always. However, if the clock's speed is small compared to that of light ($v \ll 1$), that clock will register almost the same time between the events as measured in the inertial frame: $\Delta \tau_{AB} \approx \Delta t_{AB}$. (This is true whether v is constant or not.)

When either equation R5.6 or R5.7 is applied, it is important to remember two things. First, coordinate time Δt and the proper time $\Delta \tau$ represent the time interval between two events measured in two fundamentally different ways (just as the northward displacement Δy and the pathlength ΔL represent two fundamentally different ways of measuring the spatial separation of two points on the earth's surface). The coordinate time between events is measured with a pair of *synchronized clocks* in an inertial frame, while the proper time is measured by a *clock present at both events*. One cannot use these equations to link readings on just any old clocks.

Second, the quantities $\Delta \tau$ and Δt appearing on both sides of equation R5.7 always refer to the time between the *same* pair of events, measured in these two different ways. Perhaps the most common error made by beginners in applying that equation lies in implicitly using *different* pairs of events to delimit the time intervals $\Delta \tau$ and Δt. To avoid this, think *carefully* about the events involved!

Note also that equations R5.6 and R5.7 break down if $v > 1$: in such a case, they predict that the time registered by the traveling clock is an *imaginary* number (which is even worse than being a negative number!). Remember that these equations are all based on the metric equation, whose derivation (see section R4.2) is only valid for pairs of events for which $\Delta t > \Delta d$, that is, events between which it is possible to send a clock traveling with $v < 1$. Therefore, the equations presented so far do not specify what a clock traveling faster than the speed of light would read between two events. We will see in chapter R8 that the principle of relativity in fact implies that it is *impossible* for a clock to travel faster than the speed of light in any reference frame. The failure of these equations for the case where $v > 1$ is our first indication of this.

> Any proper time between two events is smaller than the coordinate time between the same events

> Two things to remember

> The equations for proper time fail when $v > 1$

Example R5.1

Problem   Imagine that you are at rest in an inertial frame (the Home Frame) and you are whirling a clock around your head at a constant rate on the end of a string 3.0 m long. A friend compares the reading of the whirling clock as it speeds by with readings from a stationary clock. Find out how long it takes the whirling clock to go once around its circular path if its reading for one cycle is 0.01 percent smaller than the period read by your friend's clock.

Model   The first step is to rephrase the problem in terms of *events*. Let event A be the whirling clock passing by the stationary clock. Let event B be the next such passage event. The whirling clock measures a proper time between these events. The stationary clock measures a *different* proper time between the events (because it is also present at both events). Since the stationary clock is at rest in the Home Frame, the time that it registers is the same as the coordinate time between the events in the Home Frame. The whirling clock, on the other hand, is noninertial (its velocity is constantly changing direction as it goes around the circle). But since its *speed* is constant, we can use equation R5.7 to find the proper time that it measures between events A and B.

Solution   We are given that the result is 99.99% of the time measured by the stationary clock, so

$$\Delta\tau_{AB,\text{whirl}} = \sqrt{1 - v^2}\,\Delta t_{AB} = 0.9999\,\Delta t_{AB} \tag{R5.8}$$

implying that $\sqrt{1 - v^2} = 0.9999$, or $1 - v^2 = 0.9999^2$, implying that

$$v^2 = 1 - 0.9999^2 \quad\Rightarrow\quad v = \sqrt{1 - 0.9999^2} \approx 0.014 \tag{R5.9}$$

The radius of the circle in seconds is $(3.0\text{ m})(1\text{ s}/3.00 \times 10^8\text{ m}) = 1.0 \times 10^{-8}\text{ s} = 10\text{ ns}$. The coordinate time that a clock traveling at $v = 0.014$ would take to go once around this circle is

$$\Delta t_{AB} = \frac{2\pi R}{v} = \frac{2\pi(1.0 \times 10^{-8}\text{ s})}{0.014} \approx 4.40 \times 10^{-6}\text{ s} \tag{R5.10}$$

This implies frequency of revolution of $1/\Delta t_{AB} \approx 225{,}000$ Hz.

Evaluation   This answer makes it clear that the scenario presented in this problem is completely unrealistic. Yet this speed is what would be necessary to get even a 0.01% difference in the rate of the whirling clock relative to a stationary clock. It is no wonder that we think of time as being universal and absolute!

Exercise R5X.1

If the clock in this situation has a mass of 100 g, estimate the tension force that the string would have to exert on it. Is this realistic?

Example R5.2

Problem   Imagine that the speed of a certain spaceship relative to an inertial frame fixed to the sun is given by $v = at$, where $a = 10\text{ m/s}^2$. How long does it take the ship to accelerate from rest to a speed of 0.5 (in SR units) relative to the sun, as measured by clocks on the *ship*?

Model   Again, the first step is to rephrase the problem in terms of events. Let event A be the event of the ship starting to accelerate and event B the event of its passing $v = 0.5$ (in SR units). Since the ship is present at both events, its clock measures the proper time between them. Since the speed of the ship is *not* constant in this case, we must use equation R5.6 to compute this proper time:

$$\Delta\tau_{AB} = \int_{t_A}^{t_B} \sqrt{1 - v^2}\,dt = \int_{t_A}^{t_B} \sqrt{1 - a^2t^2}\,dt \tag{R5.11}$$

Solution   In spite of the simple form of the equation $v = at$, this is not a simple integral to evaluate. We can put it in a somewhat simpler form by doing the following. First, note that $t_A = 0$, because if $v = at$, then $t = 0$ when the ship is at rest but beginning to accelerate. Second, note that $dv = a\,dt$ in this case, so we can change the variable in the integral from t to v as follows:

$$\Delta\tau_{AB} = \frac{1}{a}\int_0^{t_B}\sqrt{1-a^2t^2}\,a\,dt = \frac{1}{a}\int_0^{0.5}\sqrt{1-v^2}\,dv \qquad (R5.12)$$

This integral now has a simple enough form that we can try to look it up in a table of integrals. My table of integrals says that

$$\int\sqrt{1-v^2}\,dv = \frac{v\sqrt{1-v^2}}{2} + \frac{\sin^{-1}v}{2} \qquad (R5.13)$$

Therefore,

$$\Delta\tau_{AB} = \frac{1}{2a}\left(0.5\sqrt{1-0.5^2} + \sin^{-1}0.5 - 0\sqrt{1-0^2} - \sin^{-1}0\right)$$

$$= \frac{1}{2a}(0.433 + 0.524 - 0 - 0) = \frac{0.478}{a} \qquad (R5.14)$$

To finish the derivation, we have to know what a is in SR units (since equation R5.6 and everything we have done presumes that we are working in SR units). To change $a = 10\text{ m/s}^2$ to SR units, we must convert the meters appearing in this expression to seconds:

$$a = 10\frac{\text{m}}{\text{s}^2}\left(\frac{1\text{ s}}{3.0\times10^8\text{ m}}\right) = \frac{1}{3.0\times10^7\text{ s}} \qquad (R5.15)$$

Plugging this into equation R5.14, we get

$$\Delta\tau_{AB} = 0.478\,(3.0\times10^7\text{ s}) = 1.43\times10^7\text{ s}\left(\frac{1\text{ y}}{3.16\times10^7\text{ s}}\right) \approx 0.453\text{ y} \quad (R5.16)$$

Evaluation   The units come out right, and the answer seems reasonable.

Because integrals in cases where $v \neq$ constant are so difficult, we will generally stick to cases where $v =$ constant in this course. This example is intended mainly to illustrate how a calculation where $v \neq$ constant can be done: you will not often be asked to do anything as difficult as example R5.2.

Exercise R5X.2

Show that $\Delta\tau_{AB}$ found in example R5.2 is 0.957 times the coordinate time Δt_{AB} measured between the events in the sun's frame.

Exercise R5X.3

Imagine that electrons in a certain particle accelerator ring travel in a circular path at a constant speed of 0.98. What is the time that would be measured by a clock traveling with the electrons for one complete cycle around the ring, as a multiple of the time measured by a clock at rest in the laboratory?

R5.3   The Binomial Approximation

The square root that appears in equations R5.6 and R5.7 is rather difficult to evaluate for very small speeds ($v \ll 1$). The speeds of objects we encounter on an everyday basis are on the order of $v = 10^{-8}$ in SR units, meaning

Motivation for the binomial approximation

$v^2 \approx 10^{-16}$. When one tries to evaluate the square root $(1 - v^2)^{1/2}$ in such a case, one's calculator usually simply returns just 1.0, since few calculators keep track of enough decimal places to accurately register the subtraction of 10^{-16} from 1. Such an answer is not really helpful. In such cases, however, there is an approximation we can use to help us convert the square root to a more usable form.

It can be shown (see problems R5S.6 and R5A.2) that

General statement of the
binomial approximation

$$(1 + x)^a \approx 1 + ax \qquad \text{if } x \ll 1 \qquad \text{(R5.17)}$$

Purpose: This equation is a useful trick for simplifying calculations when we are trying to compute an arbitrary power a of a very small quantity x added to 1.
Limitations: You will not get very good results if $|ax|$ is much larger than 0.1.

Equation R5.17, called the **binomial approximation,** is a very useful approximation with a wide range of applications in physics (memorize it!). To evaluate $(1 - v^2)^{1/2}$ for very small v, we identify $x \equiv -v^2$ and $a = \frac{1}{2}$, yielding

$$\sqrt{1 - v^2} \approx 1 + \tfrac{1}{2}(-v^2) = 1 - \tfrac{1}{2}v^2 \qquad \text{(R5.18)}$$

Exercise R5X.4

Using your calculator, check the accuracy of the approximation in equation R5.18 for $v = 0.1, 0.01$, and 0.001.

Example R5.3

Problem   You and a friend stand at a street corner and synchronize your watches. You leave your friend (call this event A) and walk around the block, traveling at a constant speed of 2 m/s (about 4.5 mi/h). After a time $\Delta t_{AB} = 550$ s, as measured by your friend's watch, you return (call this event B). How much less than 550 s does your watch register between events A and B?

Model   Take your friend's frame to be inertial; your friend's watch thus measures the coordinate time between A and B in that frame (also the spacetime interval!). Since you are not moving at a constant *velocity*, you measure a proper time $\Delta\tau_{AB}$ between those events. In your friend's frame, you have a constant *speed* of $v = 2$ m/s (or in SR units, $v \approx 6.7 \times 10^{-9}$), so we can use equation R5.6 to calculate your proper time. Your speed is also extremely small compared to 1, so we can employ the binomial approximation to evaluate the square root.

Solution   The proper time that you measure between events A and B is therefore

$$\Delta\tau_{AB} = \sqrt{1 - v^2}\,\Delta t_{AB} \approx \left(1 - \tfrac{1}{2}v^2\right)\Delta t_{AB} \qquad \text{(R5.19)}$$

Now, if we were to plug $\Delta t_{AB} = 550$ s into this equation, we would *still* find that $\Delta\tau_{AB} = 550$ s to the accuracy of a typical calculator, in spite of our use of the binomial approximation. The entire *point* of the binomial approximation, however, is that it makes it much easier to calculate the *difference* between your time $\Delta\tau_{AB}$ and your friend's time Δt_{AB}, which is what the problem requests. This difference is

$$\Delta\tau_{AB} - \Delta t_{AB} \approx \left(1 - \tfrac{1}{2}v^2\right)\Delta t_{AB} - \Delta t_{AB} = \left(1 - \tfrac{1}{2}v^2 - 1\right)\Delta t_{AB}$$

$$= -\tfrac{1}{2}v^2\,\Delta t_{AB} = -\tfrac{1}{2}(6.7 \times 10^{-9})^2(550\text{ s})$$

$$\approx -1.2 \times 10^{-14}\text{ s} \qquad\qquad (R5.20)$$

meaning that the time is smaller than your friend's time by about 12 fs.

Evaluation If you really want to evaluate $\Delta\tau_{AB}$, you can subtract 1.2×10^{-14} s from 550 s *by hand* to get 549.999999999999988 s. However, since we probably don't know Δt_{AB} to 15 decimal places to begin with, the difference between $\Delta\tau_{AB}$ and Δt_{AB} is not going to be even remotely measurable. Again, it is no wonder that we all intuitively have the idea that time is universal and absolute!

Example R5.3 illustrates that the binomial approximation is most useful when we want to calculate the *difference* between some clock's proper time and coordinate time in a frame where the clock moves very slowly. While we cannot easily calculate an expression like $\Delta\tau_{AB} - \Delta t_{AB} = [(1 - v^2)^{1/2} - 1] \cdot \Delta t_{AB}$ when v is small, we can easily calculate $\Delta\tau_{AB} - \Delta t_{AB} = (1 - \tfrac{1}{2}v^2 - 1) \cdot \Delta t_{AB}$ because the 1s cancel. Try to make the 1s cancel similarly in any problems in which you use the binomial approximation.

If you really do need to calculate $\Delta\tau_{AB} = (1 - v^2)^{1/2}\,\Delta t_{AB}$ directly, though, the binomial approximation is still useful because it is much easier to calculate a simple difference by hand than it is to calculate a square root by hand (and a hand calculation is going to be necessary either way if your calculator yields one when you try to compute the square root).

R5.4   The Spacetime Interval Is the Longest Possible Proper Time

Note that equation R5.6 implies that generally the proper time measured by a clock between two events will indeed depend on the worldline that the clock follows between the events: specifically, the proper time depends on the particular way that the speed v of the clock varies with time. This is analogous to the way that the pathlength between two points on a plane depends on the curvature of the path along which it is measured.

In euclidean geometry, the straight-line distance between two points is the shortest possible pathlength between the two points. In this section, we will prove that an *inertial* clock that travels between two events (which thus measures the spacetime interval Δs between them) measures the longest *possible* proper time between those events, longer than any noninertial clock: $\Delta s \geq \Delta\tau$ *for all possible worldlines between the events.*

How might we prove such a theorem? Consider an arbitrary pair of events, A and B. The time between these events is measured by two clocks that are present at both events. Imagine that clock I follows an inertial worldline between the events, while clock NI follows a noninertial worldline. Since clock I is inertial, its proper time $\Delta\tau_I$ will be equal (by definition) to the

Along a *straight* worldline between two events $\Delta\tau$ is the *longest* possible $\Delta\tau$ between those events

Proof of this theorem

spacetime interval Δs_{AB} between the events. We will take advantage of the fact that we can calculate the proper time for *any* given worldline by using *any* inertial reference frame that we please, since the *result* is frame-independent. With this in mind, it is convenient for us to evaluate the proper times for clocks I and NI in that particular inertial frame where clock I is at *rest*. Since that clock is inertial, the frame in which it is at rest will also be inertial: let's call this the Home Frame.

To compute the proper time along *any* path from event A to event B in this frame, we calculate $\int_{t_A}^{t_B} (1 - v^2)^{1/2}\, dt$, where $v(t)$ and the endpoints t_B and t_A are all measured in the Home Frame. For clock I, $(1 - v^2)^{1/2} = 1$, since that clock is at rest in the Home Frame by construction. Since clock NI travels along a *different* worldline, it must at least *sometimes* have $v \neq 0$ in the Home Frame, so the integrand $(1 - v^2)^{1/2}$ must be *less* than 1 for at least *part* of the range of integration. So $\Delta s_{AB} = \Delta\tau_I = \int_{t_A}^{t_B} 1\, dt > \int_{t_A}^{t_B} (1 - v^2)^{1/2}\, dt = \Delta\tau_{NI}$ (note that both integrals have the same endpoints). Since the values of the proper times $\Delta\tau_I$ and $\Delta\tau_{NI}$ are frame-independent, this inequality must be true no matter *what* inertial frame we use to actually calculate $\Delta\tau_I$ and $\Delta\tau_{NI}$. Q.E.D.

Note that in a spacetime diagram based on measurements made in an inertial reference frame, an inertial clock will have a straight worldline (since it moves with constant velocity with respect to any inertial frame) whereas a noninertial clock will have a curved worldline. The theorem we have just proved thus says that *a straight worldline between any two events on a spacetime diagram is the worldline of greatest proper time between the events.*

Analogy to a straight line
being the shortest distance
between two points

That the spacetime interval between two events in spacetime represents the *longest* proper time between those events while the distance between two points on a plane represents the *shortest* pathlength between those points is direct consequence of the minus signs that appear in the metric equation where only plus signs appear in the corresponding pythagorean relation. This is another of the basic differences between spacetime geometry and the euclidean geometry of points on a plane. Even so, there remains a similarity in that a straight worldline (or path) leads to an *extreme* value for the proper time (or pathlength).

Equation R5.7 implies that the coordinate time Δt between two events measured in any inertial reference frame is greater than (or at minimum equal to) *any* proper time measured between the points (*including* the spacetime interval Δs between those events). The three kinds of time interval that you can measure between two events thus stand in the strict relation:

An important inequality
involving Δt, Δs, and $\Delta\tau$

$$\Delta t \geq \Delta s \geq \Delta\tau \qquad (\text{R5.21})$$

where the first inequality $\Delta t \geq \Delta s$ becomes an equality if the events occur at the *same place* in the inertial reference frame where Δt is measured (so that a single clock in that frame is present at both events) and the second inequality $\Delta s \geq \Delta\tau$ becomes an equality if the clock measuring the proper time follows an *inertial path* between the events (so that the proper time that it measures between the events *is* the spacetime interval between them as well).

Exercise R5X.5

Alan and Beth operate synchronized clocks at two points on the earth's surface. Clarissa rides a train that travels at a constant speed between these two points: let the event of her departure from Alan's clock be event A and her arrival at Beth's clock be event B. Dave, on the other hand, pilots a plane that takes a complicated three-dimensional path between these two points, but leaves Alan's clock at the same time as Clarissa and arrives at Beth's clock just as Clarissa does. Which registers the *greatest* time interval between the events: the difference between Alan's and Beth's clock readings, the difference between Clarissa's initial and final watch readings, or the difference between Dave's initial and final watch readings? Which measures the smallest interval? Explain.

R5.5   Experimental Evidence

Equation R5.6 implies that if two clocks are synchronized at event A and then travel to event B along different worldlines, the clocks will generally *not* be synchronized when they arrive at event B, since the clocks' speeds (as measured in some inertial frame) will not generally be the same as they follow their different worldlines. This prediction severely conflicts with our newtonian intuition about time but is a direct and testable consequence of the principle of relativity.

In one well-known experiment,[†] muons were generated using a particle accelerator and then kept in a circular storage ring around which they traveled at a constant speed of about $v = 0.99942$. Although the muons' *speed* was constant, their *velocity* was not constant because they were traveling in uniform circular motion. In fact, the worldline of such a muon is quite curved and looks something like that shown in the spacetime diagram in figure R5.3. Figure R5.4 shows a muon storage ring almost exactly like the one used in this experiment.

Imagine that a pulse of muons passes is injected into the ring: call this event A. The muons subsequently travel around the ring: let event B be the time and place where one-half of the bunch has decayed. The bunch of muons, since it is present at both events, measures a proper time $\Delta\tau_{AB}$ between events A and B. The laboratory clock measures coordinate time Δt_{AB} between those events. The muons travel along a curved worldline of constant speed ($v = 0.99942$ in the laboratory frame) but with ever-changing velocity. However, since the *speed* of the muon bunch is constant, we can use equation R5.7 to relate the proper time to the coordinate times. We find that

$$\Delta\tau_{AB} = \sqrt{1 - v^2}\, \Delta t_{AB} \qquad (R5.22)$$

As we have discussed before, the fact that muons decay with a certain fixed half-life makes them effectively little clocks, and measuring the decay rate of a batch of muons amounts to reading those clocks. Since the half-life of a muon at rest was known from other experiments to be 1.52 μs when muons are at rest, $\Delta\tau_{AB} = 1.52$ μs in this experiment. All that remains to be done is to measure the half-life Δt_{AB} of the muons in the storage ring according to the laboratory clock and their exact speed in that storage ring. In this particular experiment, muons were observed to have a half-life of (44.623 ± 0.18) μs in the laboratory frame (about 30 times longer than normal, in stark contrast to the newtonian prediction), and their speed was found to be such that $1/(1 - v^2)^{1/2} = 29.327 \pm 0.004$. Equation R5.22 predicts

A muon decay experiment

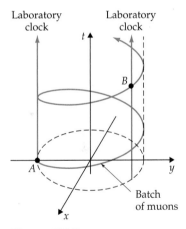

Figure R5.3
The worldlines of two laboratory clocks and a batch of orbiting muons, as drawn by an observer in the laboratory frame. A clock traveling with the muons would register less time between events A and B than the laboratory clock does. (In the experiment described, the muons actually circled the ring more than 300 times between events A and B.)

[†]J. Bailey et al., *Nature*, vol. 268, 1977.

Figure R5.4
The muon storage ring at Brookhaven National Laboratory. This ring has almost exactly the same dimensions and features as the storage ring at the CERN laboratory in Switzerland that was used for the experiment described in this section.

that $\Delta\tau_{AB} - \Delta t_{AB}(1 - v^2)^{1/2} = 0$: this experiment showed that

$$\frac{\Delta\tau_{AB} - \Delta t_{AB}(1 - v^2)^{1/2}}{\Delta\tau_{AB}} = (4 \pm 18) \times 10^{-4} \qquad (95\% \text{ confidence})$$

meaning that this experiment shows that equation R5.7 is good to at least three significant figures. As a point of interest, the acceleration of these muons as measured in the laboratory frame is about 10^{15} times the acceleration of gravity! Therefore, equation R5.7 (at least) is seen to apply even in cases of extreme acceleration.

Exercise R5X.6

Check that the laboratory frame acceleration of the muons in this experiment really is about 10^{15} times the acceleration of gravity (the radius of the muons' circular path was 7.01 m).

An experiment involving a clock flown around the world in a jet plane

In a less extreme experiment performed by J. C. Hafele and R. E. Keating in 1971,[†] a pair of very accurate atomic clocks were synchronized, and then one was put on a jet plane and sent around the world. Upon its return, the jet clock (which followed a noninertial worldline in its trip around the earth) was compared with the inertial clock that remained at rest in the frame of the earth. With suitable corrections for the effects of gravity (predicted by *general* relativity), the results were found to be in complete agreement with equation R5.6.

Exercise R5X.7

Ignoring gravitational effects and the rotation of the earth, estimate the amount by which the two clocks in the previous experiment will disagree if the plane goes around the equator at a constant speed of 300 m/s.

[†]J. C. Hafele and R. E. Keating, *Science*, vol. 117, p. 168, July 14, 1972.

Many other experiments have been performed to check this prediction of the theory of relativity, and all have been in complete agreement with the predictions of equation R5.6. Outrageous as it may seem, the idea that two clocks present at the same two events do not register the same time between those events is a well-established experimental fact.

R5.6   The Twin Paradox

As a result of misinterpreting the meaning of equation R5.7, many people (including competent professors of physics) have been unnecessarily perplexed by apparent paradoxes in the theory of relativity. Physicists call one of the most famous of these the **twin paradox** (or sometimes the *clock paradox*). This problem generated reams of journal articles (as late as the 1960s) before the inadequacy of the language and concepts commonly used to describe relativity at that time was sufficiently well understood.

Here is a statement of the apparent paradox. Andrea and Bernard are twins. When they are both 25 years old, Andrea accepts a commission to be an exobiologist on a expedition to Tau Ceti, a star nearly 8 light-years away from earth. So she flies away on a ship that is capable of near-light speeds, leaving her brother Bernard on earth. The years roll by and the world waits. Finally, hurtling out of the emptiness of space, the spacecraft returns.

As he waits for his sister to emerge from the newly arrived spacecraft, Bernard (now a distinguished man of 50) muses on the bit of relativity that he remembers from college. He recalls that "$\Delta\tau_{AB} = \int \sqrt{1 - v^2}\, dt$." Since Andrea has been moving with a large speed v for much of the trip, Andrea's clocks should measure much less time for the trip than his clocks register. This includes biological as well as mechanical clocks, and so Bernard expects to see a substantially younger sister emerge from the hatch, still displaying their once common youthful vitality. Bernard chews his lip, wondering what it will be like to have a younger "twin" sister.

Bernard expects Andrea to be younger

Similar thoughts run through Andrea's mind as she prepares to disembark. In Andrea's frame of reference, however, she and the spacecraft were motionless, and it is the earth (and thus Bernard) that has moved backward 8 light-years and returned. Andrea (who had the same course in college) thinks that since it is Bernard whose speed $v \neq 0$, it will be *Bernard* who will be younger.

Andrea expects Bernard to be younger

The paradox is clear: each expects the other to be younger from their partial recollection of relativity theory. To this confusion, we can add a third perspective. The principle of relativity states that the laws of physics are the same in every inertial frame. This means that there is no way of making a physical distinction between two inertial frames: if you perform identical experiments within each reference frame, you must get the same results. But isn't the aging process essentially a physical experiment that each person performs in his or her reference frame? If *either* twin is younger than the other, won't that distinguish between the frame of the earth and the frame of the spaceship, contrary to the principle? So perhaps they should have the same age?

Doesn't the principle of relativity imply that they should have the same age?

We have in fact *already* resolved this paradox in this chapter: we simply need to rephrase it in more appropriate language. The first task is to clarify what *events* we are talking about. Let us define event A to be the departure of the ship from earth. Let its arrival back on earth be event B. Both twins are present at both events A and B, so their clocks (including their biological clocks) measure proper time between the events. The question is, Which of

Solution to the paradox

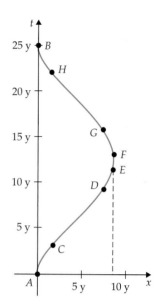

Figure R5.5

The worldline of Andrea's spaceship as drawn by an observer in the frame of reference of the sun.
A: Andrea's departure from earth;
C: ship reaches cruising speed;
D: ship begins decelerating;
E: ship arrives at Tau Ceti;
F: ship departs from Tau Ceti;
G: ship reaches cruising speed;
H: ship begins decelerating;
B: ship arrives at earth.

these twins measures the longer proper time between these events? To find out, we need to sketch their worldlines as measured in some inertial reference frame.

For our master inertial reference frame, let us choose a frame at rest with respect to the sun. Since the sun is freely falling around a very distant galactic center of mass, this will be an excellent approximation to an inertial frame.

Now let us sketch the twins' worldlines as observed in this frame. Andrea takes off from the earth at event A, which we can take to be the origin event (i.e., the event that defines $t = x = 0$ in the sun frame). Her spacecraft travels slowly at first, but gradually picks up speed as the spacecraft strains toward the speed of light. Finally, after a few years, the spacecraft reaches cruising speed. But long before it reaches Tau Ceti, it must begin to slow down, lest it flash by the star at some outrageous velocity. Finally, it coasts into the star system and lands. The process of acceleration and deceleration repeats on the way home. Event B is the event of her return to earth. The resulting worldline is drawn on the spacetime diagram of figure R5.5.

On this diagram, the earth's (and thus Bernard's) worldline is a tiny helix winding around the t axis, twisting around it about 25 times in the roughly 25-y duration of the flight. So Bernard does *not* measure coordinate time between events A and B, as his vague argument seems to suggest: rather he measures a *proper time* between those events that is somewhat different from coordinate time. Since Bernard is never more than about 8 light-minutes ($\approx 1.5 \times 10^{-5}$ y) from the sun, though, the squiggles of his little helix are far too tiny to show up on the diagram: his worldline is essentially a straight line up the t axis. Moreover, since Bernard's speed in his worldline (roughly equal to the orbital velocity of the earth) is about 30 km/s $\approx 10^{-4}$ in SR units, the proper time measured by his clocks only differs from that measured by coordinate clocks in the sun's frame by about $(1 - \sqrt{1 - v^2})\,\Delta t_{AB} \approx +\frac{1}{2}v^2 \Delta t_{AB} = (5 \times 10^{-9})(25 \text{ y}) \approx 4$ s over the time period between events A and B. We see for Bernard that the proper time he measures between events A and B is essentially the same as the time measured in the sun's frame between those events.

But for Andrea, the situation is different. In the sun's frame, Andrea spends quite a bit of time traveling at nearly the speed of light: her average speed in this frame is $16.0 \, \text{ly}/25 \, \text{y} \approx 0.64$. Therefore the factor that appears in the formula for her proper time will be a small number for major portions of the trip. As a result, Andrea's clocks (including her body's biological clock) register much less time between A and B than Bernard's clocks do. Even though 25 y passes on earth, clocks traveling with Andrea will measure a proper time of (approximately) $(1 - 0.64^2)^{1/2}(25 \, \text{y}) \approx 19 \, \text{y}$ between her departure and arrival (we would need to do the integral of equation R5.6 more carefully to get a more accurate answer). So it is indeed a young Andrea of about age 44 that bounds out of the spacecraft to greet her substantially older twin, Bernard.

But Andrea's reasoning seems perfectly logical. Why is it wrong? And what of the principle of relativity? The answer to both questions is the same: *Andrea is in a noninertial reference frame*. Every time that the spacecraft engines fire, first-law detectors in *Andrea's* frame register a violation of that law (equivalent detectors in the sun's frame would read nothing). It is *Andrea* who is pressed into her chair as the engines accelerate and decelerate the spacecraft, not Bernard who is sitting at home. Since Andrea's frame is not inertial, the *principle of relativity does not apply to her*. This exposes the error in the argument favoring the *equality* of the twins.

The core issue: Andrea's frame is *not inertial*

Andrea's mistake is to apply the proper time formula as if her own reference frame were inertial (and thus her clocks measure coordinate time Δt_{AB}). She should compute proper times between events A and B, using speeds and times measured in a *real* inertial reference frame, which her frame is clearly not.

Thus there is no paradox! Bernard's *answer* is right (although his *reasoning* is really no better than Andrea's, since he is not in an inertial frame either!). But since Bernard's frame is *nearly* inertial, when he applies the proper time formula to the times measured between events A and B, he gets an answer that is at least *approximately* correct. Andrea *really is* about 6 y younger than Bernard.

This situation should be no more paradoxical than the following (more familiar) situation. Imagine that Chris and Dana both set off from a given point A on the surface of the earth. Dana (analogous to Bernard here) takes an approximately straight path from point A to the destination point B, while Chris takes a curved path. Should they be shocked when they arrive at B and find that Chris has walked more miles than Dana? Hardly! Chris might try to claim that it was Dana who departed and then returned and so must have taken the curved path, but this is misleading. The curvature of Chris's path is an absolute physical property of that path, a property that can be displayed in any fixed coordinate system. Chris's personal coordinate system, whose axes change direction every time Chris takes a new turn, is not an appropriate coordinate system for displaying a path's curvature. We don't have trouble accepting the nonequivalence of the distance measured along Chris's curved and Dana's straight footpaths in *this* case: we should not have trouble accepting the nonequivalence of the proper times measured along Andrea's and Bernard's worldlines in the first case.

A footpath analogy

Exercise R5X.8

Why is the statement that Andrea's proper time is equal to $(1 - 0.64^2)^{1/2} \, (25 \, \text{y})$ an approximation? Do you think that it is likely to be a bit too high or a bit too low as an estimate?

TWO-MINUTE PROBLEMS

R5T.1 In figure R5.6, the clock(s) following which world-line(s) measure the spacetime interval between events A and B?
- A. ⎯⎯⎯
- B. — —
- C. - - - - -
- D. ⎯⎯⎯ and - - - - -
- E. — — and - - - - -
- F. None of these choices

R5T.2 In figure R5.6, the clock(s) following which world-line(s) measure the *shortest* time interval between events A and B?
- A. ⎯⎯⎯
- B. — —
- C. - - - - -
- D. ⎯⎯⎯ and - - - - -
- E. — — and - - - - -
- F. None of these choices

R5T.3 In figure R5.6, the clock(s) following which world-line(s) measure the *longest* time interval between events A and B?
- A. ⎯⎯⎯
- B. — —
- C. - - - - -
- D. ⎯⎯⎯ and - - - - -
- E. — — and - - - - -
- F. None of these choices

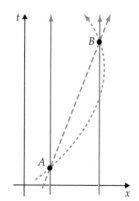

Figure R5.6

This figure is a spacetime diagram showing a pair of events A and B and the worldlines of various clocks for use in the two-minute problems R5T.1 through R5T.3.

The next four problems refer to the following situation. Two identical atomic clocks initially standing next to each other (call them P and Q) are carefully synchronized. Clock Q is then placed on a jet plane, which then flies around the world at an essentially constant speed of 300 m/s, returning 134,000 s (37.1 h) later. The two clocks are then compared

again. Assume for the sake of these problems that the earth's surface defines an inertial reference frame, and ignore possible effects of gravity.

R5T.4 Which clock measures the spacetime interval between the synchronization and comparison events?
- A. Clock P
- B. Clock Q
- C. Both
- D. Neither

R5T.5 Which clock measures the coordinate time between the synchronization and comparison events?
- A. Clock P
- B. Clock Q
- C. Both
- D. Neither

R5T.6 Which clock measures the shorter time interval between the synchronization and comparison events?
- A. Clock P
- B. Clock Q
- C. Measured times are equal

R5T.7 What is the minimum accuracy that the clocks must have to clearly display the relativistic effect?
- A. Both clocks must be accurate to the nearest 10 ms.
- B. Both clocks must be accurate to the nearest 10 μs.
- C. Both clocks must be accurate to the nearest 10 ns.
- D. Both clocks must be accurate to the nearest 10 ps.

Problems R5T.8 and R5T.9 refer to the following situation. Jennifer bungee-jumps from a bridge (event A). Since Jennifer's bungee cord is perfectly elastic, she bounces exactly back up to the bridge and lands on her feet (event B). The time between these events is measured by Jennifer's watch, a stopwatch held by Jennifer's friend Rob, who is standing on the bridge, and by two passengers (one present at event A and one present at event B) who are riding on a train traveling at a constant velocity across the bridge at the time (the passengers have synchronized watches and compare readings later).

R5T.8 Who measures the longest time interval between these events?
- A. Jennifer
- B. Rob
- C. The train passengers

R5T.9 Who measures the shortest time interval between these events?
- A. Jennifer
- B. Rob
- C. The train passengers

HOMEWORK PROBLEMS

Basic Skills

R5B.1 Two clocks are synchronized, and then one is put in a high-speed train car that subsequently runs 50 times around a circular track (radius 10.0 km) at a constant speed of 300 m/s. The two clocks are then compared. By how much do the clocks now differ?

R5B.2 Two clocks are synchronized, and then one is put in a race car that subsequently goes 50 times around a circular track (radius 5.0 km) at a constant speed of 60 m/s. The two clocks are then compared. By how much do the clocks now differ?

R5B.3 A spaceship leaves earth (event A), travels to Pluto (which is 5.0 h of distance away at the time), and then returns (event B) exactly 11.0 h later. If the spaceship's acceleration time is very short, so that it spends virtually all its time traveling at a constant speed, estimate the time measured between events A and B by the ship's clock.

R5B.4 A spaceship leaves earth (event A), travels to Alpha Centauri (which is 4.3 y of distance away), and then returns (event B) exactly 9.0 y later. If the spaceship's acceleration time is very short, so that it spends virtually all its time traveling at a constant speed, estimate the time measured between events A and B by the ship's clock.

R5B.5 Compare $(1 - v^2)^{1/2}$ and $1 - \frac{1}{2}v^2$ for the following values of v: 0.5, 0.2, 0.1, 0.05, 0.01, 0.002. Which is the largest of these values for which the difference between the two expressions is smaller than 1%? Smaller than 0.01%?

R5B.6 Imagine that a new hyperjet is constructed that can go all the way around the world in 6.235 s as measured by clocks at the airport. Assuming that the jet cruises at a constant speed, how long do the pilots' watches register for a complete circumnavigation of the globe? The radius of the earth is about 6380 km.

R5B.7 The designers of particle accelerators use electromagnetic fields to boost particles to relativistic speeds while at the same time constraining them to move in a circular path inside a donut-shaped evacuated cavity. Imagine a particle traveling in such an accelerator in a circular path of radius 7.01 m at a constant speed of 0.9994 (as measured by laboratory observers). Let event A be the particle passing a certain point on its circular path, and let event B be the particle passing the point of the circle directly opposite that point, as illustrated in figure R5.7.
(a) What are the coordinate time Δt and the distance Δd between these events in the laboratory frame? [*Hint:* $\Delta d \neq \pi (7.01 \text{ m})$! Think about it more carefully!]

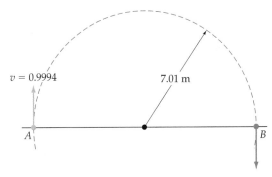

Figure R5.7
A particle in a particle accelerator ring. The light colored point and arrow show the particle's position and velocity at event A, while the darker colored point and arrow show the same at event B. (See problem R5B.7.)

(b) What is the spacetime interval Δs between the events?
(c) What is the proper time $\Delta \tau$ between the events, as measured by a clock traveling with the particle? About how many times greater than $\Delta \tau$ is Δt?

Synthetic

R5S.1 A jogger runs exactly 22 times around a 0.5-km track in 48 min, as measured by a friend sitting at rest on the side. If the jogger and friend synchronize watches before the run, how much are they out of synchronization afterward? Is this the reason that many people expect joggers to live longer than people who don't jog? Explain.

How much longer will this jogger live? See problem R5S.1.

R5S.2 Imagine that a person commutes 50 mi to work and back 250 days per year for 35 y. During the commute, the person drives almost entirely on the freeway at an approximately constant speed of 65 mi/h. How much less time has this person's watch registered in 35 y than the watch of someone who has stayed at home? Is it true that commuters live significantly longer, or does it just seem longer?

R5S.3 Alice is driving a race car around an essentially circular track at a constant speed of 60 m/s. Brian, who is sitting at a fixed position at the edge of the track, measures the time that Alice takes to complete a lap by starting his watch when Alice passes by his position (call this event E) and stopping it when Alice passes his position again (call this event F). This situation is also observed by Cara and Dave, who are passengers in a train that passes very close to Brian. Cara happens to be passing Brian just as Alice passes Brian the first time, and Dave happens to pass Brian just as Alice passes Brian the second time (see figure R5.8). Assume that the clocks used by Alice, Brian, and Cara are close enough together that we can consider them all to be "present" at event E; similarly, that those used by Alice, Brian and Dave are "present" at event F. Assume that the ground frame is an inertial reference frame.

(a) Who measures the shortest time between these events? Who measures the longest?
(b) If Brian measures 100 s between the events, how much less time does Alice measure between the events?
(c) If the train carrying Cara and Dave moves at a speed of 30 m/s, how much larger or smaller is the time that they measure compared to Brian's time? Explain carefully.

R5S.4 The half-life of a muon at rest is 1.52 μs. One can store muons for a much longer time (as measured in the laboratory) by accelerating them to a speed very close to that of light and then keeping them circulating at that speed in an evacuated ring. Assume that you want to design a ring that can keep muons moving so fast that they have a laboratory half-life of 0.25 s (about the time that it takes a person to blink).

(a) How fast will the muons have to be moving? (*Hint:* Define $u \equiv \Delta\tau/\Delta t$, write an equation that connects u to v, then solve for v and use the binomial approximation. You will probably have to do the final calculation for v by hand.)
(b) If the ring is 7.01 m in diameter, how long will it take a muon to go once around the ring (as measured in the laboratory)?

R5S.5 Here is a simple way to understand the binomial approximation. Consider $(1 + x)^2 = (1 + x)(1 + x)$. If you multiply this out, you get $(1 + x)^2 = 1 + 2x + x^2$. Now, if $|x| \ll 1$, then $x^2 \ll x$, so we have $(1 + x)^2 \approx 1 + 2x$, just as the binomial approximation states. Apply the same kind of reasoning to $(1 + x)^3$ and $(1 + x)^4$, and show that the binomial approximation works in these cases, too, when x is small enough that we can ignore x^2 (and higher powers of x) compared to x. (While this is not a proof, it may help you understand the basic issues involved.)

R5S.6 It is possible to *prove* the binomial approximation more generally as follows. According to the definition of the derivative, we have for any function $f(x)$

$$\frac{f(x) - f(0)}{x} \approx \left[\frac{df}{dx}\right]_{x=0} \qquad \text{if } x \text{ is very small}$$

(R5.23)

where $[df/dx]_{x=0}$ tells us to evaluate the derivative at $x = 0$. Now consider the function $f(x) = (1 + x)^a$. Apply equation R5.23 and the chain rule to this function to arrive at equation R5.17.

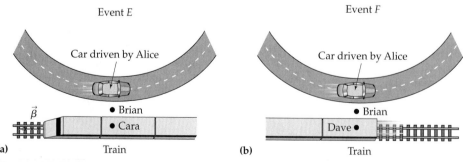

Event E — Car driven by Alice — Brian — Cara — $\vec{\beta}$ — Train — (a)

Event F — Car driven by Alice — Brian — Dave — Train — (b)

Figure R5.8
(a) Event E in the situation described in problem R5S.3. (b) Event F in the situation described in problem R5S.3.

R5S.7 In section R5.6, we proved that the coordinate time, the spacetime interval, and the proper time between a given pair of events stand in the relationship $\Delta t \geq \Delta s \geq \Delta \tau$, no matter what inertial frame is used to measure Δt and no matter what worldline is followed by the clock measuring $\Delta \tau$. It was asserted there that $\Delta t = \Delta s$ if and only if Δt is measured in a frame where the distance Δd between these events is zero. It was also *asserted* that $\Delta s = \Delta \tau$ if and only if the clock measuring $\Delta \tau$ is inertial. Write a short argument supporting each of these statements (for both directions of the "if and only if").

R5S.8 Imagine that the shuttle astronauts go into near-earth orbit for a period of 14 days 3.45 h, as measured by a clock at rest at Cape Canaveral (where shuttles are launched and return). About how much less time passed between the departure and arrival of the shuttle according to the astronauts' clocks than passed on the ground? Assume for the sake of argument that the surface of the earth defines an inertial reference frame. (*Hint:* You might review chapter N12 if you don't remember how fast an object moves in low earth orbit.)

R5S.9 Integrating equation R5.6 is a lot less tricky if v is always small enough that we can use the binomial approximation. Imagine that a spaceship starts from rest from Space Station Alpha floating in deep space and accelerates at a constant rate of $a = 10 \text{ m/s}^2$ (relative to the station) for 1.0 Ms (≈ 12 days), decelerates for the same amount of time to arrive at rest at Space Station Beta (which floats at rest relative to Alpha), and then repeats the same acceleration and deceleration processes in the opposite direction to get back to space station Alpha.
(a) Find the spaceship's top speed. Is it small compared to the speed of light?
(b) About how much less time has elapsed on the spaceship compared to clocks on Station Alpha? (*Hint:* Divide the trip into four pieces, and use the binomial approximation to convert the integral in equation R5.6 to two simpler integrals for each piece.)

Rich-Context

R5R.1 Consider the Hafele-Keating experiment discussed in section R5.6. In this experiment, two atomic clocks were synchronized, one was put on a jet and flown around the world, and then the clocks were compared. Our task in this problem is to make a more realistic prediction of the discrepancy that we would expect between the clocks. Imagine that the plane starts at a point on the equator and flies around the world at a speed of 290 m/s relative to the ground. Assume the plane cruises at about 35,000 ft ≈ 10.7 km.

(a) Using this information, make a prediction (accurate to about $\pm 1\%$) of how much the clocks will disagree when they are compared at the end of the experiment if the plane flies east around the equator (do not ignore the earth's rotation). Describe any assumptions or approximations that you are compelled to make. (*Hint:* Do your analysis in an inertial reference frame fixed to the earth's center that does *not* rotate with the earth. You can use the *coordinate* time Δt measured in this frame between the initial synchronization and the final comparison events as a reference to compare earth clocks and plane clocks. It is tricky to calculate an *exact* number for this coordinate time, but in your final calculation, note that this time will be equal to the time measured on the earth's surface to *much* better than $\pm 1\%$.)
(b) Repeat your calculation, assuming that the plane flies west. Why is your answer different from the one you found for part a?
(c) Note also that general relativity predicts that a clock that is a distance h higher in a gravitational field than a second clock will run *faster* than the lower clock by the factor $1 + (g/c^2)h$, where g is the gravitational field strength (9.8 m/s^2). How does including this information change your answer to part a?

R5R.2 Because the earth is freely falling around the sun, its center defines a pretty good inertial reference frame. With this in mind, consider the fates of two identical twins, one living since infancy at the north pole and the other living since infancy on the equator. Imagine that both twins die after exactly the same biological time has passed (as determined by their own bodies). If this is so, about how much longer will the equatorial twin live than the northern twin, if both live for approximately normal time spans? Describe any approximations or assumptions that you have to make.

Advanced

R5A.1 Consider an inertial frame at rest with respect to the earth. An alien spaceship is observed to move along the x axis of this frame in such a way that $x(t) = [\sin(\omega t + \pi/4) - b]/\omega$, where both x and t are measured in the inertial reference frame, $\omega = \pi/2$ rad/h, and $b = \sin(\pi/4)$. Assume also that the earth is located at the origin ($x = 0$) in this frame.
(a) Argue that the ship passes the earth at $t = 0$ and again at $t = 1.0$ h. (*Hint:* The value of ωt is $\pi/2$ at this time.)
(b) Draw a quantitatively accurate spacetime diagram of the spaceship's worldline, labeling the events where and when it passes the earth as events A and B.

(more)

(c) Show that the ship's x-velocity is $v_x = \cos(\omega t + \pi/4)$ as measured in the inertial frame attached to the earth. (*Hint:* You don't need to know anything about relativity!)

(d) Find the proper time measured by clocks on the ship between the time the ship passes earth the first time and the time it passes the second time. (*Hint:* $1 - \cos^2 x = \sin^2 x$.)

R5A.2   If you know about Taylor series, you can prove the binomial approximation quite generally. Any continuous and differentiable function $f(x)$ can be ex-

pressed in terms of a Taylor series expansion as follows:

$$f(x) = f(0) + x\frac{df}{dx} + \frac{x^2}{2!}\frac{d^2 f}{dx^2} + \frac{x^3}{3!}\frac{d^3 f}{dx^3} + \cdots$$

(R5.25)

Apply this to the function $f(x) = (1 + x)^a$ and show that if you drop terms in this power series involving x^2 or higher, you end up with the binomial approximation. Also show how you would write the approximation if you were to keep terms involving x^2 but drop higher-order terms.

ANSWERS TO EXERCISES

R5X.1 $F_T = mv^2/R = 0.59 \text{ TN} \approx 130$ billion lb. This is not remotely reasonable for a normal string.

R5X.2   Since $a \equiv \text{mag}(\vec{a}) = \text{mag}(\Delta\vec{v})/\Delta t$ if a is constant, in this case

$$\Delta t_{AB} = \frac{\text{mag}(\Delta\vec{v}_{AB})}{a} = \frac{0.5}{a} \qquad \text{(R5.25)}$$

Therefore,

$$\frac{\Delta\tau_{AB}}{\Delta t_{AB}} = \frac{0.478/a}{0.5/a} = 0.956 \qquad \text{(R5.26)}$$

R5X.3   Let the proper time measured by a clock traveling with the electrons between two events A and B marking out a complete cycle be $\Delta\tau_{AB}$ and coordinate time measured between the same two events by a clock in the laboratory frame be Δt_{AB}. In this case,

$$\frac{\Delta\tau_{AB}}{\Delta t_{AB}} = \sqrt{1 - v^2} = \sqrt{1 - 0.98^2} = 0.20$$

$$\Rightarrow \qquad \Delta\tau_{AB} = 0.20\,\Delta t_{AB} \qquad \text{(R5.27)}$$

R5X.4   See the table below. All results are rounded to nine decimal places.

v	$(1 - v^2)^{1/2}$	$1 - \frac{1}{2}v^2$
0.1	0.994987437	0.995000000
0.01	0.999949999	0.999950000
0.001	0.999999500	0.999999500

R5X.5   Alan and Beth measure the longest time interval (since they measure coordinate time), and Dave measures the smallest (since he measures proper time along a curved worldline).

R5X.6   The ratio of a muon's acceleration a to g in the laboratory frame is

$$\frac{a}{g} = \frac{v^2}{gR} = \frac{0.99942^2}{(9.8 \text{ m/s}^2)(7.01 \text{ m})}\left(\frac{3.0 \times 10^8 \text{ m}}{1 \text{ s}}\right)^2$$

$$\approx 1.3 \times 10^{15} \qquad \text{(R5.28)}$$

R5X.7   Let the plane's departure be event A, and let its return be event B. If the earth is not rotating, then the time that it takes the plane to go around the world at a speed of 300 m/s is

$$\Delta t_{AB} = \frac{2\pi R}{v} = \frac{2\pi(6380 \text{ km})}{300 \text{ m/s}}\left(\frac{1000 \text{ m}}{1 \text{ km}}\right)$$

$$= 1.34 \times 10^5 \text{ s} \qquad \text{(R5.29)}$$

The plane's speed in SR units is

$$v = 300\,\frac{\text{m}}{\text{s}}\left(\frac{1 \text{ s}}{3.00 \times 10^8 \text{ m}}\right) = 1.00 \times 10^{-6} \qquad \text{(R5.30)}$$

Therefore, according to equation R5.7, the difference between the proper time $\Delta\tau_{AB}$ measured on the plane and the time Δt_{AB} measured on the ground is

$$\Delta\tau_{AB} - \Delta t_{AB} = \sqrt{1 - v^2}\,\Delta t_{AB} - \Delta t_{AB}$$

$$\approx \left(1 - \tfrac{1}{2}v^2\right)\Delta t_{AB} - \Delta t_{AB}$$

$$-\tfrac{1}{2}v^2\,\Delta t_{AB} = -\tfrac{1}{2}(1.00 \times 10^{-6})^2(1.34 \times 10^5 \text{ s})$$

$$= -0.67 \times 10^{-7} \text{ s} \qquad \text{(R5.31)}$$

So the plane's clock measures about 67 ns less time than the earth-based clocks do.

R5X.8   Andrea is not moving at a constant speed, so the use of the constant-speed formula for proper time is not appropriate, even when one uses the average speed. On the other hand, she spends most of her time traveling at a constant speed, so this will be a reasonable approximation. To see whether this is likely to yield an answer too high or too low, let us consider a specific and fairly extreme case. Imagine that Andrea travels at a speed of 0.96 for two-thirds of the time and is at rest for the remaining time, as measured in the sun-based frame (i.e., we imagine here periods of acceleration to be extremely short). Andrea's average speed in this case is 0.64. If we use this average speed to predict her elapsed time,

we get $\Delta\tau \approx (1 - 0.64^2)^{1/2}\,\Delta t = 0.768\,\Delta t$, where Δt is the trip time measured in the sun-based frame. In this case, though, we can compute the proper time more accurately by breaking the path into two segments, one where she is moving at a constant speed of 0.96 and one where she is at rest. Andrea's total elapsed proper time is

$$(1 - 0.96^2)^{1/2}\tfrac{2}{3}\,\Delta t + (1)^{1/2}\tfrac{1}{3}\,\Delta t = 0.52\,\Delta t \qquad (\text{R5.32})$$

If this example calculation is any indication, using the average speed yields an estimate that is too high.

R6

Coordinate Transformations

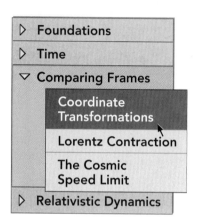
Chapter Overview

Introduction

To delve further into the implications of the principle of relativity, we need to find a way of linking an event's coordinates t, x, y, and z in one inertial frame with its coordinates t', x', y', and z' in another. This chapter develops tools for doing this, and chapters R7 and R8 use these tools to explore applications.

Section R6.1: Overview of Two-Observer Diagrams

A **two-observer spacetime diagram** is a spacetime diagram where the coordinate axes of two different observers are superimposed on events in spacetime.

Section R6.2: Conventions

We will assume in this and future chapters that the two observers are using inertial reference frames in *standard orientation* (i.e., whose frame axes point in the same directions in space) with the Other Frame moving in the $+x$ direction relative to the Home Frame, and that spatial origins of both frames coincide at $t = t' = 0$.

Section R6.3: Drawing the Diagram t' Axis

The t' axis is the set of all events that happen at $x' = 0$ (the Other Frame's origin). Since that origin moves with speed β in the $+x$ direction relative to the Home Frame, the t' axis is simply a worldline with a slope of $1/\beta$. The section argues that we should calibrate this axis either by using hyperbolas or by drawing marks whose *vertical* separation is $\gamma \equiv (1 - \beta^2)^{-1/2}$ larger than that marks on the t axis.

Section R6.4: Drawing the Diagram x' Axis

The **diagram x' axis** is the set of all events that happen at $t' = 0$ (this is to be distinguished from the Other Frame's **spatial x' axis**, which is one of the frame's three coordinate axes *in space*). The radar method implies that the diagram x' axis slopes *upward* with a slope of β. We calibrate this axis, just as we did the t' axis.

Section R6.5: Reading the Two-Observer Diagram

Since all events that lie on a line parallel to the t' axis occur at the same *place* in the Other Frame, and all events that lie on a line parallel to the x' axis occur at the same *time*, to find the Other Frame coordinates of a given event, we draw lines *parallel* to the t' and x' axes from this event until each intersects the other axis.

Figure R6.1 summarizes the process of constructing a two-observer diagram.

Section R6.6: The Lorentz Transformation

Using a two-observer spacetime diagram, we can argue that the coordinate separations between any two events in the Home Frame and Other Frame are related by

$$\Delta t' = \gamma(\Delta t - \beta \Delta x) \quad \text{(R6.12a)}$$
$$\Delta x' = \gamma(-\beta \Delta t + \Delta x) \quad \text{(R6.12b)}$$
$$\Delta y' = \Delta y \quad \text{(R6.12c)}$$
$$\Delta z' = \Delta z \quad \text{(R6.12d)}$$

$$\Delta t = \gamma(\Delta t' + \beta \Delta x') \quad \text{(R6.13a)}$$
$$\Delta x = \gamma(+\beta \Delta t' + \Delta x') \quad \text{(R6.13b)}$$
$$\Delta y = \Delta y' \quad \text{(R6.13c)}$$
$$\Delta z = \Delta z' \quad \text{(R6.13d)}$$

Purpose: These equations allow us to calculate the coordinate differences $\Delta t'$, $\Delta x'$, $\Delta y'$, and $\Delta z'$ in the Other Frame from the corresponding differences Δt, Δx, Δy, and Δz in the Home Frame or vice versa.

Symbols: β is the frame's relative speed and $\gamma \equiv (1 - \beta^2)^{-1/2}$.

Limitations: The two frames must be inertial and in standard orientation, with the Other Frame moving in the $+x$ direction relative to the Home Frame.

Notes: Equations R6.12 are called the **Lorentz transformation equations,** and equations R6.13 the **inverse Lorentz transformation equations.**

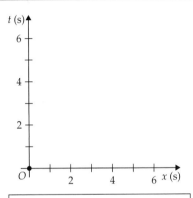

Choose one frame to be the Home Frame. Draw its axes in the usual manner, and indicate the position of the origin event O. Calibrate the axes with some convenient scale.

(a)

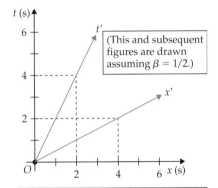

(This and subsequent figures are drawn assuming $\beta = 1/2$.)

Draw the t' axis of the Other Frame from the origin event O with a slope $1/\beta$ (where β is the x-velocity of the Other Frame with respect to the Home Frame). Draw the diagram x' axis of the Other Frame with slope β.

(b)

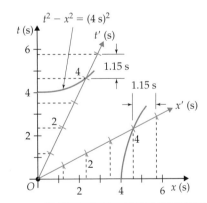

Calibrate the t' axis using hyperbolas or by drawing marks that are vertically separated by $\Delta t = \gamma \Delta t'$, where $\Delta t'$ is the intended mark separation in the Other Frame (1 s in this case). Calibrate the diagram x' axis with marks having the same spacing. (When $\beta = 1/2$, $\gamma \approx 1.15$.)

(c)

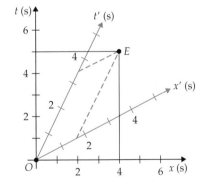

Read the coordinates (in either frame) of any event E by dropping parallels from the event to the appropriate axes. (In this case, E has the coordinates $t \approx 5$ s, $x \approx 4$ s in the Home Frame and $t' \approx 3.4$ s, $x' = 1.7$ s in the Other Frame.)

(d)

Figure R6.1
A summary of the steps in drawing and interpreting a two-observer diagram.

R6.1   Overview of Two-Observer Diagrams

In chapters R4 and R5, we learned how to use the metric equation to calculate the frame-independent spacetime interval Δs and the frame-independent proper time $\Delta \tau$ along a worldline, using frame-dependent coordinate measurements performed in some inertial frame. To delve further into the implications of the principle of relativity, we need to go a step further. We need to know how to link the coordinates t, x, y, z in one inertial reference frame to the same event's coordinates t', x', y', z' in another inertial reference frame. We need to understand such coordinate transformations to understand length contraction (see chapter R7) and find the generalization of the galilean velocity transformation equations (see chapter R8). This chapter therefore opens a three-chapter subdivision on such coordinate transformations and their implications.

Our specific task in this chapter is to understand how, given an event's spacetime coordinates t, x, y, z in one inertial reference frame, we can find the same event's coordinates t', x', y', z' in another inertial reference frame. Physicists call the equations that mathematically describe this coordinate transformation process the **Lorentz transformation equations.**

The derivation of the Lorentz transformation equations is straightforward but somewhat abstract, and that abstraction can blunt one's intuition about what is really going on. Therefore, in this text we address the same problem by using a more visual, intuitive tool called a **two-observer spacetime diagram.** In a two-observer spacetime diagram, we superimpose the coordinate axes for two different observers on the same spacetime diagram. What we end up with will be analogous to the drawing in figure R6.2, which shows two ordinary cartesian coordinate systems (one rotated with respect to the other) superimposed upon the plane. Once we have set up such a cartesian two-observer diagram, if we know the coordinates of a given point A in the xy coordinate system ($x_A = 7$ m, $y_A = 4$ m in the case shown), we can plot point A relative to point O. Then we can just *read* the coordinates of A in the $x'y'$ coordinate system from the diagram (the coordinates are $x'_A = 8$ m, $y'_A = 1$ m in this case). We do not have to use any equations or do any calculations at all!

Setting up such a diagram is straightforward for plane cartesian coordinate systems: it merely involves drawing two sets of perpendicular axes (one rotated with respect to the other), scaling the axes, and drawing coordinate grid lines for each set of axes. Setting up the two sets of coordinate axes representing different inertial frames on a spacetime diagram is a *similar*

What is a two-observer spacetime diagram?

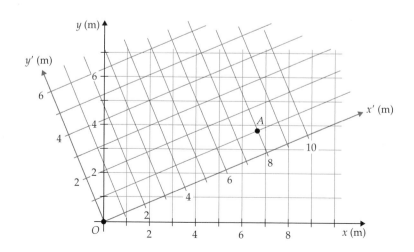

Figure R6.2

A drawing showing two sets of cartesian coordinate axes superimposed on the same plane. We can easily read the coordinates of point A in both coordinate systems from such a diagram.

process, but the peculiarities of spacetime geometry relative to plane geometry lead to some surprising dissimilarities as well. Therefore we need to develop the procedure in a careful, step-by-step manner so that we are sure to catch all these dissimilarities.

This is well worth the effort, though, because two-observer diagrams vividly illustrate how different observers see the same events differently in a way that the dry Lorentz transformation equations cannot. The Lorentz transformation equations help us *calculate* how events in one frame will be viewed in another, but two-observer diagrams help us really *understand* what is going on.

Two-observer diagrams are a powerful tool for displaying how different observers see events

R6.2   Conventions

To make the task of constructing two-observer spacetime diagrams easier, it is convenient to make several assumptions. First, *we assume that the two inertial reference frames are in standard orientation* with respect to each other, as defined in section R1.7. That is, we assume that our frames' corresponding spatial axes point in the same directions in space and that the relative motion of the frames is directed along the common x direction. Since we can choose whatever orientation desired for the axes of a spatial coordinate system, we do not really lose anything by choosing the frames to have this orientation, and we gain much in simplicity.

Standard orientation

Second, *we will work with only those events that occur along the common x and x′ axes of these frames* (i.e., those having coordinates $y = y' = z = z' = 0$). This is a substantial concession to convenience: we would really like to be able to handle any event, but plotting an event with arbitrary coordinates on a spacetime diagram would require that the diagram have four dimensions, which is impossible to represent on a sheet of paper. We choose therefore to limit our attention to events that can be easily plotted on a two-dimensional spacetime diagram. We will see that many interesting problems can still be treated within this restriction. So until you are told otherwise, you should assume that $y' = y = z' = z = 0$ for all events under discussion.

In two-observer diagrams, we can only display events that occur on the x axis

The first step in actually drawing a two-observer spacetime diagram is to pick one of the two frames to be the **Home Frame** and to call the other frame the **Other Frame**. Remember that the terms *Home Frame* and *Other Frame* are capitalized in this text to emphasize that these phrases are actually *names* of inertial frames.

It is conventional (but not absolutely necessary) to pick the Other Frame to be the frame of the two that moves in the $+x$ direction with respect to the Home Frame, as shown in figure R6.3 (alternatively, one can think of the Home Frame as the one that moves in the $-x$ direction with respect to the

The distinction between the Home Frame and Other Frame

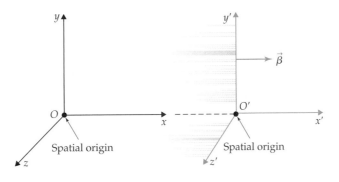

Figure R6.3
Two inertial reference frames in standard orientation. The frames are represented schematically here by bare orthogonal axes. Note that when frames are in standard orientation, the Other Frame is always taken to be the frame moving in the positive direction along the common x direction (alternatively, we can say that the Home Frame is the frame moving in the *negative* x direction).

Other Frame): some signs in equations that follow assume we are following this convention. It is also conventional to distinguish each frame's axes and coordinate measurements by using t, x, y, and z for Home Frame axes and coordinates and t', x', y', and z' for the Other Frame axes and coordinates.

Now, our choice of a Home Frame does not *necessarily* mean we are considering that frame to be at rest and the Other Frame to be moving: we still want to reserve the freedom to consider either frame as being at rest. What this choice *does* imply is simply that we will represent the Home Frame t and x axes in the *usual* manner in a spacetime diagram (i.e., we will draw its t axis as a vertical line and its x axis as a horizontal line).

Exercise R6X.1

A train moves due west relative to the ground. If the train moves at a constant velocity, the train and the ground each define an inertial reference frame. If we define the x axis to be positive eastward, which would we call the Home Frame and which the Other Frame? Does the answer change if we define the x axis to be positive westward?

R6.3   Drawing the Diagram t' Axis

The **time axis** for any frame on a spacetime diagram is by definition the line connecting all events having x coordinate $= 0$ in that frame. This means that the time axis is the worldline of the clock at the spatial origin of the reference frame (all events happening at the spatial origin of a reference frame have spatial coordinate $x = 0$ by definition). The t axis of the Home Frame is drawn as a vertical line by convention. How should we draw the t' axis of the Other Frame?

The slope of the t' axis

The Other Frame moves with speed β along the $+x$ direction with respect to the Home Frame by hypothesis. This means that as measured in the Home Frame, the Other Frame's spatial origin moves β units in the $+x$ direction every unit of time. The worldline of the Other Frame's origin as plotted on the spacetime diagram is thus a straight line of slope $1/\beta$, as shown in figure R6.4a. Note that this line goes through the origin event O since the spatial origins of both frames coincide at $t = t' = 0$ if the frames are in standard orientation.

An appropriate scale for the t' axis

The next step is to put an appropriate scale on the Other Frame's t' axis. It is (unfortunately) *not* correct to simply mark this axis using the same scale as used for the t and x axes. How can we *correctly* scale the t' axis?

Assume that the marks on the t and t' axes are to be separated by some specific time difference δ in their respective frames (in figure R6.4b, for example, this difference is $\delta = 1$ s). Imagine that we want to locate the nth mark on the t' axis. This mark is actually an event that occurs at $t' = n\delta$ and $x' = 0$. The spacetime interval between this mark event and the origin event O (where $t = x = t' = x' = 0$) is

$$\Delta s_n^2 = (t')^2 - (x')^2 = (n\delta)^2 - 0^2 = (n\delta)^2 \quad \Rightarrow \quad \Delta s_n = n\delta \quad \text{(R6.1)}$$

The nth mark on the Home Frame t axis is at $t = n\delta$ and $x = 0$, so this mark is *also* separated from the origin event by a spacetime interval of $\Delta s_n = n\delta$. Now, we learned in chapter R4 that the set of all events on a spacetime diagram that are the same spacetime interval Δs from the origin event lies on the hyperbola $t^2 - x^2 = \Delta s^2$. Therefore, to locate the nth mark on the t' axis,

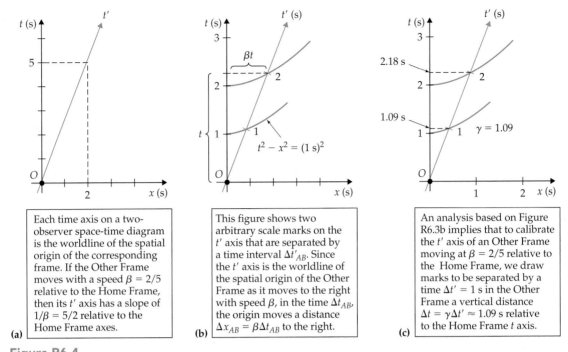

(a) Each time axis on a two-observer space-time diagram is the worldline of the spatial origin of the corresponding frame. If the Other Frame moves with a speed $\beta = 2/5$ relative to the Home Frame, then its t' axis has a slope of $1/\beta = 5/2$ relative to the Home Frame axes.

(b) This figure shows two arbitrary scale marks on the t' axis that are separated by a time interval $\Delta t'_{AB}$. Since the t' axis is the worldline of the spatial origin of the Other Frame as it moves to the right with speed β, in the time Δt_{AB}, the origin moves a distance $\Delta x_{AB} = \beta \Delta t_{AB}$ to the right.

(c) An analysis based on Figure R6.3b implies that to calibrate the t' axis of an Other Frame moving at $\beta = 2/5$ relative to the Home Frame, we draw marks to be separated by a time $\Delta t' = 1$ s in the Other Frame a vertical distance $\Delta t = \gamma \Delta t' \approx 1.09$ s relative to the Home Frame t axis.

Figure R6.4

Steps in figuring out how to construct the t' axis on a two-observer spacetime diagram.

we simply draw the hyperbola $t^2 - x^2 = \Delta s_n^2 = (n\delta)^2$ and draw the mark where the t' axis intersects that hyperbola. (Note that this hyperbola will also go through the corresponding nth mark on the t axis.) We can repeat this for different values of n to draw however many marks we need on the t' axis. This process is illustrated in figure R6.4b.

This is great if you can draw very accurate hyperbolas (or you can down-load and print the hyperbola graph paper available on the *Six Ideas* website). However, we can also pretty easily *calculate* where the mark should be math-ematically without having to draw any hyperbolas.

Because the t' axis is a line going through the origin event O with a slope of $1/\beta$, the equation describing that line in terms of Home Frame coordinates is $t = x/\beta$, or $x = \beta t$. We can find the place where this line intersects the hy-perbola $t^2 - x^2 = (n\delta)^2$ by plugging $x = \beta t$ into the equation for the hyper-bola and solving for the remaining variable t:

$$(n\delta)^2 = t^2 - x^2 = t^2 - (\beta t)^2 = (1 - \beta^2)t^2 \quad \Rightarrow \quad t = \frac{n\delta}{\sqrt{1 - \beta^2}} \quad \text{(R6.2)}$$

Note that this specifies the Home Frame t coordinate (vertical coordinate) of the nth mark on the t' axis. Now, the quantity $1/\sqrt{1 - \beta^2}$ occurs so often in rel-ativity theory that it is given its own special symbol (the Greek letter gamma):

$$\gamma \equiv \frac{1}{\sqrt{1 - \beta^2}} \quad \text{(R6.3)}$$

(Note that γ is a number that is always larger than 1.) Using this symbol, equation R6.2 becomes

$$t = n(\gamma\delta) = t \text{ coordinate of } n\text{th mark on } t' \text{ axis} \quad \text{(R6.4)}$$

This completely and correctly locates all the marks on the t' axis. Equiva-lently, we could say that since adjacent marks correspond to $\Delta n = 1$, the vertical coordinate *difference* Δt between adjacent marks on the t' axis as projected on the Home Frame t axis is

$$\Delta t = \Delta n(\gamma \delta) = \gamma \delta = \gamma\,\Delta t' \qquad \text{for events along } t' \text{ axis} \qquad (R6.5)$$

where $\Delta t' = \delta$ is the time between marks in the Other Frame.

To give a concrete example, imagine that our Other Frame moves at a speed of $\beta = \frac{2}{5}$ relative to the Home Frame, and we want to draw marks on the Other Frame t' axis that correspond to events that are $\delta = \Delta t' = 1$ s apart as measured in the Other Frame. In this case

$$\gamma \equiv \frac{1}{\sqrt{1 - \left(\frac{2}{5}\right)^2}} = 1.09 \qquad (R6.6)$$

Therefore, we should draw these marks so that the vertical coordinate of the nth mark as projected onto the Home Frame t axis is

$$t = n\gamma\,\Delta t' = n(1.09)(1\text{ s}) = n(1.09\text{ s}) \qquad (R6.7a)$$

Equivalently, we could draw the marks so that the vertical separation (as projected onto the Home Frame t axis) between adjacent marks is

$$\Delta t = \gamma\,\Delta t' = (1.09)(1\text{ s}) = 1.09\text{ s} \qquad (R6.7b)$$

This is illustrated in figure R6.4c.

Note that equation R6.5 describes the *vertical* separation between the marks, *not* their spacing along the t' axis, which is even *larger* than $\gamma\,\Delta t'$. Problem R6S.8 describes how to calculate the separation between marks measured *along* the t' axis, if you are really interested. But equation R6.4 or R6.5 is actually much more useful in practice.

Exercise R6X.2

Construct and calibrate the t' axis for an Other Frame that moves with a speed of $\beta = \frac{4}{5}$ with respect to the Home Frame.

R6.4   Drawing the Diagram x' Axis

Definition of the diagram x' axis

In section R6.3 we defined the t axis for a given frame to be the line on a spacetime diagram connecting all events that occur at the spatial origin $x = 0$ of that frame (i.e., at the same *place* as the origin event). Analogously, we define the **diagram x axis** for a given inertial reference frame to be the line on a spacetime diagram connecting all events that occur at $t = 0$ in that frame (i.e., at the same *time* as the origin event). We conventionally draw the Home Frame's diagram x axis as a horizontal line on a spacetime diagram. How should we draw the diagram x' axis for the Other Frame?

The distinction between the diagram x axis and the spatial x axis

(*Note:* In this text, the phrases "diagram x' axis" and "diagram x axis" refer to a line drawn on a *spacetime diagram* that connects all events occurring at zero time. This is to be sharply distinguished from the line in *physical space* that goes through the spatial origin and connects all points with $y = z = 0$. When talking about the latter, I will speak of the x *direction* or the **spatial x axis**.)

Intuitively, we might expect to draw the diagram x' axis perpendicular to the t' axis. Unfortunately, drawing the x' axis in this way is *not* consistent with the definition of the diagram x' axis just stated. To figure out the *right* way to draw this axis, we have to consider carefully the implications of the idea that *the diagram x' axis connects events that are simultaneous in the Other Frame.*

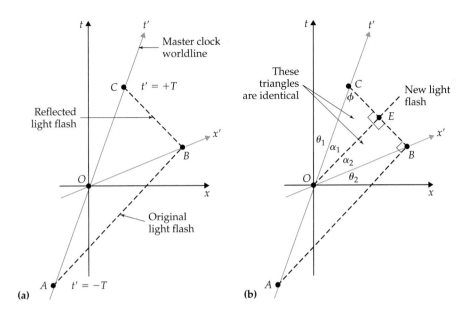

(a)

(b)

Figure R6.5
(a) This diagram shows that the *x'* axis on a two-observer diagram must tilt upward at an angle. If a flash emitted from $x' = 0$ at a certain time *T before* the origin event *O* is reflected from event *B* and returns to $x' = 0$ at the same time *T after* *O*, then event *B* must have occurred at $t' = 0$, and thus should lie on the *x'* axis, as shown. (b) This diagram shows that the *x'* axis must make the same angle with the *x* axis as the *t'* axis makes with the *t* axis. Triangles *OEC* and *OEB* are identical, so $\alpha_1 = \alpha_2 \Rightarrow \theta_1 = \theta_2$.

We begin by considering a set of events that illustrates the use of the radar method to determine coordinates in the Other Frame. Imagine that at some time $t' = -T$ (where *T* is some arbitrary number) a light flash is sent from the master clock (located at the spatial origin of the Other Frame) in the $+x$ direction (call the emission of the flash event *A*). This flash reflects from some event *B* and returns to the Other Frame's master clock at time $t' = +T$ (call the reception of the flash by the master clock event *C*). Since the light flash must take the same time to return to the origin from event *B* as it took to get there, observers in the Other Frame are forced to conclude that event *B must* have happened at a time halfway between $t'_A = -T$ and $t'_C = +T$, that is, at time $t'_B = 0$. This implies that event *B* is *simultaneous* with the origin event *O*.

Let us draw this set of events on a spacetime diagram based on the Home Frame (see figure R6.5a). Events *A*, *O*, and *C* all occur at $x' = 0$, so they all lie on the *t'* axis of the spacetime diagram. Since events *A* and *C* occur at $t' = -T$ and $t' = +T$, they must be symmetrically spaced along the *t'* axis on opposite sides of the origin event, as shown. Note that since the speed of light is 1 in every reference frame, these light-flash worldlines must have a slope of ± 1 on this diagram.

Now, by the definition of clock synchronization in the Other Frame, events *O* and *B* both occur at time $t' = 0$. The diagram *x'* axis is defined to be the line connecting all events that occur at time $t' = 0$. Therefore, the diagram *x'* axis must go through events *O* and *B*. This means that *the diagram x' axis must angle upward*, as shown.

In fact, by considering the geometric relationships implicit in figure R6.5a, we can see that *the diagram x' axis makes the same angle with the diagram x axis that the t' axis makes with the t axis*. The argument is easier if we imagine that the master clock emits a right-going light flash at the origin event *O*: let event *E* be this flash meeting the incoming reflected flash. This new light flash is shown in figure R6.5b. (This new light flash has no physical importance: it just makes the following argument simpler.) Since light flash worldlines always have a slope of ± 1, they all make a 45° angle with respect to the vertical or horizontal directions. This means that if light-flash worldlines cross at all, they always cross at right angles.

Now, I claim that triangles *ABC* and *OEC* in figure R6.5b are *similar* triangles: they are right triangles that share the common angle ϕ. Moreover, the

The diagram *x'* axis must tilt upward

The diagram *x'* makes the same angle with the *x* axis as the *t'* axis makes with the *t* axis hypotenuse of ABC is *twice* as long as that of OEC, since A and C are symmetrically placed about event O. This means that the sides of triangle ABC must be exactly twice as large as those of the triangle OEC, implying that line BC must also be twice as large as line EC (remember that if two triangles are similar, the lengths of their corresponding sides are proportional).

But if line BC is twice as large as EC, then the length of line BE must be equal to the length of line EC. This means that triangles OEC and OEB are *identical,* since they are both right triangles and their corresponding legs are equal in length. Therefore, $\alpha_1 = \alpha_2$, which in turn means that $\theta_1 \equiv 45° - \alpha_1$ is equal to $\theta_2 \equiv 45° - \alpha_2$. Thus *the diagram x' axis makes the same angle with the diagram x axis that the t' axis makes with the t axis,* as previously asserted.

Another important consequence is that the length of line OC (which represents the coordinate time interval $\Delta t' = +T$) is the same as the length of line OB (which represents the coordinate displacement $\Delta x' = T$, the distance in the Other Frame that the light signal had to travel to get to event B). This means that *the scale of both axes must be the same;* that is, the spacing of marks on the diagram x' axis will be exactly the same as the spacing of marks on the t' axis!

Exercise R6X.3

Explain why the x' coordinate difference between events O and B must be $\Delta x' = T$.

The slope of the diagram x' axis

Note that $\tan \theta_1 = \text{run}/\text{rise}$ for the t' axis $= 1/(\text{slope of } t' \text{ axis}) = \beta$. Note also that $\tan \theta_2 = \text{rise}/\text{run}$ for the diagram x' axis $=$ slope of diagram x' axis. Since we have just seen that $\tan \theta_1 = \tan \theta_2$, we have

$$\text{Slope of } x' \text{ axis} = \beta \qquad (R6.8)$$

So to be consistent with the principle of relativity, we *must* draw the Other Frame diagram t' and x' axes with slopes $1/\beta$ and β, respectively.

Exercise R6X.4

Draw and calibrate the t' and x' axes for an Other Frame moving with a speed of $\beta = \frac{4}{5}$ with respect to the Home Frame.

R6.5   Reading the Two-Observer Diagram

Summary of how to construct a two-observer spacetime diagram

In summary, what have we discovered? To construct a two-observer spacetime diagram, we proceed as follows: (1) We construct the Home Frame t axis and diagram x axis perpendicular to each other (with the t axis vertical). (2) We draw the diagram t' axis of the Other Frame with slope $1/\beta$ and the diagram x' axis of the Other Frame with slope β. (3) We calibrate the Other Frame time axis with marks that are separated vertically by $\Delta t = \gamma \, \Delta t'$ [where $\Delta t' = \delta$ is the time interval between the marks in the Other Frame and $\gamma \equiv (1 - \beta^2)^{-1/2}$]. (4) We calibrate the diagram x' axis with marks separated by the *same* distance as marks on the t' axis (i.e., the marks should be separated *horizontally* by $\Delta x = \gamma \, \Delta x'$, where $\Delta x' = \delta$ is the spatial separation between the marks along the x' axis as measured in the Other Frame).

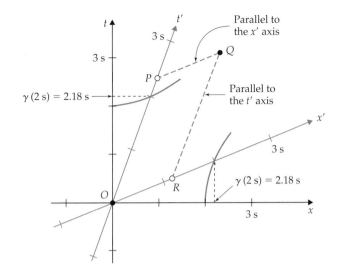

Events *Q* and *R* occur at the same place in the Other Frame, since all events that occur at the same place in that frame lie along a line parallel to the *t'* axis. Similarly, events *P* and *Q* occur at the same time in the Other Frame. Thus the time of *Q* is the same as the time of *P* in that frame (that is, $t'_Q \approx 2.3$ s in this case), and the position of *Q* is the same as the position of *P* (that is, $x'_Q \approx 1.2$ s in this case). The relative speed of the frames is $\beta = \frac{2}{5}$ in the case shown.

We can now find the *t'* and *x'* coordinates of any event on the diagram as follows. The *t'* axis is by definition the line on the spacetime diagram connecting all events that occur at $x' = 0$. The line connecting all events that have coordinate $x' = 1$ s (or any given value not equal to zero) will be a line *parallel* to the *t'* axis, because the Other Frame's lattice clock at $x' = 1$ s moves at the same velocity as the master clock at $x' = 0$, and the latter's worldline defines the *t'* axis.

Similarly, the line on the diagram connecting all events that have the same *t'* coordinate must be a line parallel to the diagram *x'* axis (the line connecting all events having $t' = 0$). Here is an argument for this statement. If the line connecting all events that occur at $t' = 1$ s, for example, were *not* parallel to the line connecting all events that occur at $t' = 0$, then these lines would intersect at some point on the diagram. The event located at the point of intersection would thus occur at both $t' = 1$ s *and* $t' = 0$, which is absurd. Thus a line connecting events having the same *t'* coordinate *must* be parallel to any other such line.

So if the line connecting all events occurring at the same *time* in the Other Frame is parallel to the diagram *x'* axis, and the line connecting all events occurring at the same *place* in that frame is parallel to the *t'* axis, we find the coordinates of an event in the Other Frame by drawing lines through the event that are *parallel* to the diagram *t'* and *x'* axes (and *not* perpendicular to them). The places where these lines of constant *x'* and *t'* cross the coordinate axes indicate the coordinates of the event in the Other Frame (see figure R6.6).

Finding the coordinate values by dropping *parallels* instead of perpendiculars may seem strange to you, and will probably take some getting used to. Nonetheless, I hope you see from the argument above that dropping parallels is the only way to read the coordinates that makes any sense in this case.

Why we have to find coordinates by drawing lines *parallel* to the axes

---

Exercise R6X.5

Use figure R6.6 to estimate the Other Frame coordinates of an event that occurs at $t = 0$ and $x = +2$ s in the Home Frame.

Exercise R6X.6

Use figure R6.6 to estimate the Home Frame coordinates of an event that occurs at $t' = 3$ s and $x' = -1$ s in the Other Frame.

Exercise R6X.7

Are events that are simultaneous in one inertial frame generally simultaneous in any other frame? (*Hint:* Consider figure R6.6.)

R6.6   The Lorentz Transformation

The two-observer spacetime diagram discussed in the preceding sections provides a very powerful visual and intuitive tool for linking the coordinates of an event measured in one inertial reference frame with the coordinates of the same event measured in another inertial frame. Because it is visual in nature, it is much more immediate and less abstract than working with equations. But this tool does not give us the quantitative precision that only equations can provide.

The purpose of this section is to develop a set of *equations* that link the coordinates of an event measured in the Home Frame with the coordinates of the same event measured in the Other Frame. These equations do *mathematically* exactly what the two-observer diagram does *visually*. Together, these two tools will enable us to discuss relativity problems with both clarity and precision.

Derivation of the Lorentz
transformation equations

Consider an arbitrary event Q, as illustrated in figure R6.7. Imagine that we know the coordinates t'_Q and x'_Q of this event in the Other Frame. This means that we can locate an event P which occurs at $t' = 0$ (that is, on the diagram x' axis) and at the same *place* as Q in the Other Frame (that is, $x'_Q = x'_P$). Let the time coordinate separation between events P and Q be Δt_{PQ} and the spatial coordinate separation between O and P be Δx_{OP} in the Home Frame. Proper calibration of the Other Frame axes requires that $\Delta t_{PQ} = \gamma t'_Q$ and $\Delta x_{OP} = \gamma x'_Q$. Also, since the line connecting events P and Q is parallel to the t' axis, its slope must be $1/\beta$, implying that the bottom leg of the triangle involving points P and Q has to have length $\beta \Delta t_{PQ}$. Similarly, the slope of the diagram x' axis is β, so the vertical leg of the triangle involving points O and P must have length $\beta \Delta x_{OP}$. All these relationships are illustrated in figure R6.7.

Now, you can see from the diagram that

$$t_Q = \Delta t_{PQ} + \beta \, \Delta x_{OP} = \gamma t'_Q + \gamma \beta x'_Q \qquad \text{(R6.9a)}$$

$$x_Q = \Delta x_{OP} + \beta \, \Delta t_{PQ} = \gamma x'_Q + \gamma \beta t'_Q \qquad \text{(R6.9b)}$$

Figure R6.7

Pick an arbitrary event Q. Then choose event P to occur at $t' = 0$ (that is, on the x' axis) and at the same place as Q in the Other Frame. Note that since the line connecting events P and Q is parallel to the t' axis, its slope must be equal to $1/\beta$. Note also that the x' axis has a slope of β. These slopes, along with the procedure for calibrating axes that we developed in section R6.2, imply the relationships displayed on the diagram.

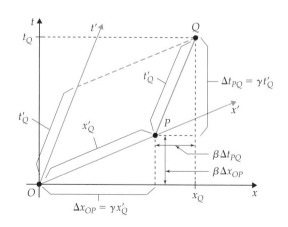

Since the event Q is purely arbitrary, we can drop the subscript and simply say that the Home Frame coordinates t and x of any event can be expressed in terms of the Other Frame coordinates t' and x' of the *same* event as follows:

$$t = \gamma(t' + \beta x') \qquad \text{(R6.10}a)$$

$$x = \gamma(\beta t' + x') \qquad \text{(R6.10}b)$$

The inverse Lorentz transformation equations

We call these equations the **inverse Lorentz transformation equations.**

The "just plain" **Lorentz transformation equations** (or LTEs for short), which express the Other Frame coordinates of an event in terms of the Home Frame coordinates, can be easily found by solving equations R6.10 for t' and x'. The results are

$$t' = \gamma(t - \beta x) \qquad \text{(R6.11}a)$$

$$x' = \gamma(-\beta t + x) \qquad \text{(R6.11}b)$$

The (direct) Lorentz transformation equations

Exercise R6X.8

Verify equations R6.11a and b.

These equations can be easily generalized to handle events having nonzero coordinates y and z. We learned in section R4.3 that if two inertial frames are in relative motion along a given line, any displacement measured perpendicular to that line has the same value in both frames. Since frames in standard orientation move relative to each other along their common x axis, this means that

Transformation of the y and z coordinates

$$y' = y \qquad \text{(R6.11}c)$$

$$z' = z \qquad \text{(R6.11}d)$$

Together, all four equations R6.11 represent the relativistic generalization of the galilean transformation equations R1.2.

Often we are not so much interested in the raw coordinates of an event as we are in the coordinate *differences* between two events. Consider a pair of events A and B separated by coordinate differences $\Delta t \equiv t_B - t_A$ and $\Delta x \equiv x_B - x_A$ in the Home Frame. What are the corresponding coordinate differences $\Delta t' \equiv t'_B - t'_A$ and $\Delta x' \equiv x'_B - x'_A$ measured in the Other Frame? Applying equation R6.11a to t'_A and t'_B separately, we get

$$\Delta t' \equiv t'_B - t'_A = \gamma(t_B - \beta x_B) - \gamma(t_A - \beta x_A)$$

The Lorentz transformation equations for coordinate *differences*

$$= \gamma(t_B - \beta x_B - t_A + \beta x_A) = \gamma[(t_B - t_A) - \beta(x_B - x_A)]$$

$$\Rightarrow \qquad \Delta t' \equiv \gamma(\Delta t - \beta\,\Delta x) \qquad \text{(R6.12}a)$$

Similarly, you can show that

$$\Delta x' \equiv \gamma(-\beta\,\Delta t + \Delta x) \qquad \text{(R6.12}b)$$

$$\Delta y' = \Delta y \qquad \text{(R6.12}c)$$

$$\Delta z' = \Delta z \qquad \text{(R6.12}d)$$

These are the Lorentz transformation equations for coordinate differences. Note that they have the same form as the ordinary Lorentz transformation equations (equations R6.11): one simply replaces the coordinate quantities with the corresponding coordinate differences.

The inverse Lorentz transformation equations for coordinate differences are completely analogous. To summarize, then, the equations that allow us to compute coordinate differences in one inertial frame from the corresponding differences in the other are

$$\Delta t' = \gamma(\Delta t - \beta\,\Delta x) \quad \text{(R6.12a)} \qquad \Delta t = \gamma(\Delta t' + \beta\,\Delta x') \quad \text{(R6.13a)}$$

$$\Delta x' = \gamma(-\beta\,\Delta t + \Delta x) \quad \text{(R6.12b)} \qquad \Delta x = \gamma(+\beta\,\Delta t' + \Delta x') \quad \text{(R6.13b)}$$

$$\Delta y' = \Delta y \quad \text{(R6.12c)} \qquad \Delta y = \Delta y' \quad \text{(R6.13c)}$$

$$\Delta z' = \Delta z \quad \text{(R6.12d)} \qquad \Delta z = \Delta z' \quad \text{(R6.13d)}$$

Purpose: These equations allow us to calculate the coordinate separations $\Delta t'$, $\Delta x'$, $\Delta y'$, and $\Delta z'$ in the Other Frame from the corresponding differences Δt, Δx, Δy, and Δz in the Home Frame, or vice versa.

Symbols: β is the frame's relative speed and $\gamma \equiv (1 - \beta^2)^{-1/2}$.

Limitations: These equations apply only to inertial frames in standard orientation, with the Other Frame moving in the +x direction relative to the Home Frame.

Notes: Equations R6.12 are called the **Lorentz transformation equations,** and equations R6.13 the **inverse Lorentz transformation equations** (for coordinate differences in both cases).

Exercise R6X.9

Using the same approach used to derive equation R6.12a, verify equations R6.12b, c, and d.

The Lorentz transformation equations and two-observer diagrams are equivalent

I hope that you can see that the previous derivation means that the inverse Lorentz transformation equations (or the "plain" Lorentz transformation equations) simply quantify more precisely what you could read from a two-observer spacetime diagram. I chose this method of deriving the Lorentz transformation equations deliberately to drive home the point that the equations simply express *mathematically* exactly the same thing that a two-observer spacetime diagram expresses *visually*. Two-observer spacetime diagrams are most useful for visually displaying and thinking about how different observers will interpret events; the Lorentz transformation equations are best for doing accurate calculations once the thinking is done. Example R6.1 illustrates how these tools complement each other.

Example R6.1

Problem Consider an event that is observed in the Home Frame to occur at time $t = 4.2$ s and position $x = 7.0$ s. When and where does this event occur according to observers in an Other Frame that is moving with speed $\beta = \frac{2}{5}$ with respect to the Home Frame?

Solution Figure R6.8 displays the solution by using a two-observer spacetime diagram. Home Frame axes were constructed and scaled, and the event (here marked with an *E*) was plotted with respect to the Home Frame axes according to the coordinates given. Then the Other Frame's t' axis was drawn with slope $\frac{5}{2}$, and the diagram x' axis was drawn with slope $\frac{2}{5}$. The Other

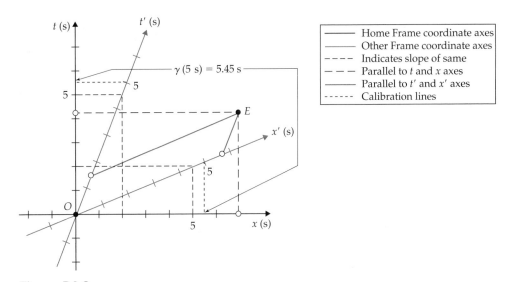

Figure R6.8
The two-observer diagram solving the question posed in this example. Event E has
coordinates $t = 4.2$ s and $x = 7$ s in the Home Frame, and about $t' = 1.5$ s and $x' = 5.7$ s
in the Other Frame.

Frame's t' and diagram x' axes were calibrated with marks separated vertically and horizontally (respectively) by γ (1 s) = 1.09 s. Finally, parallels were
dropped from event E to the t' axis and the diagram x' axis, and the coordinates of E were read from the intersection of these parallels with the axes. We
see that the coordinates of event E are about $t' \approx 1.5$ s and $x' \approx 5.7$ s.

Now let's *calculate* these coordinates, using the Lorentz transformation
equations. If $\beta = \frac{2}{5}$, then $\gamma = (1 - \beta^2)^{1/2} = 1/\sqrt{1 - \frac{4}{5}} \approx 1.09$, so

$$t' = \gamma(t - \beta x) \approx 1.09\left[4.2 \text{ s} - \tfrac{2}{5}(7.0 \text{ s})\right] \approx 1.5 \text{ s} \qquad (R6.14a)$$

$$x' = \gamma(-\beta t + x) \approx 1.09\left[-\tfrac{2}{5}(4.2 \text{ s}) + 7.0 \text{ s}\right] \approx 5.8 \text{ s} \qquad (R6.14b)$$

which are in substantial agreement with the results read from the diagram.
(One should expect an uncertainty of about 3% to 5% in results read from
even the most carefully constructed spacetime diagram.)

Exercise R6X.10

Determine (both graphically and analytically) the Other Frame coordinates
of an event with Home Frame coordinates $t = 1.0$ s, $x = 4.0$ s, where the
Home Frame and Other Frame are as described in example R6.1. (You can
use the two-observer diagram in figure R6.8 for the graphical solution.)

Exercise R6X.11

Determine (both graphically and analytically) the Home Frame coordinates
of an event whose Other Frame coordinates are $t' = 4.0$ s and $x' = 2.0$ s,
where the Home Frame and Other Frame are as described in example R6.1.
(Again, you can use the two-observer diagram provided in figure R6.8 for the
graphical solution.)

TWO-MINUTE PROBLEMS

R6T.1 The Other Frame is moving with a speed of $\beta = 0.25$ in the $+x$ direction with respect to the Home Frame. The two-observer spacetime diagram in figure R6.9 shows the diagram t and x axes of the Home Frame and the diagram t' axis of the Other Frame. Which of the choices in that figure best corresponds to the diagram x' axis?

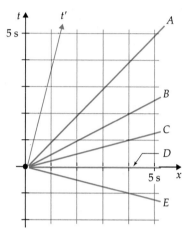

Figure R6.9
For problems R6T.1, R6T.2 and R6T.4.

R6T.2 The Other Frame is moving with a speed of $\beta = 0.25$ in the $+x$ direction with respect to the Home Frame. The two-observer spacetime diagram in figure R6.9 shows the diagram t and x axes of the Home Frame and the diagram t' axis of the Other Frame. Which of the choices in that figure would best correspond to the diagram x' axis if the newtonian concept of time were true?

R6T.3 According to our conventional frame names, the Home Frame is our frame, the frame at rest, true (T) or false (F)?

R6T.4 Imagine that the marks on the Home Frame t axis in figure R6.9 are 1.0 cm apart. What should be the *vertical* separation of the corresponding marks on the t' axis?
A.   0.94 cm
B.   0.97 cm
C.   1.0 cm
D.   1.03 cm
E.   1.07 cm
F.   Other

R6T.5 Figure R6.10 shows a two-observer spacetime diagram for an Other Frame that moves at a speed of 0.5 relative to the Home Frame. What are the coordinates of event P in the Other Frame?

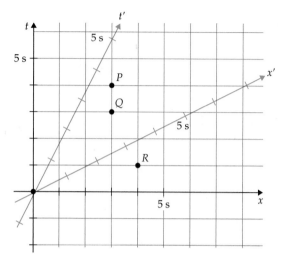

Figure R6.10
For problems R6T.5 through R6T.7.

A. $t' = 3.4$ s, $x' = 2.6$ s
B. $t' = 5.2$ s, $x' = 2.6$ s
C. $t' = 2.9$ s, $x' = 1.2$ s
D. $t' = 3.7$ s, $x' = 3.4$ s
E.   Other (specify)

R6T.6 Figure R6.10 shows a two-observer spacetime diagram for an Other Frame that moves at a speed of 0.5 relative to the Home Frame. What are the coordinates of event Q in the Other Frame?
A. $t' = x' = 5.2$ s
B. $t' = x' = 3.2$ s
C. $t' = x' = 2.6$ s
D. $t' = x' = 1.7$ s
E.   Other (specify)

R6T.7 Figure R6.10 shows a two-observer spacetime diagram for an Other Frame that moves at a speed of 0.5 relative to the Home Frame. What are the coordinates of event R in the Other Frame?
A. $t' = 1.2$ s, $x' = 4.0$ s
B. $t' = 0.9$ s, $x' = 3.4$ s
C. $t' = 6.5$ s, $x' = 1.7$ s
D. $t' = 2.2$ s, $x' = 3.2$ s
E.   Other (specify)

R6T.8 Consider two blinking warning lights 3000 m apart along a railroad track. Imagine that these beacons flash simultaneously in the ground frame. Let the event of the west light blinking be event W, and let the event of the east light blinking be event E. A train is moving eastward along the track at a known relativistic speed β. Imagine that an observer in the train passes the west light just as event W happens. By carefully measuring when the light from event E arrives and calculating the distance between the two lights (by observing how long it takes the distance between the lights to pass at the known speed β), the observer is able to infer when event E actually happened in the train frame. The observer concludes that

A. Event W happened before event E in the train frame.

B. Events W and E happened at the same time in the train frame.

C. Event E happened before event W in the train frame.

D. There is no way to determine unambiguously which event occurs first in the train frame.

HOMEWORK PROBLEMS

Basic Skills

Problems R6B.1 through R6B.6 all refer to the following situation. The spacetime diagram shown in figure R6.11 shows the worldline of an alien spaceship fleeing at a speed of $\beta = \frac{3}{5}$ in the $+x$ direction from space station DS9 after stealing some potentially destructive trilithium crystals. The departure of the ship from DS9 is event A. At event B, DS9 fires a phaser blast (which travels at the speed of light), hoping to disable the vessel. At event C, the fleeing spaceship drops a fuel tank behind, setting the tank at rest relative to DS9 while the spaceship continues on ahead of it (the tank now shields the ship from the point of view of DS9). At event D, the phaser blast hits and destroys the tank, leaving the ship unharmed.

R6B.1 The t' axis of the spaceship frame is labeled on the diagram (this is just the worldline of the spaceship itself). This axis has also been calibrated. Check that the calibration is correct (explain how you checked this). On a copy of the diagram below, draw and calibrate the diagram x' axis for the spaceship frame.

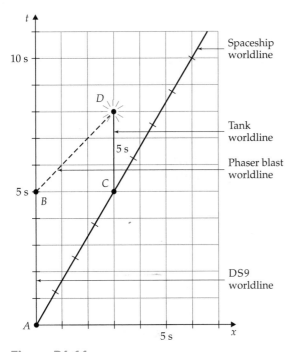

Figure R6.11
For problems R6B.1 through R6B.7.

R6B.2 When does the spaceship drop the tank according to its own clock? Explain how you arrived at your answer.

R6B.3 What is the approximate time of event D in the spaceship frame? (Do not use the Lorentz transformation equations: rather read the result from the diagram and explain how you arrived at your result.)

R6B.4 Calculate the time of event D in the spaceship's frame, using the appropriate Lorentz transformation equation.

R6B.5 Which event, B or C, occurs first in DS9's frame (or are these events simultaneous)? Which occurs first in the spaceship frame?

R6B.6 Use the appropriate Lorentz transformation equation to compute the x' coordinate of event D. Explain why the sign of your result makes sense.

R6B.7 (Do problem R6B.1 first.) Event Q is the explosion of a meteor as it collides with DS9's protective shield. According to measurements in the ship's frame, this event occurs at $t' = 3.0$ s, $x' = 2.0$ s. On the same copy of figure R6.11 that you used to do problem R6B.1, draw and label this event on the spacetime diagram.

R6B.8 Use the appropriate inverse Lorentz transformation equations to compute the Home Frame coordinates of event Q described in problem R6B.7.

Synthetic

R6S.1 An event occurs at $t = 6.0$ s and $x = 4.0$ s in the Home Frame. When and where does this event occur in an Other Frame moving with $\beta = 0.50$ in the $+x$ direction with respect to the Home Frame? Answer this question by using a two-observer spacetime diagram. Check your work, using the appropriate Lorentz transformation equations.

R6S.2 An event occurs at $t = 1.5$ s and $x = 5.0$ s in the Home Frame. When and where does this event occur in an Other Frame moving with speed $\beta = 0.60 = \frac{3}{5}$ in the $+x$-direction with respect to the Home Frame? Is the time order of this event and the origin event the same in both frames? Answer these questions, using a two-observer spacetime diagram; check your work, using the appropriate Lorentz transformation equations. (*Hint: t'* should come out to be *negative*.)

R6S.3 An Other Frame moves with speed $\beta = 0.60 = \frac{3}{5}$ in the $+x$ direction with respect to the Home Frame. In the Other Frame, an event is measured to occur at time $t' = 3.0$ s and position $x' = 1.0$ s. When and where does this event occur as measured in the Home Frame? Answer this question, using a two-observer spacetime diagram; and check your work, using the appropriate inverse Lorentz transformation equations.

R6S.4 An Other Frame moves with speed $\beta = 0.40 = \frac{2}{5}$ in the $+x$ direction with respect to the Home Frame. In the Other Frame, an event is measured to occur at $t' = -1.5$ s and $x' = 5.0$ s. What are the Home Frame coordinates of this event? Does this event occur before or after the origin event in the Home Frame? Answer these questions, using a two-observer spacetime diagram; and check your work, using the inverse Lorentz transformation equations.

R6S.5 The Federation space cruiser *Execrable* is floating in Federation territory at rest relative to the border of Klingon space, which is 6.0 min away in the $+x$ direction. Suddenly, a Klingon warship flies past the cruiser in the direction of the border at a speed $\beta = \frac{3}{5}$. Call this event *A*, and let it define time zero in both the Klingon and cruiser reference frames. At $t_B = 5.0$ min according to cruiser clocks, the Klingons emit a parting disrupter blast (event *B*) that travels at the speed of light back to the cruiser. The disrupter blast hits the cruiser and disables it (event *C*), and a bit later (according to cruiser radar measurements) the Klingons cross the border into Klingon territory (event *D*).
(a) Draw a two-observer spacetime diagram of the situation, taking the cruiser to define the Home Frame and the Klingon warship to define the Other Frame. Draw and label the worldlines of the cruiser, the Klingon territory boundary, the Klingon warship, and the disrupter blast. Draw and label events *A*, *B*, *C*, and *D* as points on your diagram.
(b) When does the disrupter blast hit, and when do the Klingons pass into their own territory, according to clocks in the cruiser's frame? Answer by reading the times of these events directly from the diagram.
(c) The Klingon–Federation Treaty states that it is illegal for a Klingon ship in Federation territory to damage Federation property. When the case comes up in interstellar court, the Klingons claim that they are within the letter of the law: according to measurements made in their

reference frame, the damage to the *Execrable* occurred *after* they had crossed back into Klingon territory: thus they were *not* in Federation territory at the time. Did event *C* (disrupter blast hits the *Execrable*) *really* happen after event *D* (Klingons cross the border) in the Klingons' frame? Answer this question by using your two-observer diagram, and check your work with the Lorentz transformation equations.

R6S.6 Fred sits 65 ns west of the east end (and thus 35 ns east of the west end) of a 100-ns-long train station at rest on earth. Sally operates a reference frame in a train racing east across the countryside at a speed $\beta = 0.50$. At a certain time (call it $t' = 0$) Sally passes Fred. At that same instant, Fred flashes a strobe lamp (call this event *F*), which sends bursts of light both east and west. Alan, who is standing at the west end of the station, receives the west-going part of the flash (call this event *A*), and a bit later (according to clocks in the station) Ellen, who is standing at the east end of the station, receives the east-going flash (call this event *E*).
(a) When do events *A* and *E* occur in the station frame? Who sees the flash first (according to clocks in the station), Alan or Ellen?
(b) Draw a two-observer spacetime diagram of the situation, showing and labeling the worldlines of Sally, Fred, Alan, Ellen, and the two light flashes. Locate and label events *F*, *A*, and *E* as points on the diagram. Carefully draw and calibrate the t' and x' axes for Sally's train frame.
(c) When and where do events *A* and *E* occur in Sally's frame? Sally claims that Ellen sees the flash first in her frame. Is this true? Verify your assertions with calculations based on the Lorentz transformation equations.

R6S.7 A Tirillian spaceship fleeing from battle passes space station DS7 at an essentially constant velocity of $\beta = \frac{3}{5}$ in the $+x$ direction as measured in DS7's frame. Let the event of the ship passing DS7 be the origin event *A* in both frames. The Tirillians have a cloaking device that they think makes them invisible to DS7's sensors. However, 40 s after passing DS7 (as measured by the Tirillian clocks) the spaceship passes through a dust cloud that emits a pulse of electromagnetic radiation when disturbed; let this be event *B*. The instant that this pulse (which travels at the speed of light) is received by DS7, the DS7 crew fires a photon torpedo (which also travels at the speed of light) toward the fleeing Tirillians; call this event *C*. The Tirillians decide 80 s after passing DS7 that they have likely been detected, so they put up their defensive shields (which involves turning off the cloaking device); call this event *D*.
(a) Use a ruler to draw a complete and carefully constructed two-observer spacetime diagram of the situation, drawing the worldlines of DS7

and the Tirillian spaceship and locating and labeling events A, B, C, and D. In the process, answer the following questions:

(b) When and where did event B occur in the Home Frame? Use an appropriate Lorentz transformation equation to check what you read from your diagram.

(c) When does event C occur in the Home Frame? Explain how you located this event on the diagram.

(d) When does event C occur in the Tirillian frame? Explain how you can read t'_C from the diagram, and use an appropriate Lorentz transformation equation to verify your result.

(e) Which event, C or D, occurs first in the DS7 frame? Which occurs first in the Tirillian frame? Explain your reasoning.

(f) Is it possible that the Tirillians could have made their decision to raise shields as a consequence of observing (somehow) that DS7 had fired a torpedo? Why or why not?

R6S.8 Imagine that the physical distance between adjacent marks on the Home Frame t and x axes is some quantity q (for example, q might be 1 cm or 0.25 in.). Argue that the physical distance q' between adjacent marks on the t' and x' axes, as measured *along* those axes, is

$$q' = q\sqrt{\frac{1 + \beta^2}{1 - \beta^2}} \tag{R6.15}$$

Rich-Context

R6R.1 When we have two inertial frames in standard orientation, our convention in preparing a two-observer spacetime diagram is to choose the frame whose relative velocity is in the common $+x$ direction to be the Other Frame in the diagram. What if we do it the other way round, that is, choose the Other Frame to be moving in the $-x$ direction with respect to the Home Frame? Go through the arguments presented in sections R6.4 through R6.6 with this change in mind, and then construct a complete and calibrated two-observer spacetime diagram for the case where the Other Frame moves with a speed of $\beta = \frac{2}{5}$ in the $-x$ direction with respect to the Home Frame. Describe why you chose to draw and calibrate the diagram as you did.

R6R.2[†] The satellites that comprise the Global Positioning System (GPS) constantly broadcast a signal that specifies what time the signal is sent (according to an atomic clock on the satellite) as well as information about its location at the time the signal is sent.

A GPS receiver uses the information sent from multiple satellites and computes its location by doing a complicated calculation that accounts for the signal travel time from the satellites, the rotation of the earth, and a variety of effects predicted by both special and general relativity.

To see just some of the effects of special relativity that are involved, consider a starkly simplified GPS system where the satellites fly at a constant speed of β a negligible distance above the x axis on a flat earth. Assume that the satellites' atomic clocks are synchronized in the satellites' frame, and that you are standing somewhere along the x axis. At a certain instant, your GPS receiver simultaneously receives a signal from somewhere along the $-x$ direction relative to you from a satellite A and another signal from somewhere along the $+x$ direction from a satellite B. The signals both state the same signal departure time ($t'_A = t'_B = 0$ in the satellite's frame) and specify that the satellites are at locations x'_A and x'_B, respectively, at that time.

(a) If the galilean transformation equations were true, and if we were to assume that the speed of both signals was the same in the *earth* frame, argue that your position would be $x = \frac{1}{2}(x'_B + x'_A)$.

(b) Use a ruler to draw a carefully constructed spacetime diagram of the situation, and argue from the diagram that $x \neq \frac{1}{2}(x'_B + x'_A)$. (*Hint:* Choose something simple for β such as $\beta = \frac{3}{5}$, and choose x'_A and x'_B to be something arbitrary.)

(c) Using this diagram to guide your thinking, develop a general equation that calculates your position along the x axis in the flat earth's frame in terms of β and the reported values of x'_A and x'_B, taking account of special relativity. Check that your proposed equation yields results consistent with your diagram for specific values of β, x'_A, and x'_B. (*Hint:* Let event C be the event of your receiver obtaining both signals. Can you find the coordinates of event C in either the spaceship frame or the earth frame?)

(d) Assume that the satellites' common speed is really about 3.9 km/s (which is roughly the orbital speed of the real GPS satellites) and that their reported positions are $x'_A = -3000$ km and $x'_B = +9000$ km. By how much would the naive calculation of part a be in error, according to your relativistic formula? Is this significant?

Advanced

R6A.1[†] We can derive the coordinate difference versions of the Lorentz transformations from scratch in the

---

[†]Adapted from an example in chapter 4 of J. B. Hartle, *Gravity: An Introduction to Einstein's General Relativity*, San Francisco: Addison-Wesley, 2002.

[†]Adapted from E. F. Taylor and J. A. Wheeler, *Spacetime Physics*, San Francisco: Freeman, 1966, p. 68.

following way (which is similar to Einstein's own derivation). For the sake of argument, we will consider events that occur only along the spatial x axis. First *assume* that the equations have to be linear; that is, they have the form

$$\Delta t' = A\,\Delta t + B\,\Delta x \qquad\qquad (R6.16a)$$

$$\Delta x' = C\,\Delta t + D\,\Delta x \qquad\qquad (R6.16b)$$

where A, B, C, and D are unknown constants that do not depend on the coordinates at all, but only on the relative orientation and velocity of the Home Frame and Other Frame. (It can be shown that only linear equations like these transform a constant-velocity worldline in the Home Frame into another constant-velocity worldline in the Other Frame. Since a free particle must be measured to move along a constant-velocity worldline in all inertial frames, this is required of any reasonable transformation equation linking two inertial frames.) Then consider the following pairs of events.

(a) Consider events E and F that both occur at the spatial origin of the Other Frame, so that $\Delta x'_{EF} = 0$. If the Home Frame and Other Frame are in standard orientation, the spatial origin of the Other Frame moves with speed β in the $+x$ direction with respect to the Home Frame, meaning that $\Delta x_{EF}/\Delta t_{EF} = \beta$. Use this to prove that the unknown constants C and D are related as follows: $C = -\beta D$.

(b) Imagine a light flash traveling in the $+x$ direction that is emitted at event G and absorbed at event H. The velocity of this light flash must be observed to be $+1$ in both frames: $\Delta x_{GH}/\Delta t_{GH} = \Delta x'_{GH}/\Delta t'_{GH} = +1$. Show that this implies that $C + D = A + B$.

(c) Imagine a different light flash emitted at event J and absorbed at event K that travels in the $-x$ direction. The velocity of this light flash must be observed to be -1 in both frames: $\Delta x_{JK}/\Delta t_{JK} = \Delta x'_{JK}/\Delta t'_{JK} = -1$. Show that this implies $C - D = -A + B$.

(d) The three equations from parts a, b, and c above, taken together, allow one to express three of the unknowns A, B, C, and D in terms of the fourth. Use these equations to find B, C, and D in terms of A.

(e) Finally, the spacetime interval Δs between two events calculated in *either* reference frame must have the same numerical value: we must have $\Delta t^2 - \Delta x^2 = \Delta t'^2 - \Delta x'^2$ (remember that we are only considering events along the x axis here). Use this condition to fix the value of A, and thus all the other constants. You should find that when you plug your results back into equation R6.16, you get the Lorentz transformation equations.

R6X.1   The train is the Home Frame in the first case and the Other Frame in the second case.

R6X.2   See the answer for exercise R6X.4.

R6X.3   In the Other Frame, the light flash is measured to have a round-trip time of $2T$ by hypothesis and thus must have taken a time T for each leg of the trip. Since light has speed 1 in all frames, and the light-flash in this case travels up and down the spatial x' axis, the reflection event must have occurred along the spatial x' axis at a distance T from the origin, implying that $x' = T$ for the reflection event.

R6X.4   The diagram t' axis in this case should have a slope of rise/run $= \frac{5}{4}$, and marks separated by $\Delta t' = 1.0$ s along the t' axis should have a vertical separation of $\Delta t = \frac{5}{3}$ s $= 1.67$ s. The diagram x' axis would have a slope of $\frac{4}{5}$, and the marks along the axis should have a horizontal separation of $\Delta x = 1.67$ s. See figure R6.12.

R6X.5 $t' \approx -0.9$ s, $x' \approx +2.2$ s

R6X.6 $t = +2.8$ s, $x \approx +0.2$ s

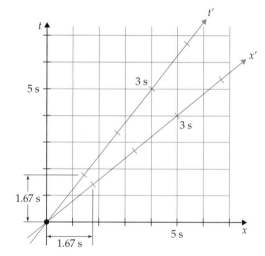

Figure R6.12

See the answer to exercise R6X.4.

R6X.7   No! The events along the diagram x' axis of figure R6.6 are by definition simultaneous in the Other Frame (they all occur at $t' = 0$ by definition), but these events are clearly *not* simultaneous in the

Home Frame, since the line connecting these events is not horizontal (indeed events farther to the right occur later in the Home Frame). Moreover, since *any* set of events simultaneous in the Other Frame must lie along lines *parallel* to the x' axis, and since such lines are not horizontal in a two-observer diagram based on the Home Frame, *no* set of events (with $y' = z' = 0$) that is simultaneous in the Other Frame will be simultaneous in the Home Frame.

R6X.8 To isolate t', multiply equation R6.10*b* by β, and then subtract this equation from R6.10*a*, yielding

$$t - \beta x = \gamma(t' + \beta x - \beta^2 t' - \beta x) = \gamma(1 - \beta^2)t'$$
$$\text{(R6.17)}$$

Remember now that $\gamma \equiv (1 - \beta^2)^{-1/2}$, so $1 - \beta^2 = 1/\gamma^2$. Substituting in for the above and multiplying both sides by γ, we get

$$\gamma(t - \beta x) = \gamma^2(1 - \beta^2)t' = t' \quad \text{(R6.18)}$$

which is equation R6.11*a*. The proof of equation R6.11*b* is entirely analogous.

R6X.9 We have

$$\Delta x' \equiv x'_B - x'_A = \gamma(-\beta t_B + x_B) - \gamma(-\beta t_A + x_A)$$

$$= \gamma[-\beta t_B + x_B + \beta t_A - x_A]$$

$$= \gamma[-\beta(t_B - t_A) + (x_B - x_A)]$$

$$= \gamma(-\beta \, \Delta t + \Delta x) \quad \text{(R6.19)}$$

The proof of equation R6.12*c* is even simpler:

$$\Delta y' \equiv y'_B - y'_A = y_B - y_A \equiv \Delta y \quad \text{(R6.20)}$$

The proof of equation R6.12*d* is analogous.

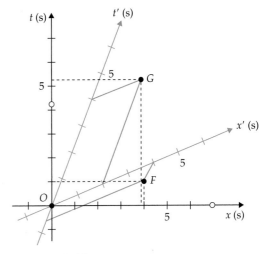

Figure R6.13
See the answers to exercises R6X.10 and R6X.11.

R6X.10 The Lorentz transformation equations imply that

$$t' = \gamma[t - \beta x] = 1.09\left[1.0 \text{ s} - \tfrac{2}{5}(4.0 \text{ s})\right] = -0.66 \text{ s}$$
$$\text{(R6.21a)}$$

$$x' = \gamma[-\beta t - x] = 1.09\left[-\tfrac{2}{5}(1.0 \text{ s}) + 4.0 \text{ s}\right] = 3.9 \text{ s}$$
$$\text{(R6.21b)}$$

The graphical solution is shown in figure R6.13 (event F).

R6X.11 By the inverse Lorentz transformation equations,

$$t = \gamma[t' + \beta x'] = 1.09\left[4.0 \text{ s} + \tfrac{2}{5}(2.0 \text{ s})\right] = 5.2 \text{ s}$$
$$\text{(R6.22a)}$$

$$x = \gamma[\beta t' + x'] = 1.09\left[\tfrac{2}{5}(4.0 \text{ s}) + 2.0 \text{ s}\right] = 3.9 \text{ s}$$
$$\text{(R6.22b)}$$

The graphical solution is shown in figure R6.13 (event G).

R7

Lorentz Contraction

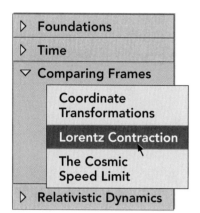
Chapter Overview

Introduction

Because two-observer spacetime diagrams concisely and vividly display the relationships between events in various inertial reference frames, problem solutions that use such diagrams often prove clearer and more compelling than mathematical solutions, even when we have to back up the diagram with mathematical calculations. In this chapter and chapter R8, we will use two-observer diagrams to explore some of the more peculiar and startling predictions of special relativity. In this chapter, we will focus on the phenomenon of *Lorentz contraction*.

Section R7.1: The Length of a Moving Object

We can precisely and operationally define an object's **length** in a given inertial frame in which it is moving to be *the distance between simultaneous events occurring at its ends in that frame*.

Section R7.2: A Two-Observer Diagram of a Stick

An object's **world-region** is the set of worldlines for all particles in the object. On a spacetime diagram, the world-region of a one-dimensional object lying along the $+x$ direction looks like a two-dimensional plane whose edges are the worldlines of the object's ends. We can use a two-observer diagram to determine an object's length in any given frame by locating two events that occur (1) along the worldlines of the object's ends and (2) simultaneously in the frame in question. The distance between those events in that frame is the object's length in that frame, by definition.

We find when we do this that the length determined in this way is always *shorter* than the object's length in its own rest frame: this is the phenomenon of **Lorentz contraction.**

We can calculate this contraction exactly as follows. Consider simultaneous events that mark the object's ends in the frame (call it the Other Frame) in which it moves with speed β. We know that $\Delta t' = 0$ between the events in that frame, and we know that these events must be separated in the object's *rest* frame (the Home Frame) by a distance Δx equal to the object's **rest length** L_R. We can then solve the Lorentz transformation equation R6.13b for the object's length $L = \Delta x'$ in the Other Frame in terms of $\Delta t'$ and Δx between these events. The result is

$$L = L_R\sqrt{1 - \beta^2} \tag{R7.2}$$

Purpose: This equation allows us to an object's measured length L in an inertial reference frame where it moves with constant speed β in terms of the object's rest length L_R (that is, its length measured in a frame where it is at rest).

Limitations: The object must have a well-defined length, and it must be at rest in an inertial frame.

Section R7.3:   What Causes the Contraction?

This contraction effect is entirely due to the fact that observers in different frames disagree about clock synchronization. Since different observers will see different pairs of events marking out the object's ends to be simultaneous, they cannot agree on the object's length.

There is an analogy for this in plane geometry. Consider a road with parallel sides. The road's east-west width depends on the orientation of one's coordinate system, even though the road's physical reality is unchanged. Similarly, an object's projection on a frame's spatial axis (which is what its length really is) is frame-dependent, even though its four-dimensional reality is unchanged.

Section R7.4:   The Contraction Is Frame-Symmetric

An object is contracted in one frame but not in another. Doesn't this contradict the principle of relativity? No! The principle only requires that the *same experiment* performed in different inertial frames yield the *same result*. Therefore if an object at rest in frame A is observed to be contracted in frame B, the principle only requires that an object at rest in frame B be similarly contracted when observed in frame A. It is easy to show that this is true.

Section R7.5:   The Barn and Pole Paradox

Consider a runner who carries a pole (whose rest length is 10 ns) through a barn (whose rest length is 8 ns) at a speed of $\beta = \frac{3}{5}$ relative to the barn. In the ground frame, the pole is 8 ns long, so there is an instant when the pole fits entirely inside the barn. In the runner's frame, though, the barn is 6.4 ns long and the pole is 10 ns long. How can a 10-ns pole fit in a 6.4-ns barn?

The solution of this "paradox" is as follows. The phrase "fits inside" really means the event B of the pole's back end entering the barn is simultaneous with the event F of the pole's front end leaving the barn. But events B and F are *not* simultaneous in the runner's frame: event B happens only after enough time has passed after event F for the 10-ns pole to move 3.6 ns beyond the front end of the barn. So the runner *never* sees the barn enclose the pole. The trick with this "paradox" (and many similar ones) is to recognize that events that are simultaneous in one frame are *not* generally simultaneous in another.

Section R7.6:   Other Ways to Define Length

We do not *have* to define length as we did in section R7.1. This section, however, argues that several reasonable and well-defined alternative definitions display the exactly the same Lorentz contraction effect.

R7.1   The Length of a Moving Object

The basic question that will concern us in this chapter can be stated simply as follows: what exactly do we mean by the *length* of a moving object?

As always, we need an *operational definition* of this word if it is to mean anything: that is, we need to describe exactly how the length of an object can be *measured* in a given inertial frame. In the particular inertial frame where the measuring stick is at rest, it is simple to compare the object to a stationary ruler. But the determination of an object's length in a frame in which it is observed to be moving presents difficulties that need to be handled carefully.

We have defined a reference frame to be an apparatus that measures the spacetime coordinates of *events*. Our first task in the problem of measuring lengths (and indeed most problems in relativity theory) is to rephrase the problem in terms of *events*. In a given reference frame, how might we characterize the length of an object in terms of events?

Let us consider a concrete example. Imagine that we are trying to measure the length of a moving train in the reference frame of the ground. A clock lattice at rest on the ground records the passage of the train through it by describing the motion in terms of events. To be specific, imagine that a certain clock in the lattice records that the *back* end of the train passed at exactly 1:00:00 p.m. (call this event O). Another clock elsewhere in the lattice records that the *front* end of the train passed at exactly 1:00:00 p.m. (call this event A). Therefore, we can say that at exactly 1:00:00 p.m., the train lies between the location of the clock registering event O and that of the clock registering event A. It therefore makes sense to define the length of the train to be equal to the distance between those events, as measured in the lattice.

With this image in mind, we will *define* the length of an object operationally as follows:

An object's **length** in a given inertial frame is defined to be the *distance* between any two *simultaneous events* that occur at its ends.

This expresses the definition of length in the language of events, enabling us to use the tools that we have been developing to describe the relationships between events to talk about the *process* of measuring an object's length.

R7.2   A Two-Observer Diagram of a Stick

Consider a measuring stick oriented along the spatial x direction and at rest in the Home Frame. How can we represent such an object on a spacetime diagram? To present the full reality of a measuring stick in spacetime, one must plot the worldline of each particle in the stick. Just as a point particle is represented on a spacetime diagram by a curve called a worldline, so a stick is represented by an infinite number of associated worldlines, which one might call a **world-region**. An example of a world-region is shown in figure R7.1.

The definition of length given in section R7.1 yields the expected result when the object in question is at rest. Consider the 4-ns measuring stick of figure R7.2, which is at rest in the Home Frame of the diagram. Events O and A lie at the ends of the measuring stick and are simultaneous in the Home Frame: both occur at $t = 0$ in that frame. According to our definition, then, the length of the measuring stick in the Home Frame is the distance between these two events, which, according to figure R7.2, is simply 4 ns.

Now consider determining the length of this same measuring stick in an Other Frame that is moving with speed β in the $+x$ direction with respect to the Home Frame. An observer in that frame will observe the stick to move in

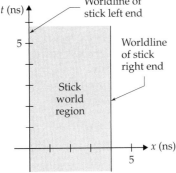

Figure R7.1
Part of the world-region of a
4-ns measuring stick at rest in
the Home Frame with one end
at $x = 0$ and the other end at
$x = 4$ ns. Because it is at rest in
the Home Frame, the worldlines
of its endpoints are vertical lines,
and the worldlines of all the
points in between fill in the region
of spacetime shown in gray.

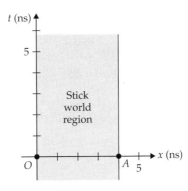

Figure R7.2
The two events O and A lie along
the worldlines of the measuring
stick's ends and also happen to
occur at the same time in the
Home Frame. The distance
between these events (which is
4 ns in this case) is the measuring
stick's length in the Home Frame,
by definition.

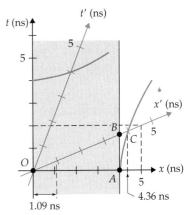

Figure R7.3
Events O and B lie along the
worldlines of the measuring stick's
ends and occur simultaneously (at
$t' = 0$) in an Other Frame moving
with $\beta = \frac{2}{5}$ with respect to the
Home Frame. According to the
diagram, the distance between
these events (which is the stick's
length in the Other Frame)
is ≈ 3.7 ns < 4 ns.

the $-x$ direction with speed β. Figure R7.3 is a two-observer spacetime diagram showing the Home Frame and Other Frame axes superimposed on the world-region of the measuring stick. (For the sake of concreteness, the diagram is constructed assuming that the relative speed of the frames is $\beta = \frac{2}{5}$.) In the Other Frame, it is not event A but B (as shown on the diagram) that is simultaneous with O and lies at the other end of the measuring stick (both O and B lie on the diagram x' axis, so both occur at $t' = 0$ in the Other Frame). This means that the length of the measuring stick as observed in the Other Frame is *defined* to be the distance between events O and B as measured in that frame.

But as the calibrated axes of the Other Frame show, the distance between O and B in that frame is *less* than 4 ns! We can see that this *must* be so as follows. Consider the 4-ns mark on the diagram x' axis (the event labeled C) on the spacetime diagram. This mark must be connected with a hyperbola to the 4-ns mark on the diagram x axis (event A). It is easy to see that the mark on the x' axis is farther along that axis at $t' = 0$ than the stick's right end is at that instant (event B). Therefore, the stick's length in the Other Frame, which is the distance between events O and B by definition, must be *smaller* than 4 ns.

We can use the other method of axis calibration to see the same thing. The 4-ns mark on the x' axis (event C) must be separated from the origin event by a horizontal displacement $\Delta x_{OC} = \gamma\, \Delta x'_{OC} = (1.09)(4\text{ ns}) \approx 4.36$ ns. This means that event B must be closer to the origin event than the mark event C at $x' = 4.0$ ns, which in turn implies that $\Delta x'_{OB} < 4$ ns!

Either way we look at it, the right edge of the measuring stick, which is always exactly 4 ns from the spatial origin of the Home Frame, intersects the diagram x' axis at an event B that is *closer* to the origin event than the 4-ns mark on the diagram x' axis, implying that the stick is measured in the Other Frame to have a length of *less* than 4 ns. In fact, you can see that in the case shown, the stick has a length of about 3.7 ns in the Other Frame.

We can also easily check this result with the help of the Lorentz transformation equations. In the Other Frame, the measuring stick's length L is *defined* to be the distance $\Delta x'$ between two *simultaneous* events that occur at the ends of the stick, that is, events for which $\Delta t' = 0$. Assuming we know that the length of the measuring stick in the Home Frame is $L_R = 4.0$ ns and that it is at rest in that frame, then the Home Frame distance between *any* pair of events that occur at the opposite ends of the measuring stick must be $\Delta x = L_R = 4.0$ ns (see figure R7.3). One of the inverse Lorentz transformation equations for coordinate differences (equation R6.13b) says that

$$\Delta x = \gamma(\beta\,\Delta t' + \Delta x') \qquad (R7.1)$$

In the case at hand, we are looking for $L = \Delta x'$, knowing $\Delta x = L_R$ and $\Delta t' = 0$ for events that simultaneously mark out the ends of the measuring stick in the Other Frame. Therefore, dropping the $\Delta t'$ term and solving for $\Delta x'$, we get

General formula for an object's length in a frame in which it is moving

$$L = \Delta x' = \frac{L_R}{\gamma} = L_R\sqrt{1 - \beta^2} \qquad (R7.2)$$

Purpose: This equation allows us to measure an object's length L in an inertial reference frame where it moves with constant speed β in terms of the object's rest length L_R (that is, its length measured in a frame where it is at rest).

Limitations: The object must have a well-defined length, and it must be at rest in an inertial frame.

Plugging in the relevant numbers in this case, we find that

$$L = (4.0\text{ ns})\sqrt{1 - \left(\tfrac{2}{5}\right)^2} = 3.7\text{ ns} \qquad (R7.3)$$

in agreement with the result displayed in figure R7.3.

We see that if we accept the definition of length given in section R7.2 (and how *else* can we define the length of a moving object?), we are confronted with the fact that an object's length is a *frame-dependent* quantity: its value depends on which inertial frame one chooses to make the measurement. Equation R7.2 implies that the length of an object measured in a frame in which the object is moving will always be *smaller* than the value of its length in the frame in which it is at rest. This phenomenon is called **Lorentz contraction.**

Definition of an object's rest length

We can use equation R7.2 quite generally to compute the magnitude of the measured contraction, as long as we remember that L_R (which we call the object's **rest length**) stands for the length of the stick measured in the frame in which it is at rest, and L stands for the contracted length of the stick measured in an inertial frame that moves with speed β with respect to the stick's rest frame.

Exercise R7X.1

By drawing appropriate axes on the diagram, estimate the length of the measuring stick shown in figure R7.3 when it is observed in a frame that is moving with $\beta = \tfrac{4}{5}$ relative to the Home Frame.

Exercise R7X.2

Use equation R7.2 to check the answer that you found for exercise R7X.1.

R7.3   What Causes the Contraction?

The discussion above shows that this contraction effect has nothing to do with some effect of motion that physically compresses a moving object (as, for example, an elastic object such as a balloon would be compressed if it were forced to move rapidly through water). The physical reality of the measuring stick (as represented by its world-region on the spacetime diagram) actually *remains the same*, no matter what reference frame one uses to describe it. The fundamental reason why observers in different inertial frames will measure the same object to have different lengths is that the observers disagree about clock synchronization, and therefore disagree about which events mark out the ends of the object "at the same time." For example, observers in the Home Frame of figure R7.3 use events O and A, while observers in the Other Frame use events O and B. We see then that the phenomenon of Lorentz contraction has its origin in the problem of clock synchronization!

Lorentz contraction has its origin in problems of clock synchronization

Nonetheless, the idea that the same object can be measured to have different lengths in different inertial frames may be hard to accept. Yet we are not at all surprised by the analogous behavior of geometric objects on a two-dimensional plane. Let me illustrate. Imagine that we want to determine the east-west width of a road running in a roughly northerly direction on the earth's surface. Two different surveyors set up differently oriented coordinate systems and make this measurement. Is it surprising that they get different results (see figure R7.4)?

A geometric analogy

The east-west width of the road shown in figure R7.4 is seen to be greater in the primed coordinate system than in the unprimed system. Has the road magically expanded for the surveyor who laid out the primed coordinate system? Hardly! The physical reality of the road does not change just because we change the coordinate system in which we measure it. But because the two surveyors cannot agree on which two points that lie along the sides of the road also lie on an east-west line, they will measure the east-west width of the road to have different values.

We do not find this problematic or even unexpected. Now, we should say that if you are going to measure the "true" width of the road, you should measure it by using a coordinate system in which the road runs parallel to the y axis, so that the x axis is *perpendicular* to the road. In that special coordinate system, the width of the road will have its "true" value (which is *shorter* than the value of the same measured in any secondary coordinate system).

The analogy between a road's "true width" and an object's rest length

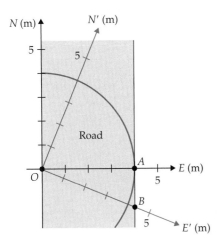

Figure R7.4

Points O and A span the east-west width of the road in the Home Frame coordinate system, while points O and B span the same in the Other Frame coordinate system. The circle shown connects all points that lie 4 m from the origin. From the picture, it is clear that the road has a greater east-west width in the Other Frame coordinate system than in the Home Frame system. Does this mean that the road has magically expanded for the surveyors in the Other Frame coordinate system?

Similarly, we might say that to measure an object's "true" length in spacetime, we should measure its length in the inertial frame in which it is at rest. This "true length" (more correctly, the object's **proper length**) will be *longer* than the value measured in any other inertial frame. For clarity's sake, let us always refer to this length as the object's **rest length.**

Exercise R7X.3

Find a formula for the east-west width of a road W in terms of its perpendicular width $W_\perp$ and the slope S of the E' axis in the Home Frame coordinate system. Compare with equation R7.2.

R7.4   The Contraction Is Frame-Symmetric

Does the Lorentz contraction provide a way of distinguishing a rest frame from a moving frame?

The principle requires that identical experiments yield identical results

But, you might ask, does this Lorentz contraction effect not violate the principle of relativity? We have seen that an object at rest in the Home Frame is measured to have a shorter length in the Other Frame. Does this not imply that there is a physically measurable distinction between the two frames, a distinction that would violate the requirement that all inertial frames be equivalent when it comes to the laws of physics?

The principle of relativity does *not* require that measurements of a specific object or of a set of events have the same values in all reference frames. What the principle *does* require is that if we do exactly the same *physical experiment* in two different inertial reference frames, we get exactly the same result (otherwise, the laws of physics that predict the outcome of the experiment will be seen to be different in the different frames). Now, we have seen that if we take a measuring stick that is at rest in the Home Frame and 4 ns long in that frame and measure its length in the Other Frame, we will find it to be Lorentz-contracted to 3.7 ns in length. The principle of relativity *does* require that if we perform the *same* experiment in the Home Frame, we get the *same* result: that is, if we take a 4.0-ns measuring stick at rest in the Other Frame and measure its length in a Home Frame moving at a speed of $\beta = \frac{2}{5}$ with respect to the Other Frame, we should find the stick to be Lorentz-contracted to 3.7 ns.

Figure R7.5 shows that this is indeed so. The worldlines of the ends of a measuring stick that is at rest in the Other Frame will be parallel to the t' axis, as shown. Events O and D mark out the two ends of the measuring stick at time $t = 0$ in the Home Frame. The distance between these events (i.e., the

Figure R7.5
This measuring stick has a rest length of 4.0 ns, because events O and E, which occur simultaneously (in the Other Frame) and at opposite ends of the measuring stick, are 4.0 ns apart in this frame. Events O and D occur at opposite ends of the stick and simultaneously in the Home Frame, so the distance between these events is defined to be the length of the stick in that frame. We can see from the diagram that this length is about 3.7 ns.

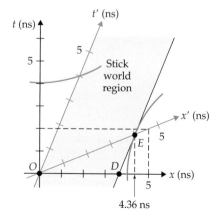

length of the measuring stick in that frame) is seen to be about 3.7 ns, as expected.

Again we can also easily check this with the help of the Lorentz transformation equations. In the Home Frame, the length L of the measuring stick is *defined* to be the distance Δx between *simultaneous* events occurring at the ends of the stick, that is, events for which $\Delta t = 0$. Assuming that we know that the length of the measuring stick in the Other Frame is $L_R = 4.0$ ns and that it is at rest in that frame, then the distance between *any* pair of events that occur at the opposite ends of the measuring stick must be $\Delta x' = L_R = 4.0$ ns in that frame. One of the Lorentz transformation equations for coordinate differences (equation R6.12b) says that

$$\Delta x' = \gamma(-\beta\,\Delta t + \Delta x) \qquad (R7.4)$$

In the case at hand, we are looking for $L = \Delta x$, knowing that $\Delta x' = L_R$ and that $\Delta t = 0$ for the events in question. Therefore, dropping the Δt term in equation R7.4 and solving for Δx, we get

$$L = \Delta x = \frac{L_R}{\gamma} = L_R\sqrt{1 - \beta^2} = (4.0\text{ ns})\sqrt{1 - \left(\tfrac{2}{5}\right)^2} = 3.7\text{ ns} \qquad (R7.5)$$

in agreement with the result displayed by figure R7.5.

In summary, we see that it doesn't matter whether the stick is at rest in the Home Frame or at rest in the Other Frame: if the stick is observed to have a length L_R in the frame in which it is at rest, it has a length $L = L_R(1 - \beta^2)^{1/2}$ in any frame that is moving with speed β relative to the stick's rest frame. If the stick is at rest in the Home Frame, it is observed to be contracted in the Other Frame. If it is at rest in the Other Frame, it is observed to be contracted in the Home Frame. This frame symmetry of the Lorentz contraction phenomenon means that it *is* consistent with the principle of relativity.

Lorentz contraction is frame-symmetric and thus consistent with the principle of relativity

R7.5   The Barn and Pole Paradox

The predictions of the theory of relativity are sufficiently counterintuitive that it is easy (as a result of fuzzy thinking) to invent situations that at first appear to be paradoxical. We dealt with one of the most famous, the *twin paradox*, in chapter R5. In this section, we will examine another famous apparent paradox, generally known as the *barn and pole paradox*, that is based on a natural but mistaken understanding of the phenomenon of Lorentz contraction.

Consider the following problem. Imagine a pole carried by a pole-vaulter who is running along the ground at a speed $\beta = \tfrac{3}{5}$. In the frame of the runner, the pole is at rest (of course): let us assume that it has a rest length of 10 ns. An observer on the ground is moving with speed β with respect to the rest frame of the pole and so will measure the pole to be Lorentz-contracted to a length of only $L = L_R(1 - \beta^2)^{1/2} = (10\text{ ns})(1 - \tfrac{9}{25})^{1/2} = (10\text{ ns})(\tfrac{16}{25})^{1/2} = (10\text{ ns})(\tfrac{4}{5}) \approx 8$ ns (see figure R7.6). As the runner presses on, she runs through a barn that also happens to be 8 ns long as measured in the ground frame. Since both the pole and the barn are 8 ns in the ground frame, there is an instant of time in that frame in which *the pole is entirely enclosed by the barn*.

Description of the barn and pole problem

But now look at the situation from the perspective of the runner. In her frame, the pole is at rest and has its normal length of 10 ns. She sees the *barn* to be moving relative to her at a speed of $\tfrac{3}{5}$, and so it is the *barn* that is Lorentz-contracted to $\tfrac{4}{5}$ of its ground frame length, that is, to $(\tfrac{4}{5})(8\text{ ns}) = (32/5)\text{ ns} = 6.4$ ns. Thus the paradox: *how can a barn that is 6.4 ns long ever enclose a 10-ns pole?*

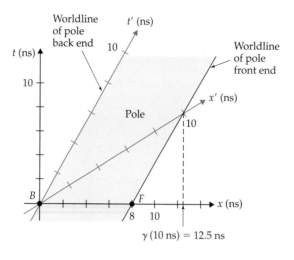

Figure R7.6
The world-region of a pole with a rest length of 10 ns being carried at a speed of $\frac{3}{5}$ relative to the ground frame (the Home Frame in this diagram). Events B and F mark out the ends of the pole at $t = 0$ in the ground frame. The pole's length in the ground frame is thus the distance between these events, or about 8 ns according to the diagram.

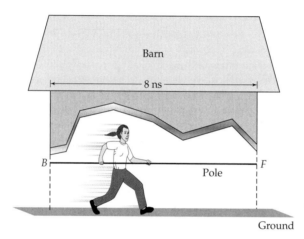

Figure R7.7
A picture of the barn and pole problem as it would be observed in the ground frame. Events F (front of the pole passes through barn's front door) and B (back of pole passes through barn's back door) are simultaneous in the ground frame: we can define the coordinate time of these events to be $t = 0$. Note that at this instant, the pole is completely enclosed by the barn.

This apparent paradox results from a naïve application of the idea that "moving objects are contracted" without really understanding *why* objects are measured to be contracted and exactly how the length of a moving object is measured. We will see that the apparent paradox is resolved if we carefully consider the precise meaning of the words used to describe it.

Restating the problem in terms of events

The first step in solving this problem (and virtually every other problem in special relativity) is to rephrase the problem in terms of *events*. Let us call the arrival of the front end of the pole at the front end of the barn event F. Call the arrival of the back end of the pole at the back of the barn event B. To say that there is an instant at which the barn encloses the pole is to say that events F and B are simultaneous in the ground frame (see figure R7.7).

For the sake of argument, let us agree to use event B as the origin of both space and time in both frames (that is, B occurs at $x = 0$ and $t = 0$ in the ground frame and $x' = 0$ and $t' = 0$ in the runner's frame). The statement that F and B are simultaneous in the ground frame then means that event F also occurs at $t = 0$ in the ground frame. But when does event F occur in the runner's frame?

Figure R7.8 shows a two-observer spacetime diagram for this problem. I have chosen the ground frame to be the Home Frame of the diagram. I have also taken the ground observer's description of the events to be truthful: events B and F *do* occur simultaneously in the ground frame, and the pole is enclosed by the barn at time $t = 0$ in the ground frame. Notice also that the diagram *does* support the runner's assertion that the barn is 6.4 ns long in her

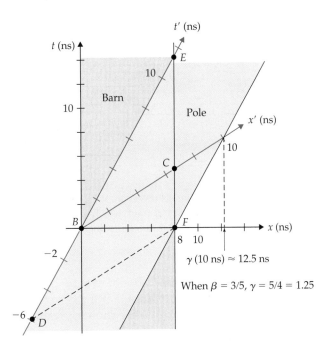

Figure R7.8
Graphical solution to the barn and pole paradox. Events *B* and *C* mark the end of the barn at $t' = 0$ in the runner's frame. These events are roughly 6.4 ns apart according to the diagram, so the barn is indeed about 6.4 ns long as measured in the runner's frame. But note that events *B* and *F* are not simultaneous in the runner's frame. Indeed, event *F* occurs at the same time as event *D*, or about 6 ns before event *B* (note that the line connecting events *F* and *D* is parallel to the *x'* axis).

frame: events *B* and *C* are simultaneous in the runner's frame and lie at the ends of the barn, so the distance between them is the length of the barn in that frame (by definition). The diagram shows that this length is indeed about 6.4 ns.

So what is the solution to the paradox? The diagram shows that *the runner never observes the pole to be enclosed by the barn.* Event *F* is *not* simultaneous with event *B* in the runner's frame: rather *F* is simultaneous with event *D* (note that the line connecting *F* and *D* is parallel to the *x'* axis). This means that event *F* (front of pole reaches front of barn) occurs about 6 ns *before* event *B* (back of pole reaches back of barn) in the frame of the runner. At the same time as event *F* occurs in the runner's frame (i.e., at $t' = -6$ ns), you can see from the diagram that the pole's back end is still sticking out behind the barn. When event *B* finally occurs (at $t' = 0$), the pole's front end is sticking out in front of the barn. (Remember that all events occurring "at the same time" as a given event in the runner's frame lie on a line parallel to the diagram *x'* axis.)

Solution to the paradox

We can use the Lorentz transformation equations to verify that event *F* does indeed occur before *B* in the runner's frame. In the Home Frame, the coordinate differences between events *F* and *B* are $\Delta t_{BF} = 0$ and $\Delta x_{BF} \equiv x_F - x_B = 8$ ns $- 0 = 8$ ns. [The factor $\gamma = 1/(1 - \beta^2)^{1/2} = 5/4$ here.] Therefore, in the runner's frame,

Check using the Lorentz transformation equations

$$\Delta t'_{BF} = \gamma(\Delta t_{BF} - \beta \, \Delta x_{BF}) = \frac{5}{4}\left[0 - \frac{3}{5}(8 \text{ ns})\right] = -\frac{3}{4}(8 \text{ ns}) = -6 \text{ ns} \quad (R7.6)$$

Since $\Delta t'_{BF} = t'_F - t'_B$ and $t'_B = 0$, we have $t'_F = -6$ ns, implying that event *F* occurs about 6 ns before *B* in the runner's frame, as we read from the diagram.

Now let us think about this for a minute. If *you* were the runner and you were told that you were about to run a 10-ns pole through a 6.4-ns barn, what would you *expect* to see? First, you would see the front end of your pole reach the front end of the barn (event *F*). At this time, your 10-ns pole would stick out 3.6 ns behind the rear of the 6.4-ns barn. After the barn moves backward relative to you another 3.6 ns, the back end of your pole will coincide with the back end of the barn (event *B*), at which time the front of the pole sticks out 3.6 ns in *front* of the barn (see figure R7.9).

To the runner, everything looks exactly as if a 10-ns pole is being carried through a 6.4-ns barn

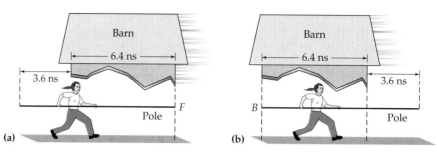

Figure R7.9

(a) The view from the runner's frame. Event *F* occurs first, at which time the pole sticks out about 3.6 ns behind the barn. (b) Event *B* occurs next, at which time the pole sticks out 3.6 ns in front of the barn. The time between events *F* and *B* is the time that it takes the barn to move a distance of 3.6 ns at its speed of $\beta = \frac{3}{5}$ relative to the runner's frame.

How long before event *B* should event *F* occur? The time between these events should be the time required for the barn to move backward a distance of 3.6 ns at a speed of $\beta = \frac{3}{5}$ (in the runner's frame), or

$$\Delta t = \frac{\Delta x}{\beta} = \frac{3.6 \text{ ns}}{\frac{3}{5}} = (3.6 \text{ ns})\left(\frac{5}{3}\right) = 6 \text{ ns} \tag{R7.7}$$

which is the time between the events indicated on the spacetime diagram in figure R7.8. In fact, you should go over that diagram very carefully and convince yourself that the description given above is indeed exactly what the runner will observe in her reference frame.

The point is that nothing strange or weird happens in either frame. In the barn frame, observed events are consistent with the interpretation that an 8-ns pole is being carried through an 8-ns barn. In the runner's frame, the time relationship between the same events is, on the other hand, consistent with the interpretation that a 10-ns pole is being carried through a 6.4-ns barn: we do not see anything like "a 10-ns pole enclosed by a 6.4-ns barn." The apparent paradox in the problem as it was stated is based on an unstated and erroneous assumption that if events *F* and *B* were simultaneous (i.e., the pole is enclosed in the barn) in the ground frame, the events will *also* be simultaneous in the runner's frame. However, when we remember that the coordinate time measured between two events will *not* in general be the same in different frames, the paradox evaporates. Excepting the phenomenon of Lorentz contraction itself, *nothing* unusual is seen to happen in either frame.

People have invented a number of apparent paradoxes analogous to the barn and pole paradox (see the homework problems for examples). Such paradoxes almost always involve a hidden assumption that two events that are simultaneous in a given inertial reference frame are simultaneous in all reference frames. We are taken in by the apparent paradox because our intuitive belief in the absolute nature of simultaneity is so natural that we hardly notice when it is assumed. But special relativity teaches us that this assumption is *not* true: because observers in different inertial frames do *not* observe clocks in other frames to be synchronized, they will disagree about whether two given events are simultaneous. Moreover, because we have defined the *length* of a moving object in a given frame in terms of simultaneous events in that frame, it follows that different observers will disagree about that length as well. The frame dependence of simultaneity and the phenomenon of Lorentz contraction are bound together, and indeed they fail to make sense without each other (as these paradoxes show).

Similar "paradoxes" hinge on the same error about simultaneous events

---

Exercise R7X.4

Using figure R7.8, describe how you could verify that at the time of event F in the runner's frame, the pole's back end is about 3.6 ns of distance behind the barn's back door.

Exercise R7X.5

Use the Lorentz transformation equations to verify that at the time of event F in the runner's frame, the pole's back end is about 3.6 ns of distance behind the barn's back door.

R7.6   Other Ways to Define Length

We have seen that if we define the length of a moving object as the distance between simultaneous events occurring at its ends, we get a frame-dependent (Lorentz-contracted) answer. Are there other ways that we might define a moving object's length? If so, do these yield different results than our original definition?

Alternative definitions of length

The answers to these questions are as follows. Yes, there are other logically reasonable ways of defining the length of a moving object, but these definitions, if they are logically reasonable, yield the *same* numerical result for the object's length as the definition involving synchronized events presented in section R7.2.

An example of such an alternative definition

For example, consider again a moving train. Instead of defining its length to be the distance between simultaneous events at its ends, we might define its length L in our frame to be the total time Δt that the train takes to pass a given point in our frame times its speed β in our frame:

$$L \equiv \beta\,\Delta t \tag{R7.8}$$

The value of $\beta\,\Delta t$ should yield the distance that the train moves as it passes the point, which is a reasonable definition of its length.

This definition, however, yields exactly the same (Lorentz-contracted) result as our original definition. To see this, let us apply this formula to determine the length in the runner's frame of the barn described in section R7.5. Event B is the event of the barn's back end passing the point $x' = 0$ (that is, the diagram t' axis) in the runner's frame, while event E is the event of the barn's front end passing this point. You can see from figure R7.8 that it takes the barn roughly $\Delta t' \approx 10.7$ ns to pass the point $x' = 0$ in the runner's frame. Since the barn is traveling backward at a speed of $\beta = \frac{3}{5}$ in the runner's frame, the barn's length according to our new definition must be

This definition yields the same result as the original

$$L \approx \tfrac{3}{5}(10.6\ \text{ns}) \approx 6.4\ \text{ns} \tag{R7.9}$$

which is roughly the same (contracted) result we found before.

We can (as usual) use the Lorentz contraction formula to calculate the exact length according to this definition. Both events B and E occur at $x' = 0$ in the runner's frame by definition, so $\Delta x' = 0$ for these events. Since these events occur at opposite ends of the barn, they are separated by $\Delta x = 8.0$ ns in the barn's frame. One of the inverse Lorentz transformation equations tells us that

$$\Delta x = \gamma(\beta\,\Delta t' + \Delta x') \tag{R7.10a}$$

so (since $\Delta x' = 0$)

$$\Delta t' = \frac{\Delta x}{\gamma \beta} = \frac{8 \text{ ns}}{(5/4)\left(\frac{3}{5}\right)} = \frac{4}{3}(8 \text{ ns}) = \frac{32}{3} \text{ ns} = 10.67 \text{ ns} \qquad (R7.10b)$$

According to equation R7.8, the length of the barn in the runner's frame is thus

$$L = \beta \, \Delta t = \frac{3}{5}\left(\frac{32}{3} \text{ ns}\right) = \frac{32}{5} \text{ ns} = 6.4 \text{ ns} \qquad (R7.11)$$

which is exactly the same as the result we found before.

... as do other definitions

This is just one example of an alternative definition of length (see problem R7A.1 for discussion of another). *All* known reasonable definitions of the length of a moving object yield the same result as equation R7.2: no matter how you compute it, the length of an object determined in a frame where it is moving is Lorentz-contracted from its rest length by the factor $(1 - \beta^2)^{1/2}$.

Exercise R7X.6

Using equation R7.8, find the length of the runner's pole in the ground frame (a) by reading Δt from figure R7.8 and (b) by computing Δt with the help of an appropriate Lorentz transformation equation.

TWO-MINUTE PROBLEMS

R7T.1   A moving object's length in a given frame is defined to be the distance between two events that occur at opposite ends of the object and that are simultaneous in that frame. *Why* is it crucial that the events we use to define a moving object's length be *simultaneous*?
   A.   This is purely conventional: there is no other reason.
   B.   This choice makes it easier to use the Lorentz transformation equations to find the length.
   C.   If the events are *not* constrained to be simultaneous, then the length is poorly defined: its value would depend on the time interval between the events.
   D.   If the events are simultaneous, then the length will be a frame-independent quantity.
   E.   Other (specify).

R7T.2   Since the ends of an object do not move in its rest frame, the events used to mark out an object's length in that frame do not *have* to be simultaneous: the distance between them is the object's rest length whether they are simultaneous or not, true (T) or false (F)?

R7T.3   An object of rest length L_R moving at one-half the speed of light will have a length equal to
   A. $\frac{1}{2} L_R$
   B. $\frac{3}{4} L_R$
   C. $\left(\frac{1}{2}\right)^{1/2} L_R$
   D. $0.87 L_R$
   E. $\frac{1}{4} L_R$
   F.   Other (specify)

R7T.4   An object is at rest in the Home Frame. Imagine an Other Frame moving at a speed of $\beta = \frac{4}{5}$ with respect to the Home Frame. The object's length in the Other Frame is measured to be 15 ns. What is its length as observed in the Home Frame?
   A.   15 ns
   B.   12 ns
   C.   9 ns
   D.   19 ns
   E.   25 ns
   F.   Other (specify)

R7T.5   The most important reason that an object is observed to be shorter in a frame where it is moving than in a frame where it is at rest is that
   A.   The force of motion strongly compresses an object that is moving at relativistic speeds.
   B.   "Simultaneity" is not a frame-independent concept.
   C.   The measuring sticks used by the moving observer are Lorentz-contracted.
   D.   The clocks used by the moving observer run slower.

R7T.6 In the pole and barn problem, the barn never actually encloses the pole in the ground frame, T or F?

R7T.7 We can define the length of a moving object to be its speed times the time that it takes to pass a given point, T or F?

HOMEWORK PROBLEMS

Basic Skills

R7B.1 How fast does an object have to be moving in a given frame if its measured length in that frame is one-half its rest length?

R7B.2 How fast does an object have to be moving in a given frame if its measured length in that frame is to be significantly different from its rest length? (Assume that you can measure the object's length to 1 part in 10,000, that is, to four significant figures.)

R7B.3 An Other Frame moves with a speed of 0.80 relative to the Home Frame. An object at rest in the Home Frame has a length of 30 ns. What is the object's length in the Other Frame?

R7B.4 An Other Frame moves with a speed of 0.80 relative to the Home Frame. An object at rest in the Other Frame has a length of 30 ns as measured in the Home Frame. What is the object's length in the Other Frame?

R7B.5 Imagine that an object with a rest length of 10 ns is at rest in a frame that is moving with a speed of $\beta = 0.50$ relative to the Home Frame. Draw a two-observer spacetime diagram of this situation, and use it to determine the length of the object in the Home Frame. Check your result by using equation R7.2.

R7B.6 Imagine that an object with a rest length of 5 ns is at rest in the Home Frame. The Other Frame is moving with a speed of $\beta = 0.50$ relative to the Home Frame. Draw a two-observer spacetime diagram of this situation, and use it to determine the length of the object in the Other Frame. Check your result, using equation R7.2.

R7B.7 An observer at rest with respect to the sun measures the diameter of the earth (in the direction of its motion) as it swings by in its orbit. How many centimeters shorter is this diameter in this frame than its rest diameter of 12,760 km? (*Hint:* Use the binomial approximation.)

R7B.8 About how many femtometers shorter than its rest length is the length of a car measured in the ground frame if the car is traveling at 30 m/s (66 mi/h) in that frame? Assume for the sake of argument that the car's rest length is 5.0 m. Remember that 1 fm = 10^{-15} m $\approx$ approximate size of an atomic nucleus.

How much is this object contracted due to its orbital motion around the sun? (See problem R7B.7.)

Synthetic

R7S.1 Imagine an alien spaceship traveling so fast that it crosses our galaxy (whose rest diameter is 100,000 ly) in only 100 y of spaceship time. Observers at rest in the galaxy would say that this was possible because the ship's speed β was so close to 1 that the proper time it measured between its entry into and departure from the galaxy was much shorter than the galaxy frame coordinate time ($\approx$ 100,000 y) between those events. But how does this look to the aliens? To them, their clocks are running normally, but the galaxy, which moves backward relative to them at speed $\beta \approx 1$, is Lorentz-contracted to a bit less than 100 ly across: *this* is what makes it possible for the whole galaxy to fly by them in only 100 y. Find the exact value of the speed β that the aliens must achieve to cross the galaxy in 100 y. Then find the diameter of the galaxy in the aliens' frame, and verify that it is possible for a galaxy moving at speed β with this diameter to pass the aliens' ship completely in 100 y.

R7S.2 In the experiment described in problem R4S.7, particles travel at $v = 0.866$ between detectors 2.08 km apart. This takes 8.0 μs as measured by laboratory clocks. Since the half-life of the particles involved is 2.00 μs, we might naively expect only about one-sixteenth of the particles to survive the trip. But it was shown in that problem that the particles' clocks only measure 4.00 μs for the trip between the detectors, and thus about one-fourth of the particles survive.

But now consider how this all looks to an observer traveling with one of the particles. In the particle's frame, the laboratory and the detectors appear to be moving past at a speed of 0.866. In 4.00 μs (as measured by the particle's clock), the laboratory will only move by a distance of 1.04 km at that speed, so the particles will only see *one-half* of the distance between the detectors go by. But laboratory observers claim that by the time that the particles' clock read 4.00 μs, they have covered the *full* distance between the detectors. Is this not a paradox?

Resolve the apparent paradox by considering the phenomenon of Lorentz contraction. How far apart are the detectors in the particle frame? How does this resolve the paradox? (*Hint:* See problem R7S.1 for a similar situation.)

R7S.3 *A useful approximation.* Consider a speed β that is very close to the speed of light: $\beta = 1 - \delta$, where δ is a very small number. Use the binomial approximation to show that under these circumstances $1/\gamma = (1 - \beta^2)^{1/2} \approx (2\delta)^{1/2}$. How accurate is this approximation when $\beta = 0.9$? When $\beta = 0.99$?

R7S.4 How fast would you have to be moving relative to our Milky Way galaxy so that the galaxy (whose rest diameter is 100,000 ly) is only 3.65 light-days across in your reference frame (enabling you to cross it in about 3.65 days according to your clock)? Express your answer in the form $\beta = 1 - \delta$. (*Hint:* See problem R7S.3.)

R7S.5 Imagine a cube 30 cm on a side. About how fast would this cube have to be moving relative to you if in your frame it was as thin as a sheet of paper in the direction of its motion? (*Hint:* You may find the approximation discussed in problem R7S.3 useful. Also, can you think of an easy way to estimate the thickness of a sheet of paper?)

R7S.6 As discussed in section R4.4, muons created in the upper atmosphere are sometimes able to reach the earth's surface. Imagine that one such muon travels the 60 km from the upper atmosphere to the ground (in the earth's frame) in one muon half-life of 1.52 μs (in the muon's frame). How thick is that part of the earth's atmosphere from the muon's creation point to the ground in the *muon's* frame?

R7S.7 We can use the metric equation to derive equation R7.2 as follows. An object with rest length L_R moving with a speed β passes a clock at rest in your inertial frame. Let event F be the object's front end

passing that clock, and let event B be the object's rear end passing that clock.
(a) Argue that in the object's frame, the coordinate time between these events is equal to L_R/β.
(b) What is the *distance* between these events in the object's frame?
(c) Define the length of the object in *your* reference frame to be the distance that the object travels in the time that it takes to pass by your clock, that is, $L = \beta\,\Delta t$, where Δt is the time measured between events F and B by your clock. Use the metric equation and the information in parts a and b above to arrive at equation R7.2. (*Hint:* Your clock is present at both events.)

R7S.8 Prove mathematically that the alternative definition of length given in section R7.7 *always* yields the same result as equation R7.2, as follows. Consider an object of rest length L_R at rest in a frame that we can choose to call the Other Frame. Let β be the speed with which the Other Frame (and thus the object) is moving with respect to the Home Frame. Let Δt be the Home Frame time between the events of the object's front end passing a certain point in the Home Frame and its rear end passing the same point. Since these events occur at the same place in the Home Frame, $\Delta x = 0$ between these events in that frame. But since these events occur at opposite ends of the object, the distance between them in the Other Frame is $\Delta x' = L_R$. Use an appropriate Lorentz transformation equation to determine Δt in terms of Δx, $\Delta x'$, and β; and then use the result to prove that $L \equiv \beta\,\Delta t = L_R(1 - \beta^2)^{1/2}$.

R7S.9 *Transformation of angles.* Consider a meterstick at rest in a given inertial frame (make this the Other Frame) oriented in such a way that it makes an angle of θ' with respect to the x' direction in that frame. In the Home Frame, the Other Frame is observed to move with a velocity of β in the $+x$ direction.
(a) Keeping in mind that the distances measured *parallel* to the line of relative motion are observed to be Lorentz-contracted in the Home Frame while distances measured perpendicular to the line of motion are not, show that the angle θ that this meterstick will be observed to make with the x direction in the Home Frame is given by

$$\theta = \tan^{-1}\left(\frac{\tan\theta'}{\sqrt{1 - \beta^2}}\right) \qquad (R7.12)$$

(b) What would the length of the meterstick be as measured in the Home Frame?
(c) Assume that the meterstick makes an angle of 30° with the x' direction in the Other Frame. How fast would that frame have to be moving with respect to the Home Frame for the meterstick to be observed in the Home Frame to make an angle of 45° with the x direction?

Rich-Context

R7R.1 *The space wars paradox.*[†] Two spacecraft of equal rest length $L_R = 100$ ns pass very close to each other as they travel in opposite directions at a relative speed of $\beta = \frac{3}{5}$. The captain of ship O has a laser cannon at the tail of her ship. She intends to fire the cannon at the instant that her bow is lined up with the tail of ship O'. Since ship O' is Lorentz-contracted to 80 ns in O's reference frame, she expects the laser burst to miss the other by 20 ns, as shown in figure R7.10a (she intends the shot to be "across the bow"). However, to the observer in ship O', it is ship O that is contracted to 80 ns. Therefore, the observer on O' concludes that if the captain of O carries out her intention, the laser burst will strike ship O' about 20 ns *behind* the bow, with disastrous consequences (figure R7.10b).

Assume that the captain of O carries out her intentions exactly as described, according to measurements in her own frame, and analyze what *really* happens as follows:

(a) Construct a carefully calibrated two-observer spacetime diagram of the situation described. Define event A to be the coincidence of the bow of ship O and the tail of ship O' and event B to be the firing of the laser cannon. Choose A to define the origin event in both frames, and locate B according to the description of O's intentions above. When and where does this event occur as measured in the O' frame, according to the diagram? (You may assume that the ships pass each other so closely that the travel time of the laser burst between the ships is negligible.)

(b) Verify the coordinates of B, using the Lorentz transformation equations.

(c) Write a short paragraph describing whether the cannon burst really hits or not, according to the results you found above. Discuss the hidden assumption in the statement of the apparent paradox, and point out how one of the drawings in figure R7.10 is misleading.

R7R.2 *The bullet hole paradox.*[‡] Two guns are mounted a distance of 40 ns apart on the embankment beside some railroad tracks. The barrels of the guns project outward toward the track so that they almost brush a speeding express train as it passes by. The train moves with a speed of $\beta = \frac{3}{5}$ with respect to the ground. Suppose the two guns fire simultaneously (as measured in the ground frame), leaving two bullet holes in the train.

(a) Let event R be the firing of the rear gun and event F the firing of the front gun. These events

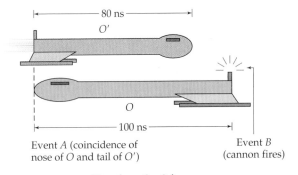

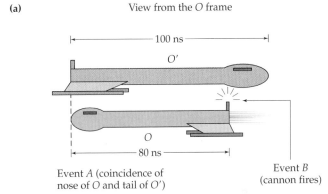

(a)

Event A (coincidence of nose of O and tail of O')

Event B (cannon fires)

View from the O frame

(b)

Event A (coincidence of nose of O and tail of O')

Event B (cannon fires)

View from the O' frame

Figure R7.10

The situation described in problem R7R.1.

occur 40 ns apart and at the same time in the ground frame. Draw a carefully constructed two-observer diagram of the situation, taking the ground frame to be the Home Frame and taking R to be the origin event. Be sure to show and label the axes of the ground and train frames, the worldlines of the guns, the worldlines of the bullet holes that they produce, and events R and F on your diagram.

(b) Argue, using your diagram, that the bullet hole worldlines are about 50 ns apart as measured in the train frame. Verify this by using the Lorentz transformation equations to show that events R and F occur 50 ns apart in the train frame.

(c) In the ground frame, the guns are 40 ns apart. In the train frame, the guns are moving by at a speed of $\beta = \frac{3}{5}$, and the distance between them will be measured to be Lorentz-contracted to less than 40 ns. Show, using the Lorentz contraction formula, that the guns are in fact 32 ns apart in the train frame. Describe how this same result can be read from your spacetime diagram.

(d) Doesn't this lead to a contradiction? How can two guns that are 32 ns apart in the train frame fire simultaneously and yet leave bullet holes 50 ns apart in the train frame? Write a paragraph in which you carefully describe the logical flaw in the description of the "contradiction" given in the last sentence. (*Hint:* Focus on the word *simultaneously.*) Describe what *really* is

---

[†]Adapted from E. F. Taylor and J. A. Wheeler, *Spacetime Physics*, San Francisco: Freeman, 1966, pp. 70–71.

[‡]Adapted from B. M. Casper and R. J. Noer, *Revolutions in Physics*, New York: Norton, 1972, pp. 363–364.

observed to happen in the train frame, and thus how it is perfectly natural for guns that are 32 ns apart to make holes 50 ns apart.

R7R.3 *The space cadets paradox.*[†] A very long measuring stick is placed in empty space at rest in an inertial frame we'll call the "stick frame." A spaceship of rest length L_R travels along the measuring stick at a speed $\beta = \frac{4}{5}$ relative to it. Two space cadets P and Q are each equipped with knives and synchronized watches and are stationed at rest on the ship frame in the ends of the spaceship. At a prearranged time, each cadet simultaneously reaches through a porthole and slices through the measuring stick.

(a) How long is the spaceship according to the cadets?

(b) How long is the spaceship according to observers along the measuring stick (i.e., observers at rest in the "stick frame")?

(c) Use the Lorentz transformation equations to show that observers along the measuring stick would conclude that the cut portion of the measuring stick had length $\frac{5}{3}L_R$.

(d) Since the cutting events occur simultaneously in the spaceship frame, they do *not* occur simultaneously in the stick frame. Use the Lorentz transformation equations to find the time separation of the two cutting events as viewed in the stick frame.

(e) Explain in a short paragraph how it is that two cadets who are only $\frac{3}{5}L_R$ apart (as measured in the stick frame) can cut a hunk of measuring stick $\frac{5}{3}L_R$ long if they really cut simultaneously according to their synchronized watches.

Advanced

R7A.1 *The radar method.* Imagine using the radar method to measure the length of an object moving at a speed β with respect to your own frame (the Other Frame). At a certain time (event A) you send forth a light flash from a clock in your frame. This flash bounces off a mirror at the far end of the object (event R) and then returns to your clock (event B). If you time this all just right so that the near end of the object passes your clock (event O) at exactly the time halfway between the emission event A and the reception event B (as measured by that clock), then you know that event O and the reflection event R are simultaneous. This means that at that instant, the object lies exactly between the clock and the light flash as it bounces off the mirror. The length L of the object in your frame is thus equal (in SR units) to the time that it takes the light to come back from event R, since light travels at a speed of 1 in all frames.

Now imagine viewing this measurement process from a Home Frame in which your frame is moving with a speed β in the $+x$ direction. The distance between the ends of the object in that frame is L_R. Draw a careful two-observer diagram of the situation as viewed by observers in the Home Frame (let O be the origin event in both frames). Argue that the coordinate distance between events O and A in the Home Frame is $\Delta x = \beta \Delta t$, where Δt is the coordinate time measured between those events in the Home Frame. Also argue (using similar triangles on the diagram) that $\Delta t = L_R$. Then use the metric equation to relate the time measured between events O and A measured in *your* frame (which is equal to L) to the coordinate time Δt measured between the events in the Home Frame, and show that you end up with the same result as that given by the Lorentz contraction equation R7.2.

R7A.2 *Light clocks.* Consider a light clock as shown in figure R4.1, except imagine the light clock to be laid on its side so that the light flash travels along the direction of motion of the clock. Show that the only way that this sideways light clock will measure the correct spacetime interval between events A and B (as any decent clock should) is if the distance between its mirrors is Lorentz-contracted by the amount stated by equation R7.2.

ANSWERS TO EXERCISES

R7X.1 When $\beta = \frac{4}{5}$, $\gamma = (1 - \frac{16}{25})^{-1/2} = 5/3 = 1.67$. A redrawn version of figure R7.3 would thus look as shown in figure R7.11. Events O and A occur simultaneously in the Other Frame and thus mark out the length of the stick in that frame. From the diagram, it looks as if the stick is about 2.4 ns long in the Other Frame.

R7X.2 A 4-ns measuring stick moving at a speed $\beta = \frac{4}{5}$ has a length of $L = L_R(1 - \beta^2)^{1/2} = (4 \text{ ns})(1 - \frac{16}{25})^{1/2} = (4 \text{ ns})(\frac{3}{5}) = 2.4$ ns.

R7X.3 Let the distance between points A and B in figure R7.4 be D. Since the slope S of the E' axis is the rise over the run of that axis in the Home Frame coordinate system, we have $S = -D/W_\perp$, implying that $D = -S(W_\perp)$. By the pythagorean theorem, then, the length of the road's projection on the E'

[†]Thanks to W. F. Titus of Carleton College.

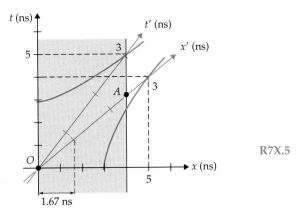

Figure R7.11
See the answer to exercise R7X.1.

axis is

$$W = \sqrt{W_\perp^2 + S^2 W_\perp^2} = W_\perp \sqrt{1 + S^2}$$

Note that except for the plus sign in the square root, this formula is quite similar in form to equation R7.2.

R7X.4   Event P shown in figure R7.12 (a slightly modified version of figure R7.8) occurs at the barn's back end

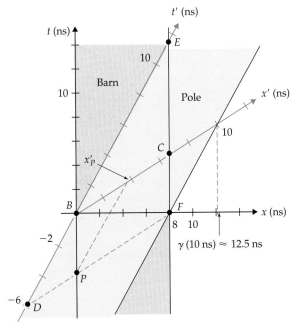

Figure R7.12
See the answer to exercise R7X.4.

at the same time as event F. To find out *where* this event occurs in the runner's frame, we draw a line parallel to the t' axis up to the x' axis, as shown. We see from this line that the back end of the barn is at $x' = 3.6$ ns at the time of event F, while the back end of the pole is (always) at $x' = 0$. Therefore the pole's back end is indeed 3.6 ns behind the barn's back end in the runner's frame at this time.

R7X.5   Event P in figure R7.12 locates the position of the barn's back end at the time of event F, and so occurs at $t'_P = -6$ ns in the runner's frame. The back end of the barn is always at $x_P = 0$ in the Home Frame. So, according to one of the inverse Lorentz transformation equations, we have

$$x_P = \gamma(\beta t'_P + x'_P) \tag{R7.13}$$

$$\Rightarrow \quad x'_P = \frac{x_P}{\gamma} - \beta t'_P = 0 - \frac{3}{5}(-6 \text{ ns}) = 3.6 \text{ ns} \tag{R7.14}$$

Since the pole's back end is always at $x' = 0$, it is sticking out 3.6 ns behind the barn at this time.

R7X.6   (a) Figure R7.8 implies that it takes about 13 ns (the time between events F and E) in the Home Frame for the pole to pass the point $x = 8$ ns in the Home Frame. Since the pole has a speed of $\beta = \frac{3}{5}$ in this frame, this implies that the pole has a length $L = \beta \, \Delta t = (\frac{3}{5})(13 \text{ ns}) = 39/5 \text{ ns} = 7.8$ ns.

(b) Events F and E occur at the same position ($x = 8$ ns) in the Home Frame, so they are separated by $\Delta x = 0$ in that frame. Since these events occur at opposite ends of the pole in the runner's frame, they are separated by $\Delta x' = -10$ ns in that frame (*negative* because $\Delta x' \equiv x'_E - x'_F = 0$ ns $- 10$ ns $= -10$ ns). So using the Lorentz transformation equation $\Delta x' = \gamma(-\beta \, \Delta t + \Delta x)$, we find that

$$\Delta t = \frac{\Delta x}{\beta} - \frac{\Delta x'}{\gamma \beta} = 0 - \frac{-10 \text{ ns}}{(\frac{5}{4})(\frac{3}{5})}$$

$$= \frac{40}{3} \text{ ns} = 13.3 \text{ ns} \tag{R7.15}$$

This means that $L = \beta \, \Delta t = \frac{3}{5}(40/3 \text{ ns}) = 8$ ns exactly. The discrepancy between this answer and the answer for part a is a result of inevitable uncertainties that arise in reading the diagram: note that the answers are close (within 2.5%).

R8

The Cosmic Speed Limit

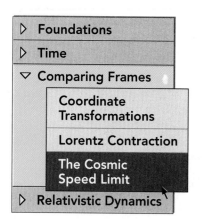

Chapter Overview

Introduction

One of the most fundamental and surprising consequences of the principle of relativity is that nothing can travel faster than the speed of light in a vacuum. In this chapter, with the help of some two-observer diagrams, we will explore why this must be.

This chapter also provides an appropriate context for discussing the Einstein velocity transformation equations that replace the galilean velocity transformation equations (equations R1.3), since the latter allow for faster-than-light speeds. The Einstein velocity transformation equations provide important background for chapters R9 and R10.

Section R8.1:   Causality and Relativity

A **causal influence** is any effect (particle, object, wave, or message) produced by one event that can cause another. For causality to make sense, the **temporal order** of the events in question must be preserved (the caused event cannot precede the event that causes it). The problem with faster-than-light travel is that it *violates causality*. We can see this as follows. Imagine that event P causes event Q and that the causal influence moving from P to Q moves with a speed faster than that of light. In such a case, we can always find an inertial frame moving slower than the speed of light where event Q occurs before event P. This is inconsistent with the principle of relativity, since it violates the concept of causality for an effect to precede its cause. The only way to make relativity consistent with causality is to insist that *nothing* (not even a message) can travel faster than light, making it a real **cosmic speed limit.**

The concept of causality expresses in colloquial terms what the law of increase of entropy (the **second law of thermodynamics**) expresses more formally. To say that we must preserve causality is really to say that the second law of thermodynamics must obey the principle of relativity.

Section R8.2:   Timelike, Lightlike, and Spacelike Intervals

Because of the minus signs in the metric equation, there are three distinct categories of spacetime interval:

1. **Timelike:** when $\Delta s^2 > 0$
2. **Lightlike:** when $\Delta s^2 = 0$
3. **Spacelike:** when $\Delta s^2 < 0$

These categories are frame-independent, since the value of Δs^2 is frame-independent.

How can we measure the spacetime interval between the two events? If Δs^2 is *timelike*, we can find an inertial frame where the events occur at the same *place:* we then measure Δs with the frame's clock at that location. If Δs^2 is *spacelike*, we can find a frame where the events occur at the same *time:* the distance that we measure with a ruler between the events in that frame is the **spacetime separation** $\Delta\sigma \equiv (-\Delta s^2)^{1/2}$ between the events.

Section R8.3: The Causal Structure of Spacetime

Events separated by a spacelike interval *cannot* be causally connected, since the causal influence would have to travel faster than the speed of light. Since the categories of spacetime interval are frame-independent, *all* observers will agree about (1) which events have a timelike (or lightlike) interval with a given event P and occur *after* it in all frames, (2) which events have a timelike (or lightlike) interval with P and occur *before* it in all frames, and (3) which events have a spacelike interval with P. Events in the first category may be causally influenced by P: we say that such events lie in the **future** of P. Events in the second category may causally influence P: we say that such events lie in the **past** of P. Events in the third category cannot be causally related to P.

In a spacetime diagram with two spatial axes, P's past and future look like cones whose points touch at P and whose surfaces are rings of light converging on or expanding from P: this structure is P's **light cone.**

Section R8.4: The Einstein Velocity Transformation

We can use a two-observer spacetime diagram to determine a particle's velocity in one inertial frame, knowing its velocity in another, and the relative velocity of the two frames as follows: (1) We set up a calibrated two-observer spacetime diagram showing axes for both frames; (2) we draw the particle's worldline with the correct slope relative to the axes of the frame in which we know its velocity; and (3) we measure that worldline's slope according to the other frame's axes.

Alternatively, one can use the Lorentz transformation equations to compute the velocity of the object in either frame, using its velocity components measured in the other. The result is

$$v'_x = \frac{v_x - \beta}{1 - \beta v_x} \qquad v'_y = \frac{v_y\sqrt{1 - \beta^2}}{1 - \beta v_x} \qquad v'_z = \frac{v_z\sqrt{1 - \beta^2}}{1 - \beta v_x} \qquad \text{(R8.14)}$$

$$v_x = \frac{\beta + v'_x}{1 + \beta v'_x} \qquad v_y = \frac{v'_y\sqrt{1 - \beta^2}}{1 + \beta v'_x} \qquad v_z = \frac{v'_z\sqrt{1 - \beta^2}}{1 + \beta v'_x} \qquad \text{(R8.8)}$$

Purpose: The first equation describes how to calculate an object's velocity components v'_x, v'_y, and v'_z measured in the Other Frame from its velocity components v_x, v_y, and v_z measured in the Home Frame; the second tells you how to do the reverse.

Symbols: β is the relative speed of the frames.

Limitations: These equations assume that the two frames are inertial, that they are in standard orientation with respect to each other, and that the Other Frame moves in the $+x$ direction relative to the Home Frame.

Notes: Equations R8.14 are the **(direct) Einstein velocity tranformation equations,** and equations R8.8 are the **inverse Einstein velocity transformation equations.** These equations replace the galilean transformation equations (equations R1.3).

R8.1   Causality and Relativity

"Nothing can go faster than the speed of light." This statement is a well-known consequence of special relativity. But why *must* this statement be true? Are there any loopholes that might make faster-than-light travel possible?

In sections R4.2 and R5.3, we saw how the metric equation and the proper time equation fail if we apply them to a clock moving faster than light: both equations imply that the time registered between two events by such a clock would be an imaginary number, which is absurd. In both cases, this absurdity results from the violation of the $\Delta t^2 > \Delta d^2$ restriction necessary for the derivation of the metric equation. Thus neither equation really says anything useful about what a clock traveling at faster-than-light speed would measure.

What is causality?

In this section, we will see that there is a deeper problem with traveling faster than light: *it violates causality*. What do I mean by *causality*? In physics (and more broadly, in daily life), we know that certain events *cause* other events to happen (see figure R8.1). For example, even couch potatoes know that if you press the appropriate button on the remote control, the TV channel will change. **Causally connected** events *must* happen in a certain order in time: the event being caused must follow the event that causes it (e.g., we would be deeply disturbed if the TV channel changed just *before* we pressed the remote control button!).

Consider two distinct events (call them P and Q) such that event P *causes* event Q, or more precisely, Q happens as a direct consequence of the reception of some kind of information that P has occurred. This information can be transmitted from P to Q in any number of ways: via some mechanical effect (such as the movement of an object or the propagation of a sound wave), via a light flash, via an electric signal, via a radio message, etc. Basically, the information can be carried by *any* object or effect that can move from place to place and is detectable.

Let's consider the TV remote control again as a specific example. Imagine that you press a button on your TV remote control handset (event P). The information that the button has been pressed is sent to the TV set in some manner, and in response, the TV set changes channels (event Q). Keep this basic example in mind as we go through the argument that follows.

Why faster-than-light causal influences are absurd

Now let us pretend that the **causal influence** that connects event P to event Q *can* flow between them at a constant speed v_{ci} *faster* than the speed

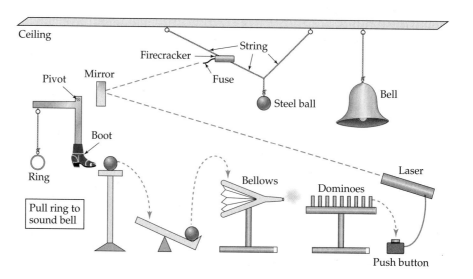

Figure R8.1

Some causal connections.

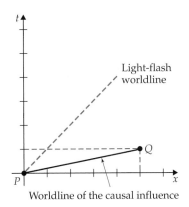

Figure R8.2

Imagine that events P and Q are connected by a hypothetical causal influence traveling with a speed v_{ci} faster than the speed of light. (In the drawing above, I have assumed for the sake of concreteness that the causal influence travels in the Home Frame at 5 times the speed of light.)

of light as measured in your inertial frame, which we will call the Home Frame. (Perhaps the TV manufacturer has found some way to convey a signal from the remote to the TV using "Z waves" that travel faster than light.) We will show that this leads to a logical absurdity. Choose event P to be the origin event in that frame, and choose the spatial x axis of the frame so that both events P and Q lie along it. (We can always do this: it is just a matter of choosing the origin and orientation of our reference frame. Choosing the frame to be oriented in this way is just a matter of convenience.)

Figure R8.2 shows a spacetime diagram (drawn by an observer in the Home Frame) of a pair of events P and Q fitting the description above. Note that if the causal influence flows from P to Q faster than the speed of light, its worldline on the diagram will have a slope $1/v_{ci} < 1$, that is, *less* than the slope of the worldline of a light flash leaving event P at the same time (which is also shown on the diagram for reference).

Now consider figure R8.3. In this two-observer spacetime diagram, I have drawn the t' and x' axes for an Other Frame that travels with a speed $\beta = \frac{2}{5}$ relative to the Home Frame. Note that according to section R6.5, the slope of the diagram x' axis in such a diagram is β. Note also that since the slope of the causal influence worldline is $1/v_{ci} < 1$, it is *always* possible to find a value of β such that $1/v_{ci} < \beta < 1$, meaning that it is always possible to find a reference frame moving slower than the speed of light relative to the Home Frame whose x' axis lies *between* the light-flash worldline and the causal influence worldline, as shown. In such a frame, event Q will be measured to occur *before* event P, as one can see by reading the time coordinates of these events from the diagram.

Thus in such an Other Frame, event P is observed to occur *after* event Q does. But this is absurd: event P is supposed to *cause* event Q. How can an event be measured to occur before its cause? This is not merely a semantic issue, nor is it mere appearance. According to any and every physical measurement that one might make in the Other Frame, event Q will really be observed to occur *before* its "cause" P.

To vividly illustrate the absurdity, consider our TV remote example. If the signal could go from your remote control to the TV faster than light, in certain inertial reference frames, you would observe the TV set changing channels *before* the button was pushed. If this were to happen in your reference frame, you would consider this a violation of the laws of physics (presuming your TV set was not broken). But the laws of physics are supposed to hold in *every* inertial reference frame. Therefore this observed inversion of cause and effect violates the principle of relativity! Causally connected events must have the same **temporal order** in *all* inertial reference frames if we are to preserve the meaning of the concept.

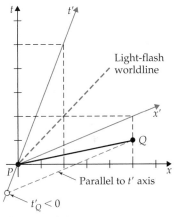

Figure R8.3

This two-observer spacetime diagram shows the same situation as in figure R8.2, but with added axes for an Other Frame moving at $\beta = \frac{2}{5}$ with respect to the Home Frame. Note that in the Other Frame, event Q occurs *before* event P.

Faster-than-light influences imply reversal of cause and effect in some frames

We have only three options at this point. We can reject the principle of relativity and start over at square one. We can radically modify our conception of causality in a way that is yet unknown. Or we can reject the assumption that got us into this trouble in the first place, namely, that a causal influence can flow from P to Q faster than light ($v_{ci} > 1$).

Consistency of causality and relativity thus implies the cosmic speed limit

The last option is clearly the least drastic. If information can only flow from P to Q with a speed $v_{ci} \leq 1$ in the Home Frame, then the worldline of the causal influence connecting event P to event Q will have a slope $1/v_{ci} > 1$. Any Other Frame must travel with $\beta < 1$, by this hypothesis (since the parts of the reference frame, like any material object, could in principle be the agent of a causal influence). Therefore, the slope β of the Other Frame's diagram x' axis on a spacetime diagram must always be less than the slope $1/v_{ci}$ of the causal connection worldline connecting P and Q, and thus Q will occur after P in *every* other inertial reference (figure R8.4).

So, if the speed of reference frames and causal influences is limited to some $v_{ci} < 1$, then effects will occur *after* their causes in every inertial reference frame, which is necessary if the idea of causality is to be consistent with the principle of relativity.

> **THEOREM: The Cosmic Speed Limit:** In order for causality (i.e., the idea that one event can *cause* another event to happen) to be consistent with the principle of relativity, information (i.e., *any* effect representing a causal connection between two events) *cannot* travel between two events with a speed v_{ci} greater than that of light.

Since anything movable and detectable can carry information (i.e., cause things to happen), this consequence of the principle of relativity applies not only to all physical objects (waves, particles, and macroscopic objects) but indeed to *any* trick or means of conveying a message that exists or might be imagined (e.g., instantaneous changes in a gravitational field, telepathy, magic).

Causality and the second law of thermodynamics

Now, at the most basic physical level, the physical law that defines the temporal order of cause and effect is the **second law of thermodynamics,** which requires that the entropy of the universe always increase (or at least remain the same) during any physical process. This law thus implies that events in certain physical processes can occur in one temporal order but *cannot* occur in the reverse order. Therefore, if the second law of thermodynamics is to be true in all inertial frames (as required by the principle of relativity), then the temporal order of all events that might be linked by that law must be preserved in all inertial frames. Thus the cosmic speed limit really follows from the assumption that the second law of thermodynamics is

Figure R8.4

Assume that the speed v_{ci} of the causal influence satisfies the restriction $v_{ci} < 1$ in the Home Frame. Then consider *any* Other Frame whose speed β relative to the Home Frame satisfies the same restriction: $\beta < 1$. Since the Other Frame's x' axis has slope β, it is less than 1 and thus less than the slope $1/v_{ci}$ of the worldline of the causal influence. This means that event Q can never be *below* the x' axis, which in turn implies that Q will occur after P in *every* inertial reference frame.

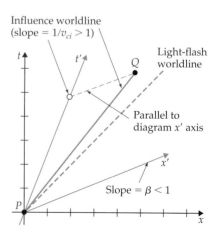

consistent with the principle of relativity. "Cause and effect" is really just an intuitive and colloquial way to talk about the invariant temporal order imposed on events by the second law.

So, with a straightforward argument using two-observer spacetime diagrams, we have proved the existence of a cosmic speed limit, an idea having profound physical and philosophical implications. As usual, this prediction is amply supported by experiment. No particle, object, or signal of any kind has ever been definitely observed to travel at faster than the speed of light in a vacuum. Science fiction fans and space travel buffs who hope for the discovery of faster-than-light travel may hope in vain: both the argument (based as it is on the firmly accepted and fundamental ideas of the principle of relativity and the physical reality of causality) and the experimental evidence present a pretty ironclad case for this cosmic speed limit.[†]

Summary

Exercise R8X.1

Imagine that a causal influence moves between events P and Q at 3 times the speed of light relative to the Home Frame. How fast would an Other Frame have to move relative to the Home Frame for event Q to occur *before* event P in the Other Frame?

R8.2   Timelike, Lightlike, and Spacelike Intervals

We are now in a position to understand more fully the true physical nature of the spacetime interval between *any* two events in spacetime. In section R4.2, we saw that for two events whose coordinate differences in a given inertial reference frame are Δt, Δx, Δy, and Δz, the quantity $\Delta s^2 = \Delta t^2 - \Delta x^2 - \Delta y^2 - \Delta z^2$ has a frame-independent value that is equal to the time registered by an inertial clock present at both events. But to make the proof of the metric equation (equation R4.5) work, we had to assume that $\Delta d = (\Delta x^2 + \Delta y^2 + \Delta z^2)^{1/2}$ was *smaller* than Δt so that there was more than sufficient time for a light flash to travel from one event to the other along the length of the light clock. The purpose of this section is to investigate the meaning of the spacetime interval when $\Delta d > \Delta t$, that is, when this condition is violated.

We have exploited the analogy between spacetime geometry and euclidean plane geometry extensively in the last few chapters. We have noted, though, that the minus signs in the metric equation (which do not appear in the corresponding pythagorean relation) lead to some subtle differences between spacetime geometry and euclidean geometry. One of these differences is the following. In euclidean geometry the squared distance Δd^2 between two points on a plane is necessarily positive:

$$\Delta d^2 = \Delta x^2 + \Delta y^2 \geq 0 \tag{R8.1}$$

Δs^2 between two arbitrary events can be positive, negative, or zero

---

[†]This does not mean that interstellar travel is out of the question. Remember that the ship time measured in a spaceship traveling close to the speed of light (relative to the galaxy) is much shorter than the coordinate time we would measure (at rest in the galaxy). Therefore, a trip to distant stars can be made as short as desired for the *passengers* by simply constraining the ship's speed to be sufficiently close to the speed of light. But (at least if special relativity is true) there appears to be no way to make a trip of 1000 ly in our galaxy in less than 1000 y *in the frame of the galaxy*, no matter how short this might seem to the passengers. This does put some severe limits on the possibilities of interstellar commerce! We will examine other difficulties associated with interstellar travel in chapter R10.

But taken at face value, the metric equation allows the squared spacetime interval between two events to be positive, zero, or negative, depending on the relative sizes of the coordinate separations Δd and Δt between those events:

$$\Delta s^2 = \Delta t^2 - \Delta d^2$$

Therefore,

$$\text{If} \quad \Delta d > \Delta t \quad \text{then} \quad \Delta s^2 < 0! \tag{R8.2}$$

We see that while there is only *one* kind of distance between two points on a plane, the possible spacetime intervals between two events in spacetime fall into *three* distinct categories depending on the sign of Δs^2. These categories are as follows:

The three categories for the spacetime interval

1. If $\Delta s^2 > 0$, we say that the interval between the events is **timelike.**
2. If $\Delta s^2 = 0$, we say that the interval between the events is **lightlike.**
3. If $\Delta s^2 < 0$, we say that the interval between the events is **spacelike.**

The reasons for these names will become clear shortly.

The peculiar category here is the *spacelike* category—there is nothing corresponding to it in ordinary plane geometry (where the squared distance between two events is always positive). What does it mean for two events to have a *spacelike* spacetime interval between them?

Spacelike spacetime intervals exist, but cannot be measured with a clock

First, note that events separated by spacelike spacetime intervals certainly do exist. Consider, for example, the case of two events that are measured in a certain inertial frame to occur at the same *time* but at different *locations*. Since the time separation between these events is zero in that reference frame, we have $\Delta s^2 = 0 - \Delta d^2 = -\Delta d^2 < 0$, so the interval between these events is necessarily spacelike. Therefore, the spacelike interval classification is needed if we are to be able to categorize the spacetime interval between arbitrarily chosen events.

As we have already discussed, the squared spacetime interval between two events Δs^2 that appears in the metric equation $\Delta s^2 = \Delta t^2 - \Delta d^2$ has been linked with the frame-independent time measured by an inertial clock present at both events *only* in the case where $\Delta t^2 > \Delta d^2$. For two events for which $\Delta t^2 < \Delta d^2$, it is not clear how one can directly measure the squared spacetime interval between the events at all. For example, for an inertial clock to be present at both events where $\Delta d > \Delta t$, it would have to travel at a speed $v > 1$ in that frame. We have just seen that this is impossible; thus a spacelike spacetime interval *cannot* be measured by a clock or anything else that travels between the events. Since the proof of the metric equation given in section R4.2 does not handle the case of spacelike intervals, it is not even obvious that the squared spacetime interval $\Delta s^2 = \Delta t^2 - \Delta d^2$ is frame-independent when it is less than zero.

Δs^2 is frame-independent even when it is negative

In fact, the squared spacetime interval Δs^2 *does* have a frame-independent value, no matter what its sign is. This can easily be demonstrated by using the Lorentz transformation equations for coordinate differences, given by equations R6.12. The argument goes like this. Let Δt, Δx, Δy, Δz be the coordinate separations of two events measured in the Home Frame, and let $\Delta t'$, $\Delta x'$, $\Delta y'$, $\Delta z'$ be the coordinate separations of the same two events measured in an Other Frame moving with speed β in the $+x$ direction with respect to the Home Frame. Then equations R6.12 imply that

$$(\Delta t')^2 - (\Delta x')^2 - (\Delta y')^2 - (\Delta z')^2$$
$$= [\gamma(\Delta t - \beta\,\Delta x)]^2 - [\gamma(-\beta\,\Delta t + \Delta x)]^2 - \Delta y^2 - \Delta z^2$$

$$
\begin{aligned}
&= \gamma^2(\Delta t^2 - 2\beta\,\Delta t\,\Delta x + \beta^2\,\Delta x^2) - \gamma^2(\beta^2\,\Delta t^2 - 2\beta\,\Delta t\,\Delta x + \Delta x^2)\\
&\quad - \Delta y^2 - \Delta z^2\\
&= \gamma^2(\Delta t^2 + \beta^2\,\Delta x^2 - \beta^2\,\Delta t^2 - \Delta x^2) - \Delta y^2 - \Delta z^2\\
&= \gamma^2(1-\beta^2)(\Delta t^2 - \Delta x^2) - \Delta y^2 - \Delta z^2\\
&= \Delta t^2 - \Delta x^2 - \Delta y^2 - \Delta z^2 \tag{R8.3}
\end{aligned}
$$

where in the last step, I used $\gamma \equiv 1/\sqrt{1-\beta^2}$. The sign of $\Delta s^2 = \Delta t^2 - \Delta d^2$ is irrelevant to this derivation, so we will find that the squared interval Δs^2 has the same frame-independent value in *every* inertial reference frame, whether Δs^2 is spacelike, timelike, or lightlike.

How can we measure the value of the spacetime interval between two events separated by a spacelike interval? We cannot use a clock, as we have noted already. In fact, we measure a spacelike spacetime interval with a *ruler*, as we will shortly see.

Let us define the **spacetime separation** $\Delta\sigma$ of two events in this way:

The spacetime separation

$$
\Delta\sigma^2 \equiv \Delta d^2 - \Delta t^2 = -\Delta s^2 \tag{R8.4}
$$

The spacetime separation, so defined, is conveniently real whenever the interval between the events is spacelike. Now note that *if* we can find an inertial reference frame where the events are simultaneous ($\Delta t = 0$), we have

$$
\Delta\sigma^2 = \Delta d^2 \quad \Rightarrow \quad \Delta\sigma = \Delta d \quad \text{(in a frame where } \Delta t = 0) \tag{R8.5}
$$

Now, I claim that we can *always* find a frame in which $\Delta t = 0$ if the events are separated by a spacelike interval. Imagine that two events occur with coordinate differences Δt and $\Delta d\ (> \Delta t)$ as measured in the Home Frame. Reorient and reposition the axes of the Home Frame so that the events in question both occur along the spatial x axis, with the later event located in the $+x$ direction relative to the earlier event. (This can be done without loss of generality: we are always free to choose the orientation of our coordinate system to be whatever we find convenient.) Once this is done, $\Delta d = \Delta x$ in the Home Frame.

Now, consider an Other Frame in standard orientation with respect to the Home Frame and traveling in the $+x$ direction with speed β with respect to the Home Frame. According to equation R6.12a, the time-coordinate difference between these events in the Other Frame is

$$
\Delta t' = \gamma(\Delta t - \beta\,\Delta x) \tag{R8.6}
$$

These events will be simultaneous in the Other Frame (that is, $\Delta t'$ will equal zero) if and only if the relative speed of the frames is chosen to be $\beta = \Delta t/\Delta x = \Delta t/\Delta d$. This relative speed β will be less than 1 since $\Delta d > \Delta t$ for our events by hypothesis. In short, given *any* pair of events that are separated by a spacelike interval in some inertial frame (which we are calling the Home Frame), it is possible to find an inertial Other Frame moving with speed $\beta < 1$ with respect to the Home Frame in which observers will find the two events to be simultaneous (see figure R8.5).

In short, if the spacetime interval between two events is *spacelike*, then

1. It is possible to find an inertial frame where these events are observed to occur at the *same time*.
2. The spacetime separation $\Delta\sigma$ is the *distance* between the events in that special frame. We can measure this with a ruler stretched between the events in that frame.
3. If observers in any other inertial frame use equation R8.4 to *calculate* $\Delta\sigma$, they will get the same value measured directly in the special frame.

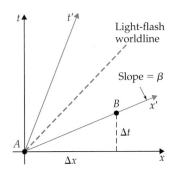

Figure R8.5

Given any pair of events A and B separated by a spacelike interval ($\Delta x > \Delta t$) in the Home Frame, we can find an Other Frame where the two events are simultaneous. The speed of this Other Frame simply must have the right value so that the diagram x' axis can connect both points. Since this axis has slope β, this means that β must equal $\Delta t/\Delta x$, where Δt and Δx are the coordinate separations of the events in the Home Frame.

How to measure a spacelike spacetime interval

Comparison with measuring
a timelike interval

These statements are directly analogous to statements that can be made about events separated by a timelike spacetime interval. If the spacetime interval between two events is *timelike*, then

1. It is possible to find an inertial frame in which these events occur at the *same place* (this is the frame of the inertial clock that is present at both events).
2. The *time* between the events in this special frame is Δs. We can measure this with a clock present at both events and at rest in this frame.
3. If observers in any other inertial frame use the ordinary metric equation to calculate Δs, they will get the same value as measured directly in the special frame.

Thus there is a fundamental symmetry between spacelike and timelike spacetime intervals, a symmetry that arises because both reflect the same underlying physical truth: it is possible to describe the separation of *any* two events in space and time with a frame-independent quantity Δs^2 (which we will call the **squared spacetime interval**) analogous to the *squared distance* between two points in plane geometry. It is simply a peculiarity of the geometry of spacetime that the quantity in spacetime that corresponds to ordinary (unsquared) distance on the plane comes in three distinct flavors (the spacetime *interval* Δs if $\Delta s^2 > 0$ for the events, the spacetime *separation* $\Delta \sigma$ if $\Delta s^2 < 0$, and the lightlike *interval* $\Delta s = \Delta \sigma = 0$ when $\Delta s^2 = 0$) which are measured in different ways using different tools. But it is important to realize that these three quantities are only different aspects of the same basic frame-independent concept Δs^2.

Why the categories of
spacetime interval have the
names that they do

We see that timelike intervals are directly measured with a *time*-measuring device (an inertial *clock* present at both events), while spacelike intervals are directly measured with a *space*-measuring device (a *ruler* stretched between the events in the particular inertial frame where Δt between the events is zero). This is why these interval classifications have the names *timelike* and *spacelike*: the names are intended to tell us whether we have to measure the interval with a clock (because the interval is timelike) or with a ruler (because it is spacelike). The lightlike interval classification stands between the other classifications. When the interval between two events is lightlike, we have $\Delta d = \Delta t$, which implies that these events could be connected by a flash of light.

Exercise R8X.2

Consider the events shown in the drawing to the right. Classify the spacetime interval between each *pair* of events as being timelike, spacelike, or lightlike.

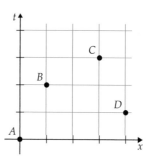

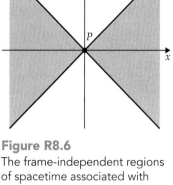

Exercise R8X.3

Imagine that event B happens 3.0 ns after and 5.0 ns east of event A in the Home Frame. In the Other Frame, the events happen at the same time. How fast is the Other Frame moving relative to the Home Frame? What is the distance between the events in that frame?

R8.3   The Causal Structure of Spacetime

Now, *because* it is true that the value of the squared spacetime interval Δs^2 is frame-independent no matter what its sign, all inertial observers will agree as to whether the interval between a given pair of events is timelike, lightlike, or spacelike (since if they all agree on the *value* of Δs^2, they will surely all agree on its sign). This means that the spacetime around any event P can be divided up into the distinct regions shown in figure R8.6, and every observer will agree about which events in spacetime belong to which region.

Because these regions can be defined in a frame-independent manner, it is plausible that they reflect something absolute and physical about the geometry of spacetime. In fact, *these regions distinguish those events that can be causally connected to P from those that cannot.* Remember that in section R8.1, we found that two events can be causally linked only if $\Delta d \leq \Delta t$ between them: otherwise the causal influence would have to travel between the events faster than the speed of light. This means that every event that can be causally linked with P must have a timelike (or perhaps lightlike) interval with respect to P: such events will lie in the white regions shown in figure R8.6.

We can be more specific yet. Since the temporal order of events is preserved in all inertial frames if $\Delta d > \Delta t$ (see figure R8.3 and the surrounding text), all events in the *upper* white region in figure R8.6 will occur *after P* in every frame (and thus could be caused by P), and all events in the lower white region of figure R8.6 will occur *before P* in every frame (and thus could cause P). We refer to these regions as the **future** and **past** of P, respectively.

Events whose spacetime interval with respect to P is spacelike ($\Delta d > \Delta t$) cannot influence P or be influenced by it. We say that these events (which inhabit the shaded region of figure R8.6) are causally *unconnected* to event P.

With this in mind, we can relabel the regions in figure R8.6 as shown in figure R8.7. Remember that every observer agrees on the value of the spacetime interval between event P and any other event, so every observer agrees as to which event belongs in which classification. The structure illustrated is thus an intrinsic, frame-independent characteristic of the geometry of spacetime.

Now, the boundaries of the regions illustrated in figure R8.7 are light-flash worldlines. If we consider two spatial dimensions instead of one, an

Figure R8.6
The frame-independent regions of spacetime associated with event P. The spacetime interval between P and any event in the white region is *timelike*, any event in the gray region is *spacelike*, and any event along the black diagonal lines is *lightlike*.

Understanding the causal structure of spacetime

The light cone associated with an event

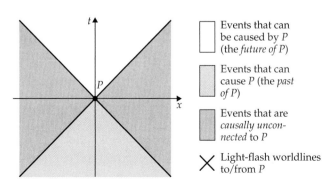

	Events that can be caused by P (the *future of P*)
	Events that can cause P (the *past of P*)
	Events that are *causally unconnected* to P
✕	Light-flash worldlines to/from P

Figure R8.7
The causal structure of spacetime relative to event P.

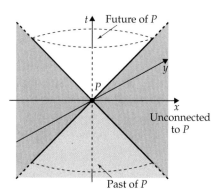

Figure R8.8

The *light cone* of P (shown on a spacetime diagram having two spatial dimensions).

omnidirectional light flash is seen as an ever-expanding ring, like the ring of waves formed by the splash of a stone into a still pool of water. If we plot the growth of such a ring on a spacetime diagram, we get a cone. The boundaries between the three regions described are then two tip-to-tip cones, as shown in figure R8.8. This boundary surface is often called the **light cone** for the given event P.

To summarize, the point of this section is that the spacetime interval classifications, which are basic, frame-independent features of the geometry of spacetime, have in fact a deeply physical significance: the sign of the squared spacetime interval between two events unambiguously describes whether these events can be causally connected or not. The light cone shown in figure R8.8 effectively illustrates this geometric feature of spacetime.

Exercise R8X.4

A meteor strikes the moon (event *A*), causing a large and vivid explosion. Exactly 0.47 s later (as measured in an inertial reference frame attached to the earth) the radiotelescope receiving signals from the moon goes on the fritz. Could these events be causally related? (*Hint:* The distance between the earth and the moon is roughly 384,000 km.)

R8.4   The Einstein Velocity Transformation

In this section we turn our attention to the relativistic generalization of the galilean velocity transformation equations R1.3. Imagine a particle that is observed in the Other Frame to move along the spatial x' axis with a constant x-velocity v'_x. The Other Frame, in turn, is moving with a speed β in the positive x direction with respect to the Home Frame. What is the particle's x-velocity v_x as observed in the Home Frame?

Transforming velocities with a two-observer diagram

Figure R8.9 shows how to construct a two-observer spacetime diagram that we can use to answer this question. After drawing both sets of coordinate axes, we simply draw the particle's worldline so that its slope in the Other Frame is $1/v'_x$. We can then find its x-velocity v_x in the Home Frame by taking the inverse slope of that line in the Home Frame, which we can do by picking an arbitrary "rise" (5 ns in figure R8.9), determining the worldline's "run" for that rise, and then taking the inverse rise/run = run/rise. In the case shown in figure R8.9, where $\beta = \frac{3}{5}$ and $v'_x = \frac{3}{5}$, we find that the value of v_x is about 0.86, and *not* $\frac{3}{5} + \frac{3}{5} = \frac{6}{5}$ that the galilean velocity transformation equations would predict.

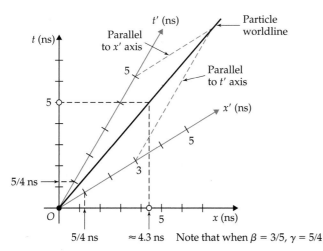

Figure R8.9

We can use a two-observer diagram to find a particle's x-velocity in one frame, given its x-velocity in another and the two frames' relative velocity. For example, if we know that the particle's Other Frame x-velocity is $v'_x = \frac{3}{5}$ and that the Other Frame moves at $\beta = \frac{3}{5}$ relative to the Home Frame, then we can construct an appropriate two-observer diagram for this β, draw the particle's worldline with a slope of $= \frac{5}{3}$ relative to the Other Frame axes, and read the slope of this line relative to the Home Frame axes (about $4.3/5 \approx 0.86$ in this case). If we knew the particle's Home Frame x-velocity, we could just reverse the process.

Now let us see if we can derive an exact equation that (like the diagram) allows us to find v_x in terms of v'_x and β. Imagine two infinitesimally separated events along the particle's worldline (which we will assume is moving along the spatial x axis). Let the coordinate differences between these events as measured in the Home Frame be dt and dx. Let the coordinate differences between the same two events as measured in the Other Frame be dt' and dx'. The x component of the particle's velocity as it travels between these events is

$$v_x \equiv \frac{dx}{dt} = \frac{\gamma(\beta\, dt' + dx')}{\gamma(dt' + \beta\, dx')} \tag{R8.7a}$$

How to derive the inverse Einstein velocity transformation equations

where I have used the difference version of the inverse Lorentz transformation equations (equations R6.13). Dividing the right side top and bottom by dt' and using $dx'/dt' \equiv v'_x$, we get the following relativistically exact equation:

$$v_x = \frac{\beta + v'_x}{1 + \beta v'_x} \tag{R8.7b}$$

In a similar fashion, you can derive the y and z component equations as well.

Exercise R8X.5

Use the same method to show that

$$v_y = \frac{v'_y\sqrt{1 - \beta^2}}{1 + \beta v'^2_x} \tag{R8.7c}$$

The complete set of equations for v_x, v_y, and v_z are

$$v_x = \frac{\beta + v'_x}{1 + \beta v'_x} \qquad v_y = \frac{v'_y \sqrt{1 - \beta^2}}{1 + \beta v'_x} \qquad v_z = \frac{v'_z \sqrt{1 - \beta^2}}{1 + \beta v'_x} \qquad \text{(R8.8)}$$

Purpose: These equations describe how to compute an object's velocity components v_x, v_y, and v_z measured in the Home Frame from its velocity components v'_x, v'_y, and v'_z measured in the Other Frame.

Symbols: β is the relative speed of the frames.

Limitations: These equations assume that the two frames are inertial, that they are in standard orientation with respect to each other, and that the Other Frame moves in the $+x$ direction relative to the Home Frame.

An example application

As an example of the use of equation R8.8a, consider the particular problem illustrated by figure R8.9, where we have $v'_x = \beta = \frac{3}{5}$. The final speed of the particle in the Home Frame (according to equation R8.8a) is

$$v_x = \frac{3/5 + 3/5}{1 + (3/5)(3/5)} = \frac{6/5}{34/25} = \frac{30}{34} \approx 0.88 \qquad \text{(R8.9)}$$

This is close to the result that we read from figure R8.9.

These equations reduce to the galilean equations at low speeds . . .

We call equations R8.8 the **inverse Einstein velocity transformation equations:** they express algebraically what a two-observer diagram like figure R8.9 expresses graphically. Note that the result is different from what the *galilean* velocity transformation predicts: solving equation R1.3a for v_x yields

$$v_x = \beta + v'_x \qquad \text{from galilean velocity transformation} \qquad \text{(R8.10)}$$

Note that when the velocities β and v'_x are very small, the factor $\beta v'_x$ that appears in the denominator of equation R8.8 becomes *very* small compared to 1. In this limit, then, equation R8.8a reduces to the galilean equation R8.10:

$$v_x = \frac{\beta + v'_x}{1 + \beta v'_x} \approx \frac{\beta + v'_x}{1} = \beta + v'_x \qquad \text{in low-velocity limit} \qquad \text{(R8.11)}$$

The same kind of argument applies to the other two component equations as well. The galilean transformation equations are therefore reasonably accurate for everyday velocities, but only represent an *approximation* to the true velocity transformation law expressed by equations R8.8.

. . . and are consistent with the cosmic speed limit and the invariant speed of light

Equation R8.8 never yields a Home Frame x-velocity that exceeds the speed of light: even if both β and v'_x are 1 (their maximum possible value), then

$$v_x = \frac{1 + 1}{1 + 1 \cdot 1} = \frac{2}{2} = 1 \qquad \text{(R8.12)}$$

Moreover, equation R8.8a (unlike the galilean equation R1.3a) is consistent with the idea that the speed of light is equal to 1 in all frames: if the x-velocity of a light flash in the Other Frame is $v'_x = 1$, its speed in the Home Frame will be

$$v_x = \frac{\beta + 1}{1 + \beta \cdot 1} = \frac{\beta + 1}{1 + \beta} = 1 \qquad \text{(R8.13)}$$

independent of the value of β.

In short, the inverse Einstein velocity transformation equations (equations R8.8) provide the answer to the question that we raised in chapter R2 about how the galilean velocity transformation equations can be modified to be consistent with the principle of relativity. Equations R8.8 reduce to the galilean transformation equations at low velocities (where the galilean transformation is known by experiment to be very accurate); but at relativistic velocities, they are consistent with both the assertion that nothing can be measured to go faster than light and the assertion that light itself has the same speed in all inertial frames.

Equations R8.8 convert Other Frame velocity components v'_x, v'_y, v'_z to Home Frame components v_x, v_y, v_z. The (direct) **Einstein velocity transformation equations** transform the velocity components the other way.

$$v'_x = \frac{v_x - \beta}{1 - \beta v_x} \qquad v'_y = \frac{v_y\sqrt{1 - \beta^2}}{1 - \beta v_x} \qquad v'_z = \frac{v_z\sqrt{1 - \beta^2}}{1 - \beta v_x} \quad \text{(R8.14)}$$

Purpose: These equations describe how to compute an object's velocity components v'_x, v'_y, and v'_z measured in the Other Frame from its velocity components v_x, v_y, and v_z measured in the Home Frame.
Symbols: β is the relative speed of the frames.
Limitations: These equations assume that the two frames are inertial, that they are in standard orientation with respect to each other, and that the Other Frame moves in the $+x$ direction relative to the Home Frame.

These equations can be derived either by solving equations R8.8 for the Other Frame components or by using the direct Lorentz transformation equations R6.12 as we used equations R6.13 in the derivation of equations R8.8.

Exercise R8X.6

Verify the first component of equation R8.14, using the direct Lorentz transformation equations.

Exercise R8X.7

An object moves at 0.80 in the $+x$ direction as measured in the Home Frame. What is its x-velocity in an Other Frame that is moving at 0.60 in the $+x$ direction, also as measured in the Home Frame?

TWO-MINUTE PROBLEMS

R8T.1 How would you classify the spacetime interval between each pair of events shown in figure R8.10 (see p. 148)?
  A. Timelike
  B. Lightlike
  C. Spacelike

R8T.2 Which pairs of events in figure R8.10 on page 148 could be causally connected? Which could not?

(For each pair, answer T if they could be causally connected, F if not.)

R8T.3 Imagine that event A is the origin event in the Home Frame and that event B occurs at $t = 1$ ns and $x = 10$ ns in that frame. What would be the minimum speed β that an Other Frame would have to have relative to the Home Frame if B occurred first in that frame?   *(more)*

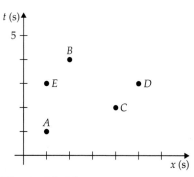

Figure R8.10

For problems R8T.1 and R8T.2.

A.   10
B.   1
C.   0.60
D.   0.40
E.   0.10
F.   B can't occur first

R8T.4   Two blinking warning lights are 3000 m apart along a railroad track. Imagine that in the ground frame, the west light blinks (event W) 5 μs before the east light blinks (event E). Now imagine that both lights are observed by a passenger on the train, who passes the west light just as it blinks. It is possible for the train to be moving fast enough that the two blinks are *observed* (not necessarily *seen*) by the passenger to be simultaneous, true (T) or false (F)?

R8T.5   Imagine that an explosion occurs at $x = 0$ in the Home Frame (event A). Light from the explosion is detected by a detector at position $x = -100$ ns (event B) and by a detector at position $x = +50$ ns (event C). Events B and C are causally connected, T or F?

R8T.6   A laser beam is emitted on earth (event A), bounces off a mirror placed on the moon by Apollo astronauts (event B), and then returns to a detector on earth (event C). The detector is 12 ns of distance from the laser.
a.   The spacetime interval between events A and B is
    A.   Timelike
    B.   Lightlike
    C.   Spacelike
b.   The spacetime interval between events A and C is
    A.   Timelike
    B.   Lightlike
    C.   Spacelike
c.   Events A and C are definitely causally connected, T or F?

R8T.7   If the spacetime interval between two events is timelike, then the temporal order of the two events is the same in *every* inertial reference frame, T or F?

R8T.8   If the spacetime interval between two events A and B is spacelike and event A occurs before event B in some Home Frame, then it is *always* possible to find an Other Frame where the events occur in the other order, T or F?

R8T.9   An object moves with speed $v_x = +0.9$ in the Home Frame. In an Other Frame moving at $\beta = 0.60$ relative to the Home Frame, the object's x-velocity v_x' is
A. $v_x' = 1.5$
B. $1 < v_x' < 1.5$
C. $0.9 < v_x' < 1$
D. $0.3 < v_x' < 0.9$
E. $v_x' < 0.3$
F. $v_x' < 0$

HOMEWORK PROBLEMS

Basic Skills

R8B.1   At 11:00:00 a.m. a boiler explodes in the basement of the Museum of Modern Art in New York City (call this event A). At 11:00:00.0003 a.m., a similar boiler explodes (call this event B) in the basement of a soup factory in Camden, New Jersey, a distance of 150 km from event A.
(a) Why is it impossible for the first event to have caused the second event?
(b) An alien spaceship cruising in the direction of Camden from New York measures the Camden event to occur at the same time as the New York event. What is the approximate speed of the spaceship relative to earth?

R8B.2   Two balls are simultaneously ejected (event A) from the point $x = 0$ in some inertial frame. One rolls in the $+x$ direction with speed 0.80 and eventually hits a wall at $x = 8.0$ ns (event B). The other rolls in the $-x$ direction with speed 0.40, eventually hitting a wall at $x = -8.0$ ns (event C). Is the spacetime interval between B and C spacelike or timelike? Could these events be causally connected?

R8B.3   Derive the first of equations R8.14 by solving equation R8.7b for v_x'.

R8B.4   An object moves with velocity $v_x' = \frac{2}{5}$ in an inertial frame attached to a train, which in turn moves with velocity $\beta = \frac{4}{5}$ in the $+x$ direction with respect to the ground. What is the object's velocity v_x with respect to the ground? Evaluate this by reading the velocity from a carefully constructed two-observer spacetime diagram. Check your answer by using the appropriate Einstein velocity transformation equation.

R8B.5 Rocket A travels to the right and rocket B to the left at speeds of $\frac{3}{5}$ and $\frac{4}{5}$, respectively, relative to the earth. What is the velocity of A measured by observers in rocket B? Answer by reading the velocity from a carefully constructed two-observer diagram. Check your answer, using an appropriate Einstein velocity transformation.

R8B.6 Two trains approach each other from opposite directions along a linear stretch of track. Each has a speed of $\frac{3}{4}$ relative to the ground. What is the speed of one train relative to the other? Answer this question by using an appropriate Einstein velocity transformation equation. As part of your solution, explain carefully which object you are taking to be the Home Frame, which object to be the Other Frame, and the object whose speed you are measuring in both frames.

What is the speed of one relative to the other?
(See problem R8B.6.)

R8B.7 Two cars travel in the same direction on the freeway. Car A travels at a speed of 0.90, while car B can only muster a speed of 0.60. What is the relative speed of the cars? As part of your solution, explain carefully which object you are taking to be the Home Frame, which object to be the Other Frame, and which is the object whose speed you are measuring in both frames.

Synthetic

R8S.1 Here is a quick argument that no material object can go faster than the speed of light. Consider an object traveling in the $+x$ direction with respect to some inertial frame (call this the Home Frame)

at a speed $v > 1$ in that frame. Show, using a two-observer spacetime diagram, that it is possible to find an Other Frame moving in the $+x$ direction at a speed $\beta < 1$ with respect to the Home Frame in which the object's worldline lies along the diagram x' axis; and find the value of β (in terms of v) that makes this happen. Why is it absurd for the worldline of any object to coincide with the diagram x' axis?

R8S.2 It is the year 2048, and you are on a jury in a terrorism trial. The facts of the case are these. On June 12, 2047, at 2:25:06 p.m. Greenwich Mean Time (GMT), the earth-Mars shuttle *Ares* exploded as it was being refueled in low earth orbit. (Fortunately, no passengers were aboard, and the refueling was handled by robots.) At 2:27:18 p.m., police video CD a raid on a hotel room on Mars conducted on the basis of an unrelated anonymous tip. In the video, the defendant is shown with a radio control transmitter in hand. Forensic experts have testified that the *Ares* was blown up by remote control, and that the reconstructed receiver was consistent with the transmitter in the defendant's possession. Just before the explosion, a caller predicted the blast and took responsibility on behalf of the Arean Liberation Army; the defendant has known links to that organization. Phone records show that the defendant spoke with someone near the earth less than 10 min before. Hall monitors in the hotel showed that the defendant entered the hotel room at 2:23:12 p.m. and was not carrying the transmitter at that time. The defendant has taken the Fifth Amendment and has offered no defense other than a plea of not guilty. A fragment of the trial transcript follows.

Prosecutor: The time shown at the bottom of the video sequence taken of the raid is that from a clock internal to the camera?

Police witness: No, I am told that that time is computed from signals originating from the master clock on earth that is part of the Solar System Positioning System (SSPS) that defines a solar-system-wide inertial frame fixed on the sun. According to the manual (*pulls out the manual*), "The signal from the earth master clock is suitably corrected in the camera for the motions of the earth and Mars and the light travel time from the earth, so that the time displayed is exactly as if it were from the clock at the camera's location, at rest in the solar system frame, and synchronized with the earth-based master clock." We do this deliberately so as to be able to correlate events on a solar-system-wide basis.

Prosecutor: There is no chance that *this* time (*freezes video display*), which shows the police yanking the device from defendant here at exactly 2:27:20 p.m. GMT, is in error?

Police witness: No, the camera was checked two days previously as part of a normal maintenance program.

Prosecutor: We have been told that the *Ares* blew up at exactly 2:25:06 p.m. GMT. Was that time determined using the SSPS also?

Police witness: Yes.

Prosecutor: You have testified that the hall monitors show the defendent entering the room at exactly 2:23:12 p.m. This time was also determined using the SSPS?

Police witness: Yes.

Prosecutor: So the defendant was alone in the room at the time that the *Ares* exploded?

Police witness: Yes, in the SSPS frame.

Prosecutor: So this video CD shows you capturing the defendant red-handed just after the destruction of the *Ares,* with the incriminating transmitter still in hand . . .

Defense: Objection, your Honor!

Judge: Sustained.

Guilty or not guilty? Write a paragraph justifying your reasoning very carefully.

R8S.3[†] Imagine spinning a laser at a speed of 100 rotations per second. How far away must you place a screen so that the spot of light on the screen produced by the laser beam sweeps along the screen at a speed faster than that of light? Would a spot speed greater than that of light violate the cosmic speed limit? (*Hint:* If you stand at one edge of the screen and a friend stands at the other, can you send a message to your friend, using the sweeping spot? If so, how? If not, why not?)

R8S.4 Starbase Alpha coasts through deep space at a speed of $\beta = 0.60$ in the $+x$ direction with respect to earth. Let the event of the starbase traveling by the earth define the origin event in both frames. Imagine that at $t = 8.0$ h, a giant accelerator on earth launches you toward the starbase at 10 times the speed of light, relative to earth. After you get to the starbase, you use a similar accelerator to launch you back toward earth at 10 times the speed of light, relative to the starbase. Using a carefully constructed (full-page) two-observer diagram, show that in such a

case you will return to the earth before you left. (This is another way to illustrate the absurdity of faster-than-light travel.)

R8S.5 A flash of laser light is emitted by the earth (event A) and is absorbed on the moon (event B). Is the space-time interval between events A and B spacelike, lightlike, or timelike? Now assume that the light flash is reflected at event B by something shiny on the moon and returns to earth, where it is absorbed at event C. Is the spacetime interval between B and C spacelike, lightlike, or timelike? What about the interval between events A and C? Support your answers by describing your reasoning.

R8S.6 A solar flare bursts through the surface of the sun at 12:05 p.m. GMT, as measured by an observer in an inertial frame attached to the sun. At 12:11 p.m., as measured by the same observer, the Macdonald family's radio fries a circuit board. Could these events be causally connected?

A solar flare. (See problem R8S.6.)

R8S.7 The first stage of a multistage rocket boosts the rocket to a speed of 0.1 relative to the ground before being jettisoned. The next stage boosts the rocket to a speed of 0.1 relative to the final speed of the first stage, and so on. How many stages does it take to boost the payload to a speed in excess of 0.95?

[†]Adapted from E. F. Taylor and J. A. Wheeler, *Spacetime Physics*, San Francisco: Freeman, 1966, p. 62.

R8S.8 Imagine that in the Home Frame, two particles of equal mass m are observed to move along the x axis with equal and opposite speeds $v = 0.60$. The particles collide and stick together, becoming one big particle which remains at rest in the Home Frame. Now imagine observing the same situation from the vantage point of an Other Frame that moves with speed $\beta = 0.60$ in the $+x$ direction with respect to the Home Frame. Find the velocities of all the particles as observed in the Other Frame, using the appropriate Einstein velocity transformation equations. Check your results, using a two-observer spacetime diagram of the situation.

R8S.9 *Do problem R8S.8 first.* In the situation described in problem R8S.8, show that while the momentum of the system is conserved in the Home Frame, it is *not* conserved in the Other Frame if momentum is defined as mass times velocity. (We'll deal with this problem in chapter R9.)

R8S.10 A train travels in the $+x$ direction with a speed of $\beta = 0.80$ with respect to the ground. At a certain time, two balls are ejected, one traveling in the $+x$ direction with x-velocity of $+0.60$ with respect to the train and the other traveling in the $-x$ direction with x-velocity of -0.40 with respect to the train.
   (a) What are the x-velocities of the balls with respect to the ground?
   (b) What is the x-velocity of the first ball with respect to the second? [*Hint:* The frames you will choose to be the Home Frame and Other Frame for part (a) will not be the same as your choices for part (b).]

R8S.11 Show, using the Einstein velocity transformation equations R8.8, that a particle traveling in any arbitrary direction at the speed of light will be measured to have the speed of light in all other inertial frames.

R8S.12 A particle moves with a speed of $\frac{4}{5}$ in a direction $60°$ away from the spatial $+x'$ axis toward the spatial y' axis, as measured in a frame (the Other Frame) that is moving with speed $\beta = 0.50$ in the $+x$ direction of the Home Frame. What are the magnitude and direction of the particle's velocity in the Home Frame?

R8S.13 A train travels in the $+x$ direction with a speed of $\beta = 0.80$ with respect to the ground. At a certain time, two balls are ejected so that they travel with a speed of 0.60 (as measured in the train frame) in opposite directions *perpendicular* to the direction of motion of the train.
   (a) What are the speeds of the balls with respect to the ground?
   (b) What is the angle that the path of each ball makes with the x axis in the ground frame?

R8S.14 Show that when $\beta \ll 1$, the Lorentz transformation equations R6.11a and R6.11b reduce to the galilean transformation equations R1.2a and R1.2b. (*Discussion:* The problem is trivial except for equation R6.11a. In that equation, how can we justify dropping the βx term when we need to keep the $-\beta t$ term in equation R6.11b? Think about typical magnitudes of quantities in an everyday experiment.)

R8S.15 Imagine that in a particle physics experiment, a particle of mass m_0 moving at a speed $v_0 = \frac{3}{5}$ in the $+x$ direction suddenly decays into a particle of mass m_1 moving at a speed of $v_1 = \frac{4}{5}$ in the $+x$ direction and a particle of m_2 at rest.
   (a) If newtonian momentum is conserved in this decay process, what must the masses m_1 and m_2 be?
   (b) Now let's look at this decay process in a frame where the initial particle is at rest. What are the final x-velocities of the decay products in this frame?
   (c) Is newtonian momentum conserved in this frame? Is this a problem?

Rich-Context

R8R.1 You are the captain of a spaceship that is moving through an asteroid belt on impulse power at a speed of $\frac{4}{5}$ relative to the asteroids. Suddenly you see an asteroid dead ahead a distance of only 24 s away, according sensor measurements in your ship's reference frame. You immediately shoot off a missile, which travels forward at a speed of $\frac{4}{5}$ relative to your ship. The missile hits the asteroid and detonates, pulverizing the asteroid into gravel. However, you learned in Starfleet Academy that it is not safe to pass through such a debris field (even with shields on full) sooner than 8 s (measured in the asteroid frame) after the detonation. Are you safe?

R8R.2 A spaceship travels at a speed of $\beta = 0.90$ along a straight-line path that passes 300 km from a small asteroid. Exactly 1.0 ms (in the asteroid frame) in time before reaching the point of closest approach, the ship fires a photon torpedo which travels at the speed of light. This torpedo is fired perpendicular to the ship's direction of travel (as measured in the ship's frame) on the side closest to the asteroid. Will this torpedo hit the asteroid? If not, does it pass the asteroid on the near side or far side (relative to the ship)? Carefully explain your reasoning.

Advanced

R8A.1 Imagine that in its own reference frame, an object emits light uniformly in all directions. Imagine also that this object moves with a speed β in the $+x$ direction with respect to the Home Frame. Show that the portion of the light that is emitted in the forward hemisphere in the object's own frame is, in the Home Frame, observed to be concentrated in a cone that makes an angle of

$$\phi = \sin^{-1}\frac{1}{\gamma} \tag{R8.15}$$

with respect to the x axis. Show that if $\beta = 0.99$, the angle where this portion (which amounts to one-half of the object's light) is concentrated is only $8.1°$. (This forward concentration of the radiation emitted by a moving object is called the *headlight effect*.)

R8A.2 Consider a very long pair of scissors. If you close the scissor blades fast enough, you might imagine that you could cause the *intersection* of the scissor blades (i.e., the point where they cut the paper) to travel from the near end of the scissors to the far end at a speed faster than that of light, without causing any *material part* of the scissors to exceed the speed of light. Argue that (1) this intersection can indeed travel faster than the speed of light in principle, but that (2) if the scissors blades are originally open and at rest and you decide to send a *message* to a person at the other end of your scissors by suddenly closing them, you will find that the intersection (and thus the message) *cannot* travel faster than the speed of light. (*Hint:* The information that the handles have begun to close must travel through the metal from the handles to the blades and then down the blades to cause the intersection to move forward. What effect carries this information? With what speed does this information travel?)[†]

ANSWERS TO EXERCISES

R8X.1 Let us take event P to be the origin event in the Home Frame. Now, consider figure R8.3. For event Q to occur below the x' axis in the Other Frame, the x' axis must have a slope greater than that of the worldline of the causal influence connecting events P and Q. If the causal influence moves at a speed of 3, then its slope on a two-observer diagram like figure R8.3 will be $\frac{1}{3}$. The slope of the Other Frame x' axis must be therefore greater than $\frac{1}{3}$. Since the slope of this axis is equal to β, the speed at which the Other Frame moves with respect to the Home Frame, the speed of the Other Frame must be greater than $\frac{1}{3}$.

R8X.2 The spacetime interval between A and B is timelike, between A and C is lightlike, and between A and D is spacelike. The interval between B and C is spacelike. The interval between D and B is spacelike, but the interval between D and C is timelike.

R8X.3 Let event A be the origin event in the Home Frame. If we look at figure R8.5, we see that if A and B are to be simultaneous in the Other Frame, the x' axis in the Other Frame has to go through both A and B and so has to have slope of $\frac{3}{5}$ in this case (since B occurs 5 units to the right and 3 units above the origin in the two-observer diagram). Since the slope of this axis is equal to the speed β of the Other Frame with respect to the Home Frame, we see that this speed must be $\beta = \frac{3}{5}$. Since the spacetime interval

between the events must have the same value in all frames and since $\Delta t' = 0$ by definition in the Other Frame, the squared distance $\Delta d'^2$ between A and B in that frame is

$$\begin{aligned}
\Delta d'^2 &= \Delta d'^2 - \Delta t'^2 = -\Delta s^2 \\
&= \Delta d^2 - \Delta t^2 = (5\text{ ns})^2 - (3\text{ ns})^2 \\
&= (25 - 9)\text{ ns}^2 \\
&= 16\text{ ns}^2 = (4\text{ ns})^2
\end{aligned} \tag{R8.16}$$

So $\Delta d' = 4$ ns.

R8X.4 The distance to the moon in seconds is

$$384{,}000\text{ km}\left(\frac{1\text{ s}}{300{,}000\text{ km}}\right) = 1.28\text{ s} \tag{R8.17}$$

Since $\Delta d > \Delta t$ in this case, the spacetime interval between the events is spacelike. Thus the events cannot be causally connected in any way.

R8X.5 According to the inverse Lorentz transformation equations, we have

$$\begin{aligned}
v_y &= \frac{dy}{dt} = \frac{dy'}{\gamma(dt' + \beta\,dx')} = \frac{dy'/dt'}{\gamma[1 + \beta(dx'/dt')]} \\
&= \frac{v_y'}{\gamma(1 + \beta v_x')} = \frac{v_y'\sqrt{1 - \beta^2}}{1 + \beta v_x'}
\end{aligned} \tag{R8.18}$$

[†]See M. A. Rothman, "Things that Go Faster than Light," *Sci. Am.,* vol. 203, p. 142, July 1960.

R8X.6 According to the direct Lorentz transformation equations, we have

$$v'_x = \frac{dx'}{dt'} = \frac{\gamma(-\beta\,dt + dx)}{\gamma(dt - \beta\,dx)} = \frac{-\beta\,dt + dx}{dt - \beta\,dx}$$

$$= \frac{-\beta + dx/dt}{1 - \beta(dx/dt)} = \frac{-\beta + v_x}{1 - \beta v_x} = \frac{v_x - \beta}{1 - \beta v_x} \quad \text{(R8.19)}$$

R8X.7 Our aim here is to convert a velocity in the Home Frame to a velocity in the Other Frame, so we want to use the direct Einstein velocity transformation equations. The problem implies that $v_x = 0.80 = \frac{4}{5}$ and $\beta = 0.60 = \frac{3}{5}$, so according to equation R8.14a, we have

$$v'_x = \frac{v_x - \beta}{1 - \beta v_x} = \frac{\frac{4}{5} - \frac{3}{5}}{1 - \left(\frac{3}{5}\right)\left(\frac{4}{5}\right)} = \frac{\frac{1}{5}}{\frac{13}{25}} = \frac{5}{13} \quad \text{(R8.20)}$$

which is about 0.38 (much larger than the $\frac{1}{5} = 0.20$ that we would have expected from the galilean equations).

R9

Four-Momentum

Chapter Overview

Introduction

Our goal in this final subdivision of the unit is to make the laws of conservation of momentum and energy compatible with the principle of relativity. This chapter introduces a redefinition of the concept of momentum (which turns out to include energy as an integral part), and chapter R10 explores the implications of this idea.

Section R9.1:   A Plan of Action

In this chapter, we will see that

1.   Conservation of **newtonian momentum** $\vec{p} = m\vec{v}$ violates the principle of relativity.
2.   *Four-momentum* is a natural relativistic redefinition of momentum.
3.   Conservation of four-momentum is consistent with the principle of relativity.
4.   The conserved fourth component of four-momentum is energy.

Section R9.2:   Conservation of Newtonian Momentum Is Not Frame-Independent

Consider a collision that conserves newtonian momentum in the Home Frame. If we use the Einstein velocity transformation equations to calculate the velocities of the colliding objects in another inertial frame, we find that newtonian momentum is *not* conserved in the second frame. Conservation of newtonian momentum is therefore incompatible with the principle of relativity.

Section R9.3:   The Four-Momentum Vector

An object's **four-momentum P** is a four-dimensional vector quantity (a **four-vector**) whose four components in a given inertial reference frame are

$$[P_t, P_x, P_y, P_z] = \left[m\frac{dt}{d\tau}, m\frac{dx}{d\tau}, m\frac{dy}{d\tau}, m\frac{dz}{d\tau} \right] \qquad (R9.8)$$

Purpose: This equation defines (in a given inertial reference frame) the components $[P_t, P_x, P_y, P_z]$ of a particle's four-momentum **P** in terms of its mass m and the rate of change of its spacetime coordinates $[t, x, y, z]$ (in that frame) with respect to the particle's own proper time τ.

Limitations: There are no known limitations: this is a definition.

Notes: The four-momentum treats time and space coordinates evenhandedly and points tangent to the particle's worldline in spacetime.

Because $d\tau = (1 - v^2)^{1/2}\, dt$, the spatial components of **P** are indistinguishable from the components of newtonian momentum $\vec{p}$ when the particle's speed $v \ll 1$. (Equation R9.10 in figure R9.5 shows how to calculate $[P_t, P_x, P_y, P_z]$ from m and $\vec{v}$.)

Section R9.4:   Properties of Four-Momentum

The components of the four-momentum transform as follows:

$$\begin{bmatrix} P_t' \\ P_x' \\ P_y' \\ P_z' \end{bmatrix} = \begin{bmatrix} \gamma(P_t - \beta P_x) \\ \gamma(-\beta P_t + P_x) \\ P_y \\ P_z \end{bmatrix} \qquad \text{(R9.12)}$$

Purpose: This equation describes how to calculate the components $[P_t', P_x', P_y', P_z']$ of a particle's four-momentum in the Other Frame, given its components $[P_t, P_x, P_y, P_z]$ in the Home Frame.

Symbols: β is the speed of the Other Frame relative to the Home Frame; $\gamma \equiv (1 - \beta^2)^{-1/2}$.

Limitations: This equation assumes that both frames are inertial, that they are in standard orientation, and that the Other Frame moves in the $+x$ direction relative to the Home Frame.

Note: These equations are *very* similar to the Lorentz transformation equations for coordinate differences.

The particle's mass m is the frame-independent **four-magnitude** of its four-momentum: $m = (P_t^2 - P_x^2 - P_y^2 - P_z^2)^{1/2}$ (see equation R9.29 in figure R9.5).

Section R9.5: Four-Momentum and Relativity

Equation R9.12 implies quite generally that if a system's total four-momentum is conserved in any inertial frame, it will be conserved in *all* inertial frames. Therefore, a law of conservation of four-momentum *is* compatible with the principle of relativity.

Section R9.6: Four-Momentum's Time and Space Components

This conservation law only works, however, if P_t is conserved along with P_x, P_y, and P_z. In the limit that $v \ll 1$, $P_t \approx m + \frac{1}{2}mv^2$, so conservation of P_t in that limit implies that a system's total mass plus its total kinetic energy is conserved. We call P_t the particle's **relativistic energy** E. Note that even a particle at rest has a **rest energy** $E_{\text{rest}} = m$ ($= mc^2$ in SI units), and that conservation of relativistic energy allows conversion of this form of energy into other forms. We define a particle's **relativistic kinetic energy** K to be $K \equiv E - m$: this is *approximately* equal to $\frac{1}{2}mv^2$ when $v \ll 1$.

We define a particle's **relativistic momentum** p to be the magnitude of the spatial components of its four-momentum: $p \equiv (P_x^2 + P_y^2 + P_z^2)^{1/2}$. Note that a particle's speed in a given frame is $v = p/E$. Figure R9.5 summarizes virtually everything you need to know about doing calculations with four-momentum components.

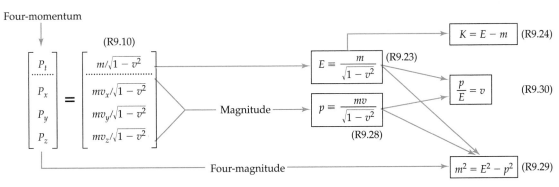

Figure R9.5
Virtually everything that you need to know about four-momentum.

R9.1   A Plan of Action

Up to this point, we have been studying **relativistic kinematics**—how we describe and measure the motion of objects in special relativity. But physicists are also interested in *dynamics*, which comprises the study of how interactions between objects *determine* the objects' motions. In this volume's final subsection, we will explore the basic principles of **relativistic dynamics.**

The laws of newtonian physics are not consistent with relativity

The basic principles of newtonian dynamics are expressed by Newton's three laws of motion, which in turn are based on the laws of conservation of momentum and energy. In section R1.6 (see also problems R1S.9 through R1S.11), we saw that various laws of newtonian dynamics were consistent with the principle of relativity *if* the galilean velocity transformation equations (equations R1.2) are true. But as we saw in chapter R8, these equations are *not* true: they only represent the low-velocity limit of the relativistically correct Einstein velocity transformation equations (equation R8.14). *This means that the laws of newtonian dynamics are **not** generally consistent with the principle of relativity;* the laws of newtonian dynamics likewise represent only low-velocity approximations to the laws of *relativistic* dynamics, laws that are the same in *all* inertial frames (as the principle of relativity requires).

Well, what are these laws of relativistic dynamics, and how can we find them? We *could* address this question by searching for a relativistic generalization of Newton's second law, then Newton's third law, then the law of universal gravitation, and so on. But it turns out that trying to do this is trickier than it looks, and the result is often ugly equations that are not really very illuminating.

To find the laws of relativistic dynamics, we start with conservation of momentum

Things work much better if we instead start with the more basic law of *conservation of momentum* and make this law consistent with the principle of relativity first. Not only is the correct adaptation of this law fairly easy to find, but also it proves to be very illuminating and rich in implications and applications. Indeed, just as we found for newtonian dynamics in unit C, we can learn virtually everything that is useful to know about relativistic dynamics by closely examining the law of conservation of momentum.

An overview of the argument to be found in this chapter

The basic argument in this chapter can be outlined as follows.

1. I will show you that conservation of **newtonian momentum** $\vec{p} = m\vec{v}$ is *not* consistent with the principle of relativity: if an isolated system's total newtonian momentum is conserved in one inertial frame, it is *not* conserved in other frames. This means that we need to redefine momentum so that its conservation law *is* consistent with the principle of relativity.

2. I will propose a natural relativistic generalization of the idea of momentum called *four-momentum,* which is a *four*-component vector having a time component as well as the usual x, y, and z components.

3. I will show you that the law of conservation of *four*-momentum *is* consistent with the principle of relativity, and thus represents a reasonable relativistic realization of the law of conservation of momentum. But for this to be true, the t component of a system's total four-momentum must *also* be conserved along with its x, y, and z components. So the law of conservation of four-momentum not only makes the idea of conservation of momentum consistent with the principle of relativity, but it tells us that something *else* is conserved as well.

4. This fourth conserved quantity turns out to be a relativistic version of the concept of *energy.* Thus the law of conservation of four-momentum actually *unifies* two of the three great conservation laws discussed in unit C.

R9.2   Conservation of Newtonian Momentum Is Not Frame-Independent

In this section, I will argue that the law of conservation of newtonian momentum is *not* consistent with the principle of relativity and thus cannot be a valid law of physics as stated. To illustrate the problem, it is sufficient to demonstrate a single instance of the inconsistency. For the sake of simplicity, I will illustrate the problem by using a simple one-dimensional collision.

Figure R9.1 shows such a collision as observed in the Home Frame. In this frame, an object with mass m is moving in the $+x$ direction with an x-velocity $v_{1x} = +\frac{3}{5}$. It then strikes an object of mass $2m$ at rest ($v_{2x} = 0$). Let's assume that the objects are isolated and the collision is elastic. If so, the newtonian equations for one-dimensional elastic collisions (see section C12.3) imply that the lighter mass will rebound from the collision with an x-velocity of $v_{3x} = -\frac{1}{5}$, while the heavier object will rebound with an x-velocity of $v_{4x} = +\frac{2}{5}$.

We can easily verify that the system's total newtonian x-momentum *is* conserved in the Home Frame for the collision as described:

A hypothetical collision

Newtonian momentum is conserved in the Home Frame for this collision

Total x-momentum before: $mv_{1x} + 2mv_{2x} = m\left(+\frac{3}{5}\right) + 2m(0) = +\frac{3}{5}m$

$$(R9.1a)$$

Total x-momentum after: $mv_{3x} + 2mv_{4x} = m\left(-\frac{1}{5}\right) + 2m\left(+\frac{2}{5}\right) = +\frac{3}{5}m$

$$(R9.1b)$$

Now consider how this collision looks when observed in an Other Frame that is moving with a speed $\beta = \frac{3}{5}$ in the $+x$ direction. Since this frame essentially moves along with the lightweight object, that object appears to be at rest in the Other Frame: $v'_{1x} = 0$. Since the larger object is at rest in the Home Frame, and the Home Frame is observed to be moving backward with respect to the Other Frame at a speed of $\frac{3}{5}$, the x-velocity of the more massive object must also be $v'_{2x} = -\frac{3}{5}$. The objects' final x-velocities are not so easy to intuit: we need to use the Einstein velocity transformation equation R8.14a to compute these velocities:

How the collision looks in the Other Frame

$$v'_{3x} = \frac{v_{3x} - \beta}{1 - \beta v_{3x}} = \frac{-\frac{1}{5} - \frac{3}{5}}{1 - \left(\frac{3}{5}\right)\left(-\frac{1}{5}\right)} = \frac{-\frac{4}{5}}{\frac{28}{25}} = \frac{20}{28} = -\frac{5}{7} \qquad (R9.2a)$$

Similarly, you can show that

$$v'_{4x} = -\frac{5}{19} \qquad (R9.2b)$$

Exercise R9X.1

Verify equation R9.2b.

View in the Home Frame

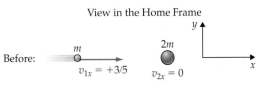

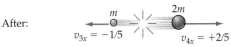

Figure R9.1
A hypothetical collision of two particles as observed in the Home Frame. The total newtonian momentum of this isolated system is conserved in this frame.

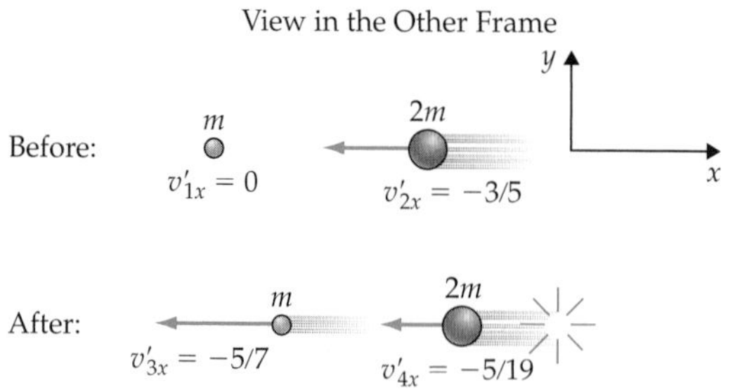

Figure R9.2
The same collision as observed in the Other Frame. Newtonian momentum is *not* conserved in this frame.

Note that these *must* be the object's final velocities if the Einstein velocity transformation equations are true and the collision actually occurs as described in the Home Frame. In the Other Frame, then, the collision is shown in figure R9.2.

Newtonian momentum is not conserved in the Other Frame

In this frame, though, the system's total x-momentum is *not* conserved:

Total x-momentum before: $mv'_{1x} + 2mv'_{2x} = m(0) + 2m\left(-\frac{3}{5}\right) = -\frac{6}{5}m$ (R9.3a)

Total x-momentum after: $mv_{3x} + 2mv_{4x} = m\left(-\frac{5}{7}\right) + 2m\left(-\frac{5}{19}\right) = -\frac{165}{133}m$
$$\text{(R9.3b)}$$

We can see that in this frame, the x component of the system's total momentum is somewhat larger after the collision than it was before the collision, since $165/133 > 6/5$. The law of conservation of newtonian momentum therefore does *not* hold in the Other Frame, even though it did hold in the Home Frame. (Note that the law of conservation of momentum requires that *each* component of the system's total momentum be conserved separately, so a violation of the conservation of even *one* component, the x component in this case, is a violation of the entire law.)

The principle of relativity requires that the laws of physics be the same in all inertial reference frames. The conclusion is inescapable: If the Einstein velocity transformation equations are true, then the law of conservation of newtonian momentum is *not* consistent with the principle of relativity.

Exercise R9X.2

The problem is the Einstein velocity transformation

It is important to understand that the root of the problem is the Einstein velocity transformation equations. Show that if the *galilean* velocity transformation equations (R1.3) were true (which they are *not*), newtonian momentum *would* be conserved in the Other Frame as well as in the Home Frame.

R9.3   The Four-Momentum Vector

Review of the definition of newtonian momentum

In unit C, we defined an object's momentum $\vec{p}$ as *mass* times *velocity:*

$$\vec{p} \equiv m\vec{v} = m\frac{d\vec{r}}{dt} \qquad \text{(R9.4)}$$

This is the *newtonian* definition of momentum, which we need to modify to make the law of conservation of momentum consistent with the principle of relativity. But *how* might we modify this definition? It helps to look at the newtonian definition very closely to try to understand its meaning.

The vector $d\vec{r}$ in equation R9.4 represents an infinitesimal displacement in space, which we divide by an infinitesimal time interval dt to get the

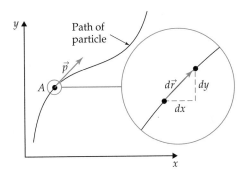

y

Path of
particle

$\vec{p}$

A

$d\vec{r}$ dy

dx

Figure R9.3

A graph showing the path of a particle through space. The particle's ordinary momentum $\vec{p}$ at point A is defined to be a vector parallel to the displacement $d\vec{r}$ that connects two infinitesimally separated points surrounding A.

object's velocity vector $\vec{v}$. The components of the infinitesimal displacement vector $d\vec{r}$ are $[dx, dy, dz]$, so the components of the newtonian momentum are

$$p_x \equiv mv_x = m\frac{dx}{dt} \qquad p_y \equiv m\frac{dy}{dt} \qquad p_z = m\frac{dz}{dt} \qquad (R9.5)$$

Notice that the momentum vector is *parallel* to the infinitesimal displacement $d\vec{r}$, and so will be tangent to the object's path through space (see figure R9.3).

How can we arrive at a relativistic generalization of this process? In special relativity, space and time are considered to be equal parts of the unitary whole that we call spacetime, so we describe the motion of an object not merely by describing its path through space but also by describing its *worldline through spacetime*. The appropriate relativistic generalization of an "infinitesimal displacement in space" $d\vec{r}$ between two infinitesimally separated points on an object's path in space is a displacement $d\mathbf{R}$ in *spacetime* between two infinitesimally separated *events* on the object's *worldline* (see figure R9.4). Note that in any given inertial reference frame, the displacement $d\mathbf{R}$ in spacetime between two events is specified by *four* numbers:

$$d\mathbf{R} = [dt, dx, dy, dz] \qquad (R9.6)$$

Including the time displacement dt on an equal footing with the spatial displacements $dx, dy,$ and dz makes the displacement $d\mathbf{R}$ a four-component vector that physicists call a **four-vector**. In this book, I will always use *boldface capital letters* (for example, $d\mathbf{R}$ and $\mathbf{P}$) to represent such four-vectors.

Given the components $[dt, dx, dy, dz]$ of the displacement four-vector $d\mathbf{R}$ that stretches between two infinitesimally separated events on the worldline of our object, how do we define its relativistic momentum? By analogy with the newtonian momentum, we want to divide $d\mathbf{R}$ by a quantity that characterizes the *time* between the events, and then multiply by the object's mass. In newtonian mechanics, time is universal and absolute, so the time between the events will be independent of who does the measuring. The most important flaw in the definition of newtonian momentum from the *relativistic* viewpoint, however, is that time is *not* really universal and absolute. The time dt measured between the events in the frame where we measure $dx, dy,$ and dz is not the same as the time dt' measured in some other inertial frame, which in turn is not generally the same as the proper time $d\tau$ between the events measured by a clock traveling with the particle. Which of these things should we choose?

Dividing by dt makes the spatial components of $m\,d\mathbf{R}/dt$ the same as the spatial components of ordinary momentum, which we know doesn't work. So let's try dividing by $d\tau$ instead. When you think about it, it makes a certain kind of sense to use the object's *own* time (which is uniquely and unambiguously linked to its motion) to characterize its momentum. Moreover, $d\tau$ has the advantage of being a frame-independent measure of time.

How we can revise this definition to make it more "relativistic"?

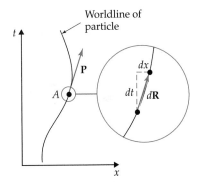

Worldline of
particle

t

$\mathbf{P}$

dx

A dt $d\mathbf{R}$

x

Figure R9.4

A spacetime diagram showing the worldline of a particle through spacetime. The object's relativistic momentum $\mathbf{P}$ at event A is defined to be an arrow parallel to the displacement arrow $d\mathbf{R}$ that connects two infinitesimally separated events surrounding event A. (The y and z dimensions are not shown here.)

So our proposed relativistic redefinition of an object's newtonian momentum $\vec{p}$ is the relativistic **four-momentum P**, defined as follows:

$$\mathbf{P} \equiv m\frac{d\mathbf{R}}{d\tau} \tag{R9.7}$$

The four-momentum is thus a four-dimensional vector having components

$$[P_t, P_x, P_y, P_z] = \left[m\frac{dt}{d\tau}, m\frac{dx}{d\tau}, m\frac{dy}{d\tau}, m\frac{dz}{d\tau} \right] \tag{R9.8}$$

Purpose: This equation defines (in a given inertial reference frame) the components $[P_t, P_x, P_y, P_z]$ of a particle's four-momentum $\mathbf{P}$ in terms of its mass m and the rate of change of its spacetime coordinates $[t, x, y, z]$ (in that frame) with respect to the particle's own proper time τ.
Limitations: There are no known limitations: this is a definition.
Notes: The four-momentum is defined to have *four* components so that the time and space coordinates are treated evenhandedly and so that the resulting four-vector always points tangent to the particle's worldline in spacetime.

Just as the newtonian momentum is a vector that is tangent to the object's *path* through space, the four-momentum (when drawn as an arrow on a spacetime diagram) is a vector tangent to the object's *worldline*, as shown in figure R9.4.

Expressing the particle's
four-momentum in terms of
its mass and velocity

We can express the components of the four-momentum in a given inertial frame in terms of the object's ordinary velocity measured in that frame. Equation R5.5 implies that the proper time between two infinitesimally separated events measured by a clock traveling between them at a speed v in a given inertial frame is related to the coordinate time dt measured in that frame between those events by

$$d\tau = \sqrt{1 - v^2}\, dt \tag{R9.9}$$

This means that

$$P_t \equiv m\frac{dt}{d\tau} = \frac{m}{\sqrt{1-v^2}}\frac{dt}{dt} = \frac{m}{\sqrt{1-v^2}} \tag{R9.10a}$$

$$P_x \equiv m\frac{dx}{d\tau} = \frac{m}{\sqrt{1-v^2}}\frac{dx}{dt} = \frac{mv_x}{\sqrt{1-v^2}} \tag{R9.10b}$$

$$P_y \equiv m\frac{dy}{d\tau} = \frac{m}{\sqrt{1-v^2}}\frac{dy}{dt} = \frac{mv_y}{\sqrt{1-v^2}} \tag{R9.10c}$$

$$P_z \equiv m\frac{dz}{d\tau} = \frac{m}{\sqrt{1-v^2}}\frac{dz}{dt} = \frac{mv_z}{\sqrt{1-v^2}} \tag{R9.10d}$$

These equations tell us how to calculate the components of the four-momentum of an object in a given frame, given the object's velocity vector $\vec{v}$ in that frame.

When the speed v of an object becomes very small compared to the speed of light ($v \ll 1$), the square roots in the denominators in equations R9.10 become almost equal to 1, and we have

$$P_t \approx m \tag{R9.11a}$$

$$P_x \approx mv_x \tag{R9.11b}$$

$$P_y \approx mv_y \qquad (R9.11c)$$

$$P_z \approx mv_z \qquad (R9.11d)$$

Thus in the limit where the velocity of an object is very small, the spatial components of the four-momentum reduce to being the same as the corresponding components of the object's newtonian momentum. Therefore, in everyday circumstances, we are unable to tell whether it is really newtonian momentum or four-momentum that is the conserved quantity.

Note also that since velocity in SR units is unitless, all four components of an object's four-momentum have units of *mass* in SR units.

Exercise R9X.3

An object with a mass of 1.0 kg moves with velocity of $[v_x, v_y, v_z] = [0, \frac{4}{5}, 0]$ in the Home Frame. What are the components of its four-momentum in this frame? What are the components of its four-momentum in the frame where it is at rest?

R9.4   Properties of Four-Momentum

Why define an object's relativistic four-momentum in this way? The definition has several attractive features. One feature has already been mentioned: *on a spacetime diagram, the four-momentum is represented by an arrow tangent to the worldline of the object through spacetime,* just as the ordinary momentum vector is an arrow tangent to the path of the object through space.

It is also nice that the definition of the four-momentum treats the time coordinate in the same manner as the spatial coordinates: all four coordinate displacements dt, dx, dy, dz appear on an equal footing in the definition of the four-momentum given by equations R9.8. We have already seen how it is important in relativity theory to treat time and space as being equal participants in the larger geometric whole that we call spacetime. The definition of four-momentum given above maintains this symmetry between the different spacetime coordinates.

All this symmetry has a certain beauty about it which an intuitive physicist like Einstein might take as corroborating evidence that we are on the right track with this definition. But we will see that the most important feature of how the four-momentum vector is defined is that given its components in one inertial frame, we can calculate its components in any other inertial frame, using a very straightforward (and already familiar) set of equations.

The components of an object's four-momentum are *frame-dependent* quantities, because the values of the coordinate differences dt, dx, dy, dz that appear in the numerators of equations R9.8 are frame-dependent. The differential proper time $d\tau$ appearing in the denominator, on the other hand, is a *frame-independent* quantity. In this text, we will also consider the mass m of the object to be a *frame-independent* measure of the amount of "stuff" in the object.[†]

A particle's four-momentum vector is tangent to its worldline in spacetime

It puts space and time on an equal footing

It has a very straightforward transformation law

[†]You may have heard in another context that special relativity implies that the mass of an object depends on its velocity. This is an old-fashioned way of looking at mass that obscures some of the simplicity and beauty of relativity theory. See C. G. Adler, "Does Mass Really Depend on Velocity, Dad?" *Am. J. Phys.,* vol. 55, no. 8, pp. 739–743, August 1987, for a careful and entertaining look at the problems with the old way of thinking about mass in relativity theory. Most modern treatments of relativity treat an object's mass as being frame-*independent*.

Let us imagine that we know the components $[P_t, P_x, P_y, P_z]$ of a given object's four-momentum in the Home Frame, and that we want to find the corresponding components in an Other Frame moving with speed β in the $+x$ direction with respect to the Home Frame. We can calculate the time component P'_t of the object's four-momentum in the Other Frame as follows:

$$P'_t = m\frac{dt'}{d\tau} = m\frac{\gamma(dt - \beta\,dx)}{d\tau} = \gamma m\frac{dt}{d\tau} - \gamma\beta m\frac{dx}{d\tau}$$

$$= \gamma P_t - \gamma\beta P_x = \gamma(P_t - \beta P_x) \tag{R9.12a}$$

where I have used the Lorentz transformation (specifically equation R6.12a) to express dt' as measured in the Other Frame in terms of dt and dx as measured in the Home Frame. Similarly, you can show that the transformation equation for the four-momentum x component is

$$P'_x = \gamma(-\beta P_t + P_x) \tag{R9.12b}$$

We also have

$$P'_y = m\frac{dy'}{d\tau} = m\frac{dy}{d\tau} = P_y \tag{R9.12c}$$

$$P'_z = m\frac{dz'}{d\tau} = m\frac{dz}{d\tau} = P_z \tag{R9.12d}$$

Exercise R9X.4

Verify that equation R9.12b is correct.

The transformation has the same form as the Lorentz transformation equations

What is nice about these equations? Compare them with the Lorentz transformation equations, given by equations R6.12. Equations R9.12 are the *same* as these Lorentz transformation equations except that the four-momentum components P_t, P_x, P_y, and P_z have been substituted for the coordinate displacement components Δt, Δx, Δy, and Δz, respectively. *Thus the components of the four-momentum transform from frame to frame according to the Lorentz transformation equations,* just as coordinate differences do!

The transformation equations for the four-momentum come out so nicely because (1) the time coordinate appears on an equal footing with the spatial components in the definition of the four-momentum, (2) we have divided the displacement by the frame-independent differential proper time $d\tau$ instead of the frame-dependent differential coordinate time dt, and (3) we define the object's mass m to be a frame-independent quantity.

Formal definition of a four-vector quantity

In fact, the technical definition of a *four-vector* requires this kind of transformation law: physicists define a **four-vector** to be a physical quantity represented by a vector whose four components transform according to the Lorentz transformation equations (i.e., just as the coordinate differences Δt, Δx, Δy, Δz do) when we go from one inertial reference frame to another.

The frame-independent four-magnitude of a four-vector

Now we know that although the coordinate differences Δt, Δx, Δy, Δz between two events are frame-*dependent* quantities, the spacetime interval Δs between the events given by

$$\Delta s^2 = \Delta t^2 - \Delta x^2 - \Delta y^2 - \Delta z^2 \tag{R9.13}$$

has a frame-*independent* value. Similarly, we can define the frame-independent **four-magnitude** of a four-vector as follows: if a four-vector **A** has components $[A_t, A_x, A_y, A_z]$, then its squared four-magnitude is defined to be

$$|\mathbf{A}|^2 \equiv A_t^2 - A_x^2 - A_y^2 - A_z^2 \tag{R9.14}$$

This is the relativistic analog to the definition (based on the pythagorean theorem) of the magnitude of an ordinary vector.

You can easily show that the frame-independent four-magnitude of an object's four-momentum vector is simply equal to its frame-independent mass:

$$m = |\mathbf{P}| = \sqrt{P_t^2 - P_x^2 - P_y^2 - P_z^2} \qquad (R9.15)$$

Exercise R9X.5

Verify that equation R9.15 is correct, using the fact that for infinitesimally separated events, there is no distinction between proper time and spacetime interval, so $d\tau = ds = (dt^2 - dx^2 - dy^2 - dz^2)^{1/2}$.

R9.5   Four-Momentum and Relativity

We now have a suitable candidate for a relativistic generalization of the concept of momentum. The final step is to verify that a law of conservation of four-momentum is in fact consistent with the principle of relativity.

Consider an arbitrary collision of two objects moving along the x axis. The law of conservation of four-momentum says that

$$\mathbf{P}_1 + \mathbf{P}_2 = \mathbf{P}_3 + \mathbf{P}_4 \qquad (R9.16)$$

Proof that four-momentum is conserved in all *inertial frames if it is in any one frame*

where $\mathbf{P}_1$ and $\mathbf{P}_2$ are the objects' four-momenta *before* the collision and $\mathbf{P}_3$ and $\mathbf{P}_4$ are the objects' four-momenta *after*. We can usefully rewrite this as follows:

$$\mathbf{P}_1 + \mathbf{P}_2 - \mathbf{P}_3 - \mathbf{P}_4 = 0 \qquad (R9.17)$$

which essentially says that the *difference* between the system's initial and final total momenta is zero. In component form, this last equation tells us that

$$\begin{bmatrix} P_{1t} \\ P_{1x} \\ P_{1y} \\ P_{1z} \end{bmatrix} + \begin{bmatrix} P_{2t} \\ P_{2x} \\ P_{2y} \\ P_{2z} \end{bmatrix} - \begin{bmatrix} P_{3t} \\ P_{3x} \\ P_{3y} \\ P_{3z} \end{bmatrix} - \begin{bmatrix} P_{4t} \\ P_{4x} \\ P_{4y} \\ P_{4z} \end{bmatrix} = \begin{bmatrix} 0 \\ 0 \\ 0 \\ 0 \end{bmatrix}$$

$$\text{or} \qquad \begin{aligned} P_{1t} + P_{2t} - P_{3t} - P_{4t} &= 0 \\ P_{1x} + P_{2x} - P_{3x} - P_{4x} &= 0 \\ P_{1y} + P_{2y} - P_{3y} - P_{4y} &= 0 \\ P_{1z} + P_{2z} - P_{3z} - P_{4z} &= 0 \end{aligned} \qquad (R9.18)$$

When expressed in component form, the single equation R9.17 thus becomes a set of *four* equations, one for each of the four-momentum's four components. Each must be *independently* satisfied if four-momentum is to be conserved.

Let us assume that we have observed a collision in the Home Frame and determined that it satisfies the law of conservation of four-momentum in that frame. The principle of relativity requires that the same law apply in *every other* inertial reference frame, that is,

$$\begin{aligned} \text{If} \qquad & \mathbf{P}_1 + \mathbf{P}_2 - \mathbf{P}_3 - \mathbf{P}_4 = 0 \qquad \text{in Home Frame} \\ \text{Then} \qquad & \mathbf{P}_1' + \mathbf{P}_2' - \mathbf{P}_3' - \mathbf{P}_4' = 0 \qquad \text{in any Other Frame} \end{aligned} \qquad (R9.19)$$

If this statement is *not* true, our proposed relativistic generalization of the idea of momentum is not any better than newtonian momentum. If the statement *is* true, then the law of conservation of four-momentum represents at least a *possible* relativistic expression of the law of conservation of ordinary momentum.

We can in fact easily show that this is true for any collision as viewed in any inertial Other Frame. Consider the x component of the conservation law in the Other Frame. According to the transformation law for the components of the four-momentum given by equation R9.12*b*,

$$P'_{1x} + P'_{2x} - P'_{3x} - P'_{4x} = \gamma(-\beta P_{1t} + P_{1x}) + \gamma(-\beta P_{2t} + P_{2x}) \\ - \gamma(-\beta P_{3t} + P_{3x}) - \gamma(-\beta P_{4t} + P_{4x}) \quad \text{(R9.20a)}$$

Collecting the terms on the right side of this equation that are multiplied by γ and those multiplied by $\gamma\beta$, we get

$$P'_{1x} + P'_{2x} - P'_{3x} - P'_{4x} = -\gamma\beta(P_{1t} + P_{2t} - P_{3t} - P_{4t}) \\ + \gamma(P_{1x} + P_{2x} - P_{3x} - P_{4x}) \quad \text{(R9.20b)}$$

But if *both* the t and x components of the four-momentum are conserved in the Home Frame, then equations R9.18 tell us that the quantities in parentheses are equal to zero. This means that

$$P'_{1x} + P'_{2x} - P'_{3x} - P'_{4x} = -\gamma\beta(0) + \gamma(0) = 0 \quad \text{(R9.20c)}$$

So *if* both the t and x components of the system's total four-momentum are conserved in the Home Frame, then the x component of the system's total four-momentum will also be conserved in the Other Frame, as hoped.

Exercise R9X.6

In the same manner, verify that the t and y components of the four-momentum are conserved in the Other Frame if all components are conserved in the Home Frame.

Thus conservation of four-momentum is consistent with the principle of relativity, but is it true?

What we have shown is that the law of conservation of four-momentum expressed by equation R9.17 is *consistent* with the principle of relativity in the sense that if it holds in one frame, it holds in all frames. That does not make the law *true*: it simply makes it *possible*. But now let me argue for the *truth* of this law. (1) We know from a multitude of experiments at low velocities that *some* quantity that reduces to newtonian momentum at such velocities is conserved. (2) Conservation of *newtonian* momentum won't work: the law is inconsistent with the principle of relativity. (3) The hypothetical law of conservation of four-momentum *is* compatible with the principle of relativity. (4) The three spatial components of the four-momentum *do* reduce to the components of newtonian momentum at low velocities. (5) Therefore, if the law of conservation of four-momentum were true, it would both explain the low-velocity experimental data and maintain compatibility with the principle of relativity. (6) Moreover, there must be *some* relativistically valid expression of the deep symmetry principle that gives rise to the law of conservation of momentum. In the absence of compelling alternatives, it makes *sense* to believe that it is the total four-momentum of a system that is conserved.

Abundant experimental evidence suggests that it is

Of course, no matter how suggestive a theoretical argument might be, there is no substitute for direct experimental evidence. Since the 1950s, physicists have been using particle accelerators to create beams of subatomic particles traveling at speeds very near the speed of light that collide with stationary targets or other particle beams. At such speeds, the distinction between newtonian momentum and four-momentum is very clear, and analysis of a typical experiment involves applying conservation of four-momentum to anywhere from thousands to millions of particle collisions. The result is that conservation of four-momentum is implicitly tested thousands of times daily in the course of such research. In spite of this enormous wealth of data, no compelling evidence of a violation of the law of conservation of four-momentum has ever been seen.

R9.6   Four-Momentum's Time and Space Components

Equations R9.20b and c make it clear that conservation of four-momentum is *only* consistent with the principle of relativity if all *four* components of the four-momentum (P_t as well as P_x, P_y, and P_z) are independently conserved. The three *spatial* components of an object's four-momentum correspond (at low velocities) to the three components of its newtonian momentum. What is the physical interpretation of the additional conserved quantity P_t?

What is the interpretation of the fourth conserved quantity P_t?

According to equation R9.10a, P_t for an object with mass m, as measured in a frame where the object moves with speed v, is given by the expression

Examination of meaning at low velocities suggests that P_t is related to energy

$$P_t = \frac{m}{\sqrt{1-v^2}} \tag{R9.21}$$

We know that this reduces to the object's mass m at low velocities, but is not exactly equal to the mass. The binomial approximation (equation R5.17) says that when $x \ll 1$, $(1+x)^a \approx 1 + ax$. Therefore, when $v \ll 1$,

$$P_t = \frac{m}{\sqrt{1-v^2}} = m(1-v^2)^{-1/2} \approx m\left[1 - \left(-\tfrac{1}{2}\right)v^2\right] = m + \tfrac{1}{2}mv^2 \tag{R9.22}$$

The first term here is the mass of the particle, and when v is zero, that is what P_t becomes. But when v is nonzero but still very small, we have an additional term in P_t that corresponds to the *kinetic energy* of the particle. If P_t is conserved in a collision at low velocities, what we are saying is that the sum of the particles' masses plus the sum of their kinetic energies is conserved. If the particles' masses remain unchanged in the collision, then conservation of P_t tells us that the kinetic energy of the particles is conserved.

Thus conservation of P_t is (at low velocities) basically the same as conservation of newtonian (kinetic) energy! As we have already generalized the concept of momentum, so now we generalize the concept of energy. We *define* the component P_t of a particle's four-momentum to be that particle's **relativistic energy** E, and we assert that it is this relativistic energy that is the fourth quantity that is conserved in an isolated system. So the relativistic energy E of an object moving at a speed v as measured in a given inertial frame is

$$E \equiv P_t = \frac{m}{\sqrt{1-v^2}} \tag{R9.23}$$

Definition of relativistic energy

Note that if the object's speed v is an appreciable fraction of the speed of light, equation R9.22 does not hold. We define an object's **relativistic kinetic energy** K (for all v) to be the difference between the object's total relativistic energy E and its mass energy m:

$$K \equiv E - m = \frac{m}{\sqrt{1-v^2}} - m = m\left(\frac{1}{\sqrt{1-v^2}} - 1\right) \tag{R9.24}$$

Definition of relativistic kinetic energy

The value of K is *approximately* equal to $\tfrac{1}{2}mv^2$ only for small values of v: in general, $K > \tfrac{1}{2}mv^2$.

In newtonian mechanics, we thought of conservation of energy and momentum as separate concepts. But just as special relativity binds space and time into a single geometry, so here it binds the laws of conservation of momentum and energy into a single statement: *the total four-momentum of an isolated system of particles is conserved.* Conservation of energy is impossible without conservation of momentum, and vice versa: in the theory of relativity, energy and momentum are indissolubly bound together as parts of the same whole.

Note, however, that an object's relativistic energy is not just its kinetic energy (even at low velocities), but includes its *mass* as well. The fact that $E = P_t$ is conserved does *not* imply that a system's total mass and its total kinetic energy are *separately* conserved: only that the *whole* (i.e., the relativistic energy) is conserved. This implies that *processes that convert mass to kinetic energy, and vice versa, do not necessarily violate the law of conservation of four-momentum* and therefore might exist. We explore this subject more fully in chapter R10.

Equation R9.22 is expressed in SR units, where velocity is unitless and both kinetic energy and mass are measured in kilograms. If we want to express a particle's relativistic energy in the SI unit of joules ($J = kg \cdot m^2/s^2$), we must multiply the energy in kilograms by two powers of the spacetime conversion factor $c = 2.998 \times 10^8$ m/s to get the units to come out right. Therefore, equation R9.22 in SI units would read

$$E \approx mc^2 + \tfrac{1}{2}mv^2 \qquad \text{when } v^2 \ll c^2 \qquad \text{SI units} \qquad \text{(R9.25)}$$

In particular, when the particle is at rest, its relativistic **rest energy** in SI units is

$$E_{\text{rest}} = mc^2 \qquad \qquad \text{(R9.26)}$$

This is the famous equation that has served as an icon representing both the essence of special relativity and Einstein's achievement. This equation really is just a special case of the more general equation R9.23. Even so, it does focus our attention on the startling new idea implicit in equation R9.23: *an object at rest has relativistic energy simply by virtue of its mass,* and this mass energy is a part of the total energy conserved by interactions inside an isolated system. Note that conservation of four-momentum does not require that a system's total mass and its total kinetic energy be separately conserved, only that their sum be conserved. Therefore, conservation of four-momentum opens the possibility that there may be processes that convert mass energy to kinetic energy, and vice versa. We will see in chapter R10 that such processes do indeed exist!

Exercise R9X.7

A rock whose mass is 1.0 kg is moving in the $-z$ direction with a speed of 0.80. What is its relativistic energy E? Its relativistic kinetic energy K? (Express both results in kilograms and joules.)

Just as special relativity teaches us that time and space are but different aspects of spacetime, we now see that energy and momentum are but different aspects of four-momentum. The fact remains, however, that just as we experience time and space very differently, so we experience energy and momentum differently. It is often convenient, therefore, to split a particle's four-momentum into time and space components.

We have seen (see equation R9.11) that at low velocities the spatial components of an object's four-momentum vector P_x, P_y, and P_z become approximately equal to the components of that object's newtonian momentum vector $\vec{p}$. Thus, they represent as close a relativistic analog to $\vec{p}$ as we have. We therefore call the three-dimensional vector $[P_x, P_y, P_z]$ the object's **relativistic three-momentum** (to distinguish it from its four-momentum). More often, we are interested in the simple vector magnitude of these three components, which we will call simply the particle's **relativistic momentum** p:

$$p \equiv \sqrt{P_x^2 + P_y^2 + P_z^2} \qquad \text{(R9.27)}$$

Definition of relativistic momentum

The relativistic momentum p is thus the relativistic generalization of the *magnitude* of an object's newtonian momentum vector.

With the help of equations R9.10, we can express an object's relativistic momentum in terms of its mass m and speed v as follows:

$$p = \sqrt{\left(\frac{mv_x}{\sqrt{1-v^2}}\right)^2 + \left(\frac{mv_y}{\sqrt{1-v^2}}\right)^2 + \left(\frac{mv_z}{\sqrt{1-v^2}}\right)^2}$$

$$= \frac{m\sqrt{v_x^2 + v_y^2 + v_z^2}}{\sqrt{1-v^2}} = \frac{mv}{\sqrt{1-v^2}} \qquad \text{(R9.28)}$$

Note that p does indeed become approximately equal to the magnitude mv of the object's newtonian momentum when v is very much smaller than 1.

We can write equation R9.15, which shows how an object's frame-independent mass can be computed by using the frame-dependent components of its four-momentum, in terms of E and p as follows:

Useful relationships between E, p, and m

$$m^2 = P_t^2 - \left(P_x^2 + P_y^2 + P_z^2\right) = E^2 - p^2 \qquad \text{(R9.29)}$$

We can also use an object's relativistic energy E and relativistic momentum p in a given frame to determine the object's speed in that frame:

$$\frac{p}{E} = \frac{mv/\sqrt{1-v^2}}{m/\sqrt{1-v^2}} = v \qquad \text{(R9.30)}$$

This relationship applies to each individual spatial component as well:

$$\frac{P_x}{E} = \frac{mv_x}{\sqrt{1-v^2}}\frac{\sqrt{1-v^2}}{m} = v_x \quad \text{similarly} \quad \frac{P_y}{E} = v_y \quad \frac{P_z}{E} = v_z$$

$$\text{(R9.31)}$$

Indeed, the spatial components of the four-momentum can be thought of as expressing the rate at which relativistic energy is transported through space:

$$\begin{bmatrix} P_t \\ P_x \\ P_y \\ P_z \end{bmatrix} = \begin{bmatrix} m/\sqrt{1-v^2} \\ mv_x/\sqrt{1-v^2} \\ mv_y/\sqrt{1-v^2} \\ mv_z/\sqrt{1-v^2} \end{bmatrix} = \begin{bmatrix} E \\ Ev_x \\ Ev_y \\ Ev_z \end{bmatrix} \qquad \text{(R9.32)}$$

Equations R9.23, R9.24, R9.28, R9.29, and R9.30 express relationships between the quantities E, p, m, K, and v that are really helpful to know when one is working with the four-momentum: they are summarized in figure R9.5.

Exercise R9X.8

Consider the rock described in exercise R9X.7. Find its relativistic momentum p, and write down its four-momentum vector.

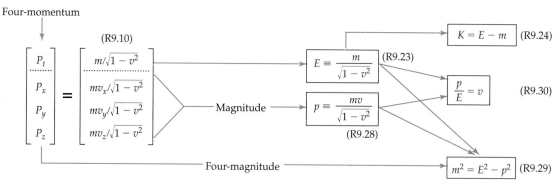

Figure R9.5

Virtually everything that you need to know about four-momentum.

TWO-MINUTE PROBLEMS

R9T.1   A particle's newtonian x-momentum p_x is always either equal to or *smaller* than the x component of its four-momentum P_x, true (T) or false (F)?

R9T.2   A particle's mass is always either equal to or smaller than the time-component of its four-momentum, T or F?

R9T.3   The absolute value of the x component of a particle's four-momentum vector is always either equal to or smaller than its t component, T or F?

R9T.4   The squared magnitude of a four-vector could be negative, T or F? The squared magnitude of a particle's four-momentum could be negative, T or F?

R9T.5   How does a particle's relativistic kinetic energy compare to its newtonian kinetic energy when the particle's speed $v \neq 0$?
A. $K > \frac{1}{2}mv^2$
B. $K < \frac{1}{2}mv^2$
C.   It depends on the value of v.

R9T.6   The components of a particle's four-momentum are the same in all inertial reference frames, T or F?

R9T.7   Imagine that a particle has a four-momentum vector whose components are $[P_t, P_x, P_y, P_z] = [5\ \text{kg}, 3\ \text{kg}, 0, 0]$. What is this particle's mass?
A.   5 kg
B.   8 kg
C.   4 kg
D.   3 kg
E. $\sqrt{34}$ kg
F.   Other (specify)

R9T.8   A dust particle has a rest energy (mass) of 1.0 μg. In joules, this rest energy is closer to
A. 10^{17} J
B. 10^{11} J
C. 10^8 J
D. 10^{-3} J
E. 10^{-23} J
F.   Other (specify)

R9T.9   As a particle's speed approaches that of light, the difference between its relativistic momentum p and its relativistic energy E becomes small (compared to E), T or F?

R9T.10   The mass of a particle is the same in all inertial reference frames (even though its speed is not), T or F?

HOMEWORK PROBLEMS

Basic Skills

R9B.1   A 2.0-kg object moves with a speed of 0.60 in the $+x$ direction. Find the components of the object's four-momentum.

R9B.2   An alien spaceship with a mass of 12,000 kg is traveling in the solar system frame at a speed of 0.80 in the $-z$ direction. Find the components of the spaceship's four-momentum vector.

R9B.3   A 12-kg rock moves with a velocity whose components $[v_x, v_y, v_z] = [\frac{4}{13}, -\frac{3}{13}, 0]$ in a certain frame. Find the components of its four-momentum in that frame.

R9B.4   Imagine that an object of mass 5.0 kg is observed in a certain inertial frame to have velocity components of $v_x = -0.866$ and $v_y = v_z = 0$. Evaluate the following in that reference frame:

(a) the object's total energy E,

(b) its relativistic momentum p,

(c) the spatial components of its four-momentum, and

(d) its relativistic kinetic energy K.

R9B.5 An object is observed in a certain frame to have a four-momentum of $[P_t, P_x, P_y, P_z] = [5.0 \text{ kg}, 4.0 \text{ kg}, 0, 0]$.

(a) Find the object's x-velocity in this frame.

(b) Find the object's mass.

(c) Find its kinetic energy in this frame.

(d) Find its relativistic momentum in this frame.

R9B.6 An object is observed in a certain frame to have a four-momentum of $[P_t, P_x, P_y, P_z] = [13 \text{ kg}, -12 \text{ kg}, 0, 0]$.

(a) Find the object's x-velocity in this frame.

(b) Find the object's mass.

(c) Find its kinetic energy.

R9B.7 A particle is observed in a certain inertial frame to have a four-momentum vector whose components are $[P_t, P_x, P_y, P_z] = [18 \text{ kg}, 9.0 \text{ kg}, 15 \text{ kg}, 1.0 \text{ kg}]$.

(a) Find the object's velocity (vector).

(b) Find the object's speed.

(c) Find the object's mass.

(d) Find the object's relativistic momentum p.

(e) Find the object's kinetic energy.

R9B.8 A particle's four-momentum in a given inertial frame is $[P_t, P_x, P_y, P_z] = [5.0 \text{ kg}, 3.0 \text{ kg}, 0, 0]$. What is its four-momentum in an inertial frame that moves in the $+x$ direction at a speed of 0.60 relative to the first frame?

R9B.9 A particle's four-momentum in a given inertial frame is $[P_t, P_x, P_y, P_z] = [5.0 \text{ kg}, -3.0 \text{ kg}, 0, 0]$. What is its four-momentum in an inertial frame that moves in the $+x$ direction at a speed of 0.80 relative to the first frame?

Synthetic

R9S.1 Verify equation R9.15 by squaring the definitions of the four-momentum components given in equations R9.10 and combining as required.

R9S.2 Prove that in the limit as $v \to 1$, $E \approx p$. What speed is required for these quantities to be equal to within 1%?

R9S.3 While solving a problem, friend claims that a particle's four-momentum vector is given by $[P_t, P_x, P_y, P_z] = [-0.10 \text{ kg}, \frac{2}{5}, -\frac{1}{5}, 0]$. Even without knowing the particulars of the problem, there are at least three things wrong with this four-vector as stated. What are they?

R9S.4 Imagine that a dust particle of mass 2.0 μg is traveling at a speed of $\frac{4}{5}$ in the xy plane at an angle of $30°$

clockwise from the x axis in a certain inertial reference frame. Evaluate the following in that frame:

(a) the particle's relativistic energy E,

(b) its relativistic momentum p,

(c) its three spatial four-momentum components, and

(d) its kinetic energy K (in joules).

R9S.5 A particle's mass is 1.0 μg, its kinetic energy is 2.0 μg, and it is moving in the $+y$ direction. Find its four-momentum vector.

R9S.6 One can use the four-momentum transformation law to do a correct relativistic transformation of velocities. Consider a 1.0-kg object moving at a speed of $\frac{3}{5}$ in the $+x$ direction in the Home Frame. What are the components of its four-momentum in this frame? Using the Lorentz transformation equations, find the components of this object's four-momentum in an Other Frame moving at a speed of $\frac{4}{5}$ in the $+x$ direction relative to the Home Frame. Then use equation R9.30 to find the object's speed in that Other Frame. Check your result with the Einstein velocity transformation equations.

R9S.7 If electrical energy can be sold at about $0.03 per 10^6 J (the approximate current price in southern California), compute how much your rest energy is worth in dollars. That is, find the amount of money that your survivors could put in your memorial fund if there was a way to convert your mass entirely to electrical energy. (It is probably a good thing that this is not easy to do.)

R9S.8 If electrical energy costs about $0.04 per 10^6 J (the approximate current price in southern California) and you have $1.5 million at your disposal to spend on energy to convert to kinetic energy, about how fast can you make a 1.0-g object travel?

R9S.9 In a number of problems in this text, we have blithely spoken about trains traveling at significant fractions of the speed of light. If electrical energy costs about $0.04 per 10^6 J (the approximate current

How much would it cost to get this train moving at a speed of $\frac{3}{5}$? (See problem R9S.9.)

price in southern California), what would it cost to accelerate an electric train with a mass of about 100,000 kg to a speed of $\frac{3}{5}$?

R9S.10 At nonrelativistic speeds ($v \ll 1$), a particle's kinetic energy is $K \approx \frac{1}{2}mv^2 = (mv)^2/2m \approx p^2/2m$, so we have $p \approx (2mK)^{1/2}$. Show that the exact relativistic expression for the particle's relativistic momentum p in terms of its relativistic kinetic energy K is

$$p = \sqrt{K(K + 2m)} \qquad \text{(R9.33)}$$

and argue that at low speeds, this reduces to $p \approx (2mK)^{1/2}$.

R9S.11 Imagine that in a certain particle physics experiment, a particle of mass m_0 moving with speed $v_0 = \frac{3}{5}$ in the $+x$ direction in the laboratory frame is observed to decay into two particles, one with mass m_1 moving with speed $v_1 = \frac{4}{5}$ in the $+x$ direction and another with mass m_2 that is essentially at rest.
   (a) Show that if total particle mass and newtonian momentum are conserved in the laboratory frame, then we must have $m_1 = \frac{3}{4}m_0$ and $m_2 = \frac{1}{4}m_0$.
   (b) Show that if four-momentum is to be conserved in the laboratory frame, we must have $m_1 = \frac{9}{16}m_0$ and $m_2 = \frac{5}{16}m_0$ (note that this decay converts some mass energy to kinetic energy if conservation of four-momentum is true).
   (c) Consider how an inertial reference frame (the Other Frame) that moves with the same speed in the $+x$ direction as the initial particle. Use the Einstein velocity transformation equations to show that the particles' x-velocities in this frame must be $v'_{0x} = 0$, $v'_{1x} = \frac{5}{13}$, and $v'_{2x} = -\frac{3}{5}$.
   (d) Show that if we use the outgoing particle masses that ensure that newtonian momentum and particle mass are conserved in the laboratory frame (see part a), newtonian momentum is *not* conserved in the Other Frame.
   (e) Show that if we use the masses that ensure that four-momentum is conserved in the laboratory frame (see part b), we find that four-momentum *is* conserved in the Other Frame.

R9S.12 Consider again the one-dimensional elastic collision discussed in section R9.2. In this collision, an object of mass m moving at an initial speed of $\frac{3}{5}$ in the $+x$ direction hits an object of mass $2m$ at rest. It turns out that conservation of *four-momentum* for this elastic collision implies that the x-velocities of the objects after the collision are $v_{3x} = -9/41$ and $v_{4x} = +39/89$ (instead of the newtonian predictions of $-\frac{1}{5}$ and $+\frac{2}{5}$, respectively) in the laboratory frame.
   (a) Show that if the objects have these final x-velocities, both the time and x components of the system's four-momentum are indeed conserved in the laboratory frame.

   (b) Use the Einstein velocity transformation equation to find the velocities in the frame in which the lighter object is initially at rest, and show that the time and x' components of the system's four-momentum are conserved in this frame also.

Rich-Context

R9R.1 To compete with e-mail, imagine that the U.S. Postal Service now offers Super Express Mail, where a letter is sent to its destination at a speed of 0.999 using a special Letter Accelerator. Assume that a typical letter has a mass of about 25 g. What will be the approximate minimum cost of the letter's stamp? If the letter misses its target and hits a nearby building, describe the consequences. (*Hint:* See problem R9S.8 for some useful information. It also may help you to know that a typical nuclear bomb releases on the order of magnitude of 5×10^{14} J of energy when it explodes.)

R9R.2 In this text we have blithely described people running at substantial fractions of the speed of light. Consider a runner who has somehow managed to accelerate to a speed of $\frac{3}{5}$. Even assuming that this highly trained athlete uses food energy very efficiently, describe the meal that the runner must have eaten before the race. (*Hint:* Remember that one food calorie is 1000 physics calories, and each physics calorie is equivalent to the thermal energy required to raise the temperature of 1.0 g of water by $1°C = 4.186$ J.)

R9R.3 A typical household might use about 2×10^{10} J of electrical energy per year. About 1 in every 5000 hydrogen atoms in a quantity of water is actually deuterium, and the fusion of two deuterium nuclei converts about 0.5% of the rest energy of the deuterium nuclei to other forms of energy. Avogadro's number of hydrogen atoms has a mass of 1 g, the same number of deuterium atoms has a mass of 2 g, and the same number of oxygen atoms has a mass of 16 g. About how long could you run a typical household on the energy that would be produced by the fusion of the deuterium in 1 gal of water?

Advanced

R9A.1 Prove that the four-magnitude of any four-vector is a frame-independent number. (*Hint:* What is the formal definition of a four-vector?)

R9A.2 Prove conclusively that a particle's relativistic kinetic energy $K \equiv E - m$ is always larger than $\frac{1}{2}mv^2$ (although the values are close for $v \ll 1$). (*Hint:* Both go to zero as v goes to zero, but you can show that the derivative of K with respect to v is always greater than the derivative of $\frac{1}{2}mv^2$ with respect to v. How does this help?)

ANSWERS TO EXERCISES

R9X.1 We have

$$v'_{4x} = \frac{v_{4x} - \beta}{1 - \beta v_{4x}} = \frac{+\frac{2}{5} - \frac{3}{5}}{1 - \left(\frac{2}{5}\right)\left(\frac{3}{5}\right)} = -\frac{\frac{1}{5}}{\frac{19}{25}} = -\frac{5}{19} \quad \text{(R9.34)}$$

R9X.2 If the galilean velocity transformations were true, then $v'_{1x} = 0$ and $v'_{1x} = -\frac{3}{5}$ as before, but

$$v'_{3x} = v_{3x} - \beta = -\frac{1}{5} - \frac{3}{5} = -\frac{4}{5} \quad \text{(R9.35a)}$$

$$v'_{4x} = v_{4x} - \beta = +\frac{2}{5} - \frac{3}{5} = -\frac{1}{5} \quad \text{(R9.35b)}$$

The system's total momentum in the Other Frame before the collision is still $-\frac{6}{5}m$, but the total momentum after the collision is now

$$mv_{3x} + 2mv_{4x} = m\left(-\frac{4}{5}\right) + 2m\left(-\frac{1}{5}\right) = -\frac{6}{5}m \quad \text{(R9.36)}$$

also. So ordinary momentum would in fact be conserved if the galilean velocity transformation equations were valid.

R9X.3 If $v = \frac{4}{5}$, then $(1 - v^2)^{1/2} = (1 - \frac{16}{25})^{1/2} = \frac{3}{5}$. So, according to equation R9.10, we have

$$\begin{bmatrix} P_t \\ P_x \\ P_y \\ P_z \end{bmatrix} = \frac{1}{\sqrt{1 - v^2}} \begin{bmatrix} m \\ mv_x \\ mv_y \\ mv_z \end{bmatrix} = \frac{5}{3} \begin{bmatrix} m \\ 0 \\ \frac{4}{5}m \\ 0 \end{bmatrix} = \begin{bmatrix} \frac{5}{3}\,\text{kg} \\ 0 \\ \frac{4}{3}\,\text{kg} \\ 0 \end{bmatrix} \quad \text{(R9.37)}$$

In the frame where the object is at rest, $(1 - v^2)^{1/2} = 1$, so

$$\begin{bmatrix} P_t \\ P_x \\ P_y \\ P_z \end{bmatrix} = \frac{1}{\sqrt{1 - v^2}} \begin{bmatrix} m \\ mv_x \\ mv_y \\ mv_z \end{bmatrix} = \begin{bmatrix} m \\ 0 \\ 0 \\ 0 \end{bmatrix} = \begin{bmatrix} 1\,\text{kg} \\ 0 \\ 0 \\ 0 \end{bmatrix} \quad \text{(R9.38)}$$

R9X.4 Using the same basic approach outlined in equation R9.12a, we get

$$P'_x = m\frac{dx'}{d\tau} = m\frac{\gamma(-\beta\,dt + dx)}{d\tau} = -\gamma\beta m\frac{dt}{d\tau} + \gamma m\frac{dx}{d\tau}$$

$$= -\gamma\beta P_t + \gamma P_x = \gamma(-\beta P_t + P_x) \quad \text{(R9.39)}$$

R9X.5 Substituting the definition of the four-momentum components into the definition of the four-magnitude of the four-momentum, we see that

$$|\mathbf{P}| \equiv \sqrt{P_t^2 - P_x^2 - P_y^2 - P_z^2}$$

$$= \sqrt{\left(m\frac{dt}{d\tau}\right)^2 - \left(m\frac{dx}{d\tau}\right)^2 - \left(m\frac{dy}{d\tau}\right)^2 - \left(m\frac{dz}{d\tau}\right)^2}$$

$$= m\frac{\sqrt{dt^2 - dx^2 - dy^2 - dz^2}}{d\tau} = m\frac{d\tau}{d\tau} = m \quad \text{(R9.40)}$$

(where in the next-to-last step, I used the provided hint).

R9X.6 We can prove conservation of P'_t as follows:

$$P'_{1t} + P'_{2t} - P'_{3t} - P'_{4t}$$
$$= \gamma(P_{1t} - \beta P_{1x}) + \gamma(P_{2t} - \beta P_{2x})$$
$$\quad - \gamma(P_{3t} - \beta P_{3x}) - \gamma(P_{4t} + \beta P_{4x})$$
$$= \gamma(P_{1t} + P_{2t} - P_{3t} - P_{4t})$$
$$\quad -\gamma\beta(P_{1x} + P_{2x} - P_{3x} - P_{4x})$$
$$= \gamma(0) - \gamma\beta(0) = 0 \quad \text{(R9.41)}$$

since the quantities $P_{1t} + P_{2t} - P_{3t} - P_{4t} = 0$ and $P_{1x} + P_{2x} - P_{3x} - P_{4x} = 0$ if four-momentum is conserved in the Home Frame. Proving conservation of P'_y is even easier:

$$P'_{1y} + P'_{2y} - P'_{3y} - P'_{4y} = P_{1y} + P_{2y} - P_{3y} - P_{4y} = 0 \quad \text{(R9.42)}$$

R9X.7 According to equation R9.23, the rock's energy is

$$E = \frac{m}{\sqrt{1 - v^2}} = \frac{1.0\,\text{kg}}{\sqrt{1 - \frac{16}{25}}} = \frac{1.0\,\text{kg}}{\sqrt{\frac{9}{25}}} = \frac{5}{3}\,\text{kg} \quad \text{(R9.43)}$$

Its kinetic energy is

$$K \equiv E - m = \frac{5}{3}\,\text{kg} - 1.0\,\text{kg} = \frac{2}{3}\,\text{kg} \quad \text{(R9.44)}$$

Converting to joules, we find that

$$E = \left(\frac{5}{3}\,\text{kg}\right)\left(\frac{3.0 \times 10^8\,\text{m}}{1\,\text{s}}\right)^2\left(\frac{1\,\text{J}}{1\,\text{kg·m}^2/\text{s}^2}\right)$$

$$= 1.5 \times 10^{17}\,\text{J} \quad \text{(R9.45)}$$

$$K = \left(\frac{2}{3}\,\text{kg}\right)\left(\frac{3.0 \times 10^8\,\text{m}}{1\,\text{s}}\right)^2\left(\frac{1\,\text{J}}{1\,\text{kg·m}^2/\text{s}^2}\right)$$

$$= 6.0 \times 10^{16}\,\text{J} \quad \text{(R9.46)}$$

For the sake of comparison, note that a typical nuclear bomb releases about 5×10^{14} J of energy when it explodes.

R9X.8 The rock's relativistic momentum is

$$p = \frac{mv}{\sqrt{1 - v^2}} = \frac{(1.0\,\text{kg})\left(\frac{4}{5}\right)}{\frac{3}{5}} = \frac{4}{3}\,\text{kg} \quad \text{(R9.47)}$$

Using equation R9.32, we see that the rock's four-momentum vector is

$$\begin{bmatrix} P_t \\ P_x \\ P_y \\ P_z \end{bmatrix} = \begin{bmatrix} E \\ Ev_x \\ Ev_y \\ Ev_z \end{bmatrix} = \begin{bmatrix} \frac{5}{3}\,\text{kg} \\ 0 \\ 0 \\ \left(\frac{5}{3}\,\text{kg}\right)\left(-\frac{4}{5}\right) \end{bmatrix} = \begin{bmatrix} \frac{5}{3}\,\text{kg} \\ 0 \\ 0 \\ -\frac{4}{3}\,\text{kg} \end{bmatrix} \quad \text{(R9.48)}$$

R10

Conservation of Four-Momentum

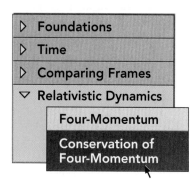

Chapter Overview

Introduction

Chapter R9 introduced the concept of conservation of four-momentum and showed that such a law could be fully consistent with the principle of relativity while reducing to the newtonian form of the law at low speeds. In this chapter, we explore some of the surprising consequences and experimental tests of this law.

Section R10.1: Energy-Momentum Diagrams

An **energy-momentum diagram** represents the four-momenta of particles or systems as arrows on a diagram whose vertical and horizontal axes are E and P_x, respectively. Such a diagram displays these four-momenta in the same way that an ordinary space-time diagram displays the coordinates of events.

A given object's four-momentum arrow on such a diagram has a slope of $1/v_x$, where v_x is its speed. We will use a flag attached to the shaft of the arrow to indicate the arrow's four-magnitude (i.e., its mass). We can qualitatively estimate the magnitude corresponding to a given four-momentum vector by projecting a hyperbola from the vector's tip back to the vertical axis. See figure R10.5 for a summary of how to read such a diagram.

Section R10.2: Working Conservation of Four-Momentum Problems

One can qualitatively solve a conservation of four-momentum problem by using an energy-momentum diagram, or quantitatively by using four-dimensional column vectors. As an example of the latter, consider a collision between incoming particles with four-momenta $\mathbf{P}_1$ and $\mathbf{P}_2$ that produces final particles with four-momenta $\mathbf{P}_3$ and

Figure R10.5
Virtually everything you need to know about four-momentum diagrams.

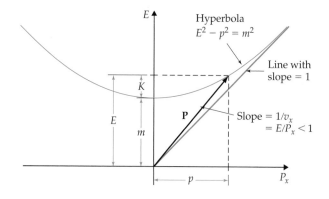

$\mathbf{P}_4$. Conservation of four-momentum requires that

$$
\begin{bmatrix} E_1 \\ P_{1x} \\ P_{1y} \\ P_{1z} \end{bmatrix} + \begin{bmatrix} E_2 \\ P_{2x} \\ P_{2y} \\ P_{2z} \end{bmatrix} = \begin{bmatrix} E_3 \\ P_{3x} \\ P_{3y} \\ P_{3z} \end{bmatrix} + \begin{bmatrix} E_4 \\ P_{4x} \\ P_{4y} \\ P_{4z} \end{bmatrix}
\tag{R10.3}
$$

As usual, each row of this column-vector equation must be satisfied independently.

This section discusses an example that illustrates both approaches. The diagrams are not often much easier than doing the algebra (although having access to hyperbola graph paper helps), but often provide a useful visual check on algebraic solutions.

Section R10.3: The Mass of a System of Particles

In chapter R9, we defined an object's mass to be the frame-independent magnitude of its four-momentum. This definition implies, however, that *mass is not additive:* the mass of a system is *not* equal to the sum of the masses of its particles. The total mass of a system at rest is in fact equal to the sum of the *energies* of all its particles, which (for noninteracting particles at least) is usually larger than the sum of the particle masses. The extra mass does not really reside anywhere that we can locate: it is rather a property of the system as a whole.

In a collision between two balls of putty of mass m that stick together afterward, for example, the mass of the final putty glob is $M > 2m$. However, the mass of the *system* is M both before and after the putty balls collide!

Section R10.4: The Four-Momentum of Light

A sufficiently short and confined flash of light is the approximate equivalent of a particle. Such a light flash carries energy and so must have a four-momentum vector. Its four-momentum arrow must be parallel to the light flash's worldline, so (1) the slope of the four-momentum arrow must be ± 1 on an energy-momentum diagram, which means that (2) the relativistic momentum p of a flash must be $p = E$, which means that (3) the flash's mass m is $m = (E^2 - p^2)^{1/2} = 0$! To put it another way, a flash of light has all its energy in the form of kinetic energy and none in the form of mass energy.

An antimatter-matter rocket would convert the mass of its fuel entirely to light that travels out the engine's nozzle with speed c. Such an engine would be the best possible rocket engine. Example R10.3 discusses such a rocket as an illustration of doing a conservation of four-momentum problem involving light.

Section R10.5: Applications to Particle Physics

Most real applications and tests of relativity involve subatomic particles, because such particles are about the only things that are light enough to be accelerated to near-light speeds without a prohibitive cost in energy. Particle physicists typically use the electronvolt (eV) as the unit of mass, energy, and momentum: $1\,\text{eV} = 1.602 \times 10^{-19}\,\text{J} = 1.782 \times 10^{-36}\,\text{kg}$. This section discusses the decay of a particle called a kaon as an example of a realistic application of relativity.

Section R10.6: Parting Comments

This section discusses a few places where interested readers can go to learn more about relativity.

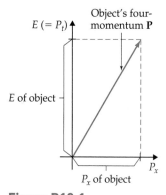

Figure R10.1

An energy-momentum diagram displaying the four-momentum of an object moving in the +x direction. The object's four-momentum is shown on the diagram as an arrow whose projections on the vertical and horizontal axes represent the values of the object's relativistic energy $E = P_t$ and its relativistic x-momentum P_x, respectively.

R10.1 Energy-Momentum Diagrams

We can visually represent an object's four-momentum as an arrow on a special kind of spacetime diagram called an **energy-momentum diagram** (see figure R10.1). We can do this conveniently only if the object is moving in the spatial x direction (so that $P_y = P_z = 0$ and $p = |P_x|$). In the rest of the chapter, we will assume that this is true (unless otherwise specified).

Just as the direction of the arrow representing an object's ordinary momentum is tangent to its path through space, the direction of the arrow representing an object's four-momentum is tangent to its worldline in spacetime (because the object's four-momentum vector $\mathbf{P} = m\,d\mathbf{R}/d\tau$ at any given point along its worldline is proportional to the object's differential displacement in spacetime $d\mathbf{R}$ along that worldline around that point: see figure R10.2).

Since the inverse slope of an object's worldline at any instant is equal to its x-velocity at that instant, the inverse slope of the object's four-momentum arrow at a given time (i.e., run/rise = P_x/E) should also be equal to its x-velocity at that time if the two vectors are to be parallel. Equation R9.31 says essentially the same thing:

$$\frac{P_x}{E} = \frac{mv_x}{\sqrt{1-v^2}}\frac{\sqrt{1-v^2}}{m} = v_x \tag{R10.1}$$

The four-magnitude of an object's four-momentum is its mass m (see equation R9.15): this value is frame-independent and independent of the object's motion. But the *length* of the arrow that represents the object's four-momentum on an energy-momentum diagram *does* depend on the object's velocity: it is *not* proportional to the four-magnitude of the corresponding four-momentum. (This is analogous to the problem with the spacetime

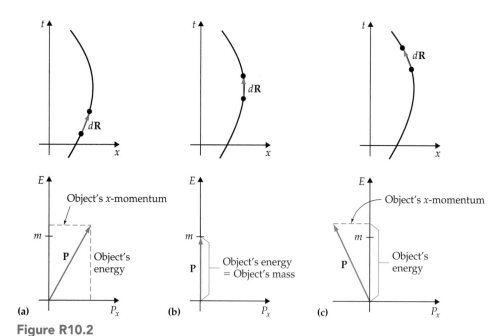

(a) **(b)** **(c)**

Figure R10.2

(a) At any given time, the arrow that represents an object's four-momentum on an energy-momentum diagram points in a direction tangent to the object's worldline, because $\mathbf{P}$ is proportional to $d\mathbf{R}$. (b) When an object is at rest (even at just an instant), its four-momentum arrow is vertical and its energy is equal to its mass (see equation R9.10a). (c) When the object moves in the −x direction, its x-momentum is negative (see equation R9.10b), but its energy remains positive.

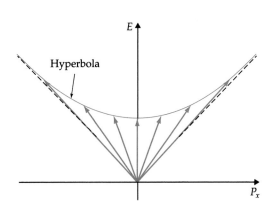

Figure R10.3

An energy-momentum diagram showing four-momentum arrows for a set of identical objects of mass m moving at different x-velocities in the Home Frame. The tips of all these arrows lie on a hyperbola whose equation is $m^2 = E^2 - p^2$. Note that as the object's x-velocity approaches ± 1 (and thus p/E approaches 1), both p and E have to become very large if the difference of their squares is to remain fixed.

interval discussed in section R4.6.) In fact, according to equation R9.29, we have

$$m^2 = E^2 - p^2 = E^2 - (P_x)^2 \qquad (R10.2)$$

This means that the tips of the four-momentum arrows for objects of identical mass m traveling at different x-velocities (or the four-momentum arrows for a single accelerating object observed at different times) lie along the hyperbola on the diagram defined by the equation $m^2 = E^2 - p^2$, as shown in figure R10.3.

If you know an object's x-velocity and its mass m, you can easily draw an energy-momentum diagram showing its four-momentum vector as follows:

How to draw the four-momentum for an object with known mass and speed

1. Set up your E and P_x axes.
2. Draw a line from the origin of those axes having the slope $1/v_x$.
3. Compute the value of $E = m/(1 - v^2)^{1/2} = m/(1 - v_x^2)^{1/2}$ for the object.
4. Draw a horizontal line from this value on the E axis until it intersects the line that you drew in step 2.
5. The arrow representing the object's four-momentum lies along the line drawn in step 2, with its tip at the intersection found in step 4.

Alternatively, if you have access to hyperbola graph paper (you can download some examples from the *Six Ideas* website), you can locate the hyperbola that intersects the vertical axis at the object's mass m and follow the hyperbola out to where it intersects the sloping line that you drew in step 2. The point of intersection will correspond to the tip of the arrow.

We can easily read an object's relativistic kinetic energy $K = E - m$ directly from an energy-momentum diagram. For example, K for an object of mass m moving at a speed $v = \frac{3}{5}$ is $m/4$, as shown in figure R10.4. [Note that

A practical reason why no object can travel at the speed of light or faster

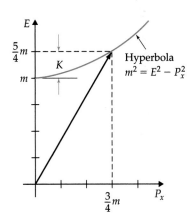

Figure R10.4

An energy-momentum diagram of an object of mass m traveling with an x-velocity $v_x = \frac{3}{5}$. The arrow representing the object's four-momentum has a slope of 5/3 and [since $(1 - v^2)^{1/2} = \frac{4}{5}$ in this case] an energy of $E = m/(1 - v^2)^{1/2} = 5m/4$. For the arrow to have the correct slope, we must have $P_x = \frac{3}{4}m$. Note that the object's relativistic kinetic energy $K = E - m = \frac{1}{4}m$ can be read directly from the diagram.

Figure R10.5

Virtually everything that you need to know about four-momentum diagrams. No matter what the x-velocity of an object of mass m might be, the tip of the arrow representing its four-momentum lies on the hyperbola $m^2 = E^2 - p^2$. The inverse slope of the arrow representing the four-momentum is equal to v_x, which always has a magnitude less than 1.

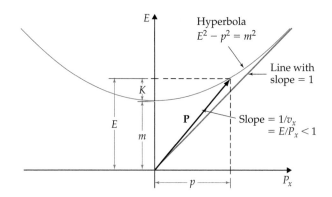

$K \neq \frac{1}{2}m(\frac{3}{5})^2$, but in fact is substantially larger!] Figures R10.3 and R10.4 together make it clear that as an object's speed v approaches 1 (the speed of light), both the object's total relativistic energy E and its kinetic energy K go to infinity. This means that *you would have to supply an infinite amount of energy to accelerate an object of nonzero mass to the speed of light.* (This is the most practical reason that no object can go faster than the speed of light: all the energy in the universe could not accelerate even a mote of dust to that speed!)

Exercise R10X.1

Check that $K = \frac{1}{4}m > \frac{1}{2}m(\frac{3}{5})^2$, as claimed above.

Virtually all that you need to know to construct and interpret an energy-momentum diagram is summarized in figure R10.5.

Exercise R10X.2

Draw an energy-momentum diagram of an object whose mass is 1.0 kg and which moves in the $-x$ direction at a speed of $\frac{4}{5}$. Compute its energy E, and then read its relativistic kinetic energy K and the x component of its four-momentum P_x from the diagram.

R10.2 Working Conservation of Four-Momentum Problems

The law of conservation of four-momentum (like the law of conservation of ordinary momentum) is most useful when applied to an isolated system of objects undergoing some type of *collision* process (i.e., a kind of sudden interaction between the objects in the system that may be strong and complicated but limited in time). In such a case, the system has a clearly defined state "before" and "after" the collision, making it easy to compute the system's total four-momentum both before and after the collision. The law of conservation of four-momentum states that the system should have the *same* total four-momentum after the collision process as it had before.

Solving conservation problems algebraically

What does this really mean mathematically? Since four-momentum is a (four-dimensional) vector quantity, conservation of four-momentum means that *each component of the system's total four-momentum is separately conserved.* For example, consider a system consisting of two objects, and let the objects' four-momenta before the collision be $\mathbf{P}_1$ and $\mathbf{P}_2$ and after the collision be $\mathbf{P}_3$

and $\mathbf{P}_4$. Conservation of four-momentum requires that

$$
\begin{bmatrix} E_1 \\ P_{1x} \\ P_{1y} \\ P_{1z} \end{bmatrix} + \begin{bmatrix} E_2 \\ P_{2x} \\ P_{2y} \\ P_{2z} \end{bmatrix} = \begin{bmatrix} E_3 \\ P_{3x} \\ P_{3y} \\ P_{3z} \end{bmatrix} + \begin{bmatrix} E_4 \\ P_{4x} \\ P_{4y} \\ P_{4z} \end{bmatrix} \qquad \text{(R10.3)}
$$

remembering that the time component of a four-momentum vector (i.e., the relativistic energy) is usually given the more evocative symbol E instead of P_t. As usual, each row of this equation has to be *separately* true for four-momentum to be conserved.

As I said before, in this chapter we will most often consider objects moving in only *one* dimension, which we can take to be the $\pm x$ direction. This simplifies the mathematics significantly, and there is not anything significant to be learned by doing more complicated cases. Even so, I typically show all four components in column-vector equations, because it reminds me visually that I am dealing with conservation of *four*-momentum, and because it reminds me that I may have to worry about these components in rare circumstances.

Example R10.1: Elastically Colliding Rocks

Problem   Imagine that somewhere in deep space a rock with mass $m_1 = 12\,\text{kg}$ is moving in the $+x$ direction with $v_{1x} = +\frac{4}{5}$ in some inertial frame. This rock then strikes another rock of mass $m_2 = 28$ kg at rest ($v_{2x} = 0$). Pretend that the first rock, instead of instantly vaporizing into a cloud of gas (as any *real* rocks colliding at this speed would), simply bounces off the more massive rock and is subsequently observed to have an x-velocity $v_{3x} = -\frac{5}{13}$. What is the x-velocity v_{4x} of the larger rock after the collision?

Solution   The first step in solving this problem is to calculate the energy E_1 and the x-momentum P_{1x} of the smaller rock before the collision. Using the definitions of these four-momentum components, we find that

$$
E_1 \equiv \frac{m_1}{\sqrt{1 - v_{1x}^2}} = \frac{m_1}{\sqrt{1 - \frac{16}{25}}} = \frac{m_1}{\sqrt{\frac{9}{25}}} = \frac{5}{3}(12\,\text{kg}) = 20\,\text{kg} \qquad \text{(R10.4a)}
$$

$$
P_{1x} \equiv \frac{m_1 v_{1x}}{\sqrt{1 - v_{1x}^2}} = \frac{m_1 \left(+\frac{4}{5}\right)}{\frac{3}{5}} = +\frac{4}{3}(12\,\text{kg}) = +16\,\text{kg} \qquad \text{(R10.4b)}
$$

Similarly, the larger rock's energy and x-momentum before the collision are

$$
E_2 \equiv \frac{m_2}{\sqrt{1 - v_{2x}^2}} = \frac{m_2}{\sqrt{1 - 0^2}} = m_2 = 28\,\text{kg} \qquad \text{(R10.5a)}
$$

$$
P_{2x} \equiv \frac{m_2 v_{2x}}{\sqrt{1 - v_{2x}^2}} = \frac{m_2(0)}{\sqrt{1 - 0^2}} = 0\,\text{kg} \qquad \text{(R10.5b)}
$$

The smaller rock's energy and x-momentum *after* the collision are

$$
E_3 \equiv \frac{m_1}{\sqrt{1 - v_{3x}^2}} = \frac{m_1}{\sqrt{1 - \left(-\frac{5}{13}\right)^2}} = \frac{m_1}{\sqrt{\frac{144}{169}}} = \frac{13}{12}(12\,\text{kg}) = 13\,\text{kg} \quad \text{(R10.6a)}
$$

$$
P_{3x} \equiv \frac{m_1 v_{3x}}{\sqrt{1 - v_{3x}^2}} = \frac{m_1 \left(-\frac{5}{13}\right)}{\frac{12}{13}} = -\frac{5}{12}(12\,\text{kg}) = -5\,\text{kg} \qquad \text{(R10.6b)}
$$

Conservation of four-momentum requires that the four-momentum vectors before the collision add up to the same value after the collision, so

$$
\begin{bmatrix} E_1 \\ P_{1x} \\ P_{1y} \\ P_{1z} \end{bmatrix} + \begin{bmatrix} E_2 \\ P_{2x} \\ P_{2y} \\ P_{2z} \end{bmatrix} = \begin{bmatrix} E_3 \\ P_{3x} \\ P_{3y} \\ P_{3z} \end{bmatrix} + \begin{bmatrix} E_4 \\ P_{4x} \\ P_{4y} \\ P_{4z} \end{bmatrix} \tag{R10.7a}
$$

which implies that

$$
\begin{bmatrix} E_4 \\ P_{4x} \\ P_{4y} \\ P_{4z} \end{bmatrix} = \begin{bmatrix} E_2 \\ P_{2x} \\ P_{2y} \\ P_{2z} \end{bmatrix} + \begin{bmatrix} E_3 \\ P_{3x} \\ P_{3y} \\ P_{3z} \end{bmatrix} - \begin{bmatrix} E_1 \\ P_{1x} \\ P_{1y} \\ P_{1z} \end{bmatrix}
$$

$$
= \begin{bmatrix} 20\ \text{kg} \\ 16\ \text{kg} \\ 0 \\ 0 \end{bmatrix} + \begin{bmatrix} 28\ \text{kg} \\ 0 \\ 0 \\ 0 \end{bmatrix} - \begin{bmatrix} 13\ \text{kg} \\ -5\ \text{kg} \\ 0 \\ 0 \end{bmatrix} = \begin{bmatrix} 35\ \text{kg} \\ 21\ \text{kg} \\ 0 \\ 0 \end{bmatrix} \tag{R10.7b}
$$

Knowledge of the energy and x-momentum of an object is sufficient to determine both its energy and x-velocity. Using equation R9.29, we see that the larger rock's mass is still

$$
m = \sqrt{E_4^2 - P_{4x}^2} = \sqrt{(35\ \text{kg})^2 - (21\ \text{kg})^2} = (7\ \text{kg})\sqrt{5^2 - 3^2} = 28\ \text{kg} \tag{R10.8a}
$$

after the collision. According to equation R9.31, its final x-velocity is

$$
v_{4x} = \frac{P_{4x}}{E_4} = \frac{+21\ \text{kg}}{35\ \text{kg}} = +\frac{3}{5} \tag{R10.8b}
$$

This completes the solution.

Exercise R10X.3

Show that the collision described in example R10.1 does *not* conserve newtonian momentum.

Example R10.2: Solving the Rock Problem Graphically

Problem   Solve the rock collision problem discussed in example R10.1, using an energy-momentum diagram.

Model   Since the sum of four-momentum arrows is defined as the sum of ordinary vectors (we simply add the components), we can add four-momenta arrows on an energy-momentum diagram just as we would ordinary vector arrows (by putting the tail of one vector on the tip of the other while preserving their directions).

Solution   Using this technique, we see in figure R10.6a that in the rock example, the system's total four-momentum *before* the collision has components $E_T = 48$ kg and $P_{Tx} = 16$ kg. The two rocks' four-momentum arrows after the collision have to add up to the *same* total four-momentum arrow;

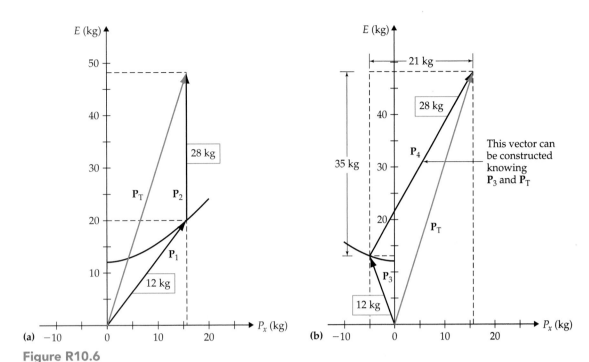

Figure R10.6

(a) The four-momenta of the rocks *before* the collision. The vector sum of these four-momenta is represented by the arrow P$_T$. Since the four-magnitudes of the individual four-momentum arrows (which equal the masses of the corresponding rocks) cannot be read directly from the diagram, I have adopted the expedient of attaching a "flag" to each four-momentum arrow that states its magnitude. (b) The four-momenta of the rocks *after* the collision. The vector sum of these four-momenta is still P$_T$ by four-momentum conservation. Since P$_3$ is also known, it is possible to construct the unknown four-momentum P$_4$, read its components from the diagram as shown, and use these to compute the corresponding rock's mass and x-velocity.

and since we know the smaller rock's four-momentum after the collision, we can *construct* the larger rock's final four-momentum arrow **P$_4$** (figure R10.6b). We can then read the components of this arrow right off the diagram, getting the same results as in equations R10.7. One can then use these results (as we did before) to compute that rock's mass and x-velocity.

This kind of graphical approach to the problem is not usually much *easier* than the algebraic approach unless one has access to hyperbola graph paper (which makes it unnecessary to calculate the object's energies). However, even in the absence of hyperbola graph paper, such a diagram does have some advantages: (1) It provides a more visual and concrete way of dealing with the problem, and may be helpful to you if you find the algebraic approach rather abstract. (2) When used in conjunction with the algebraic method, it serves as a useful check on the algebraic results: it is harder to make errors while using the graphical method. (3) In some cases, as we will see, simply *looking* at the diagram can yield qualitative information that is very difficult to get from the algebraic equations alone.

In short, the graphical method represents an alternative method for solving problems involving conservation of four-momentum that often complements the algebraic approach. Armed with both of these techniques, we are now ready to explore some of the strange and interesting consequences of the law of conservation of four-momentum.

Advantages and disadvantages of the graphical method

R10.3   The Mass of a System of Particles

Conversion of rest energy (mass) into other kinds of energy (or vice versa) is possible

As we have seen, the relativistic energy of an object is not simply equal to its kinetic energy (even at low velocities) but involves the mass of the object as well. The fact that $E = P_t$ is conserved by the internal interactions in an isolated system does not imply that the mass of an object and its kinetic energy are separately conserved—only that the sum of these two things is conserved. This implies that processes that convert mass to kinetic energy and vice versa do not necessarily violate the law of conservation of four-momentum and therefore may exist. Mass and kinetic energy are seen in the theory of special relativity to be simply two parts of the same whole (the relativistic energy). There is no reason to presuppose a barrier between these two manifestations of relativistic energy that would preclude the conversion of one to the other.

In much of the remainder of this chapter, we will be considering a variety of examples of processes that do just that. We will begin with a simple example that illustrates a crucial thing that we need to understand about "mass" before we can go further: *the mass of a system is generally **different** from the sum of the masses of its parts.*

An example showing conversion of kinetic energy to mass

Consider the collision of two identical balls of putty with mass $m = 4$ kg which in some inertial frame are observed to have x-velocities of $v_{1x} = +\frac{3}{5}$ and $v_{2x} = -\frac{3}{5}$; that is, these putty balls are approaching each other with equal speeds. Imagine that when these putty balls collide, they stick together, as shown in figure R10.7.

Note that before the collision, the x component of the system's total four-momentum is zero:

$$P_{1x} + P_{2x} = \frac{m\left(+\frac{3}{5}\right)}{\sqrt{1 - \left(\frac{3}{5}\right)^2}} + \frac{m\left(-\frac{3}{5}\right)}{\sqrt{1 - \left(\frac{3}{5}\right)^2}} = \frac{m\left(\frac{3}{5} - \frac{3}{5}\right)}{\sqrt{1 - \left(\frac{3}{5}\right)^2}} = 0 \qquad \text{(R10.9)}$$

so conservation of four-momentum implies that the x component of the final mass's four-momentum is zero as well, meaning that it must be at rest.

What of relativistic energy conservation in this case? A newtonian analysis of this collision would speak of the kinetic energy being converted to thermal energy in this inelastic collision. Such an analysis would also assert that the mass of the coalesced particle is $M = m + m = 2m$. But we have more constraints to consider in a relativistic solution to this problem. If the spatial components of the four-momentum are conserved in this collision, the time component must also be conserved, whether the collision is elastic or not. But how can we think of the relativistic energy being conserved in this case, since no mention has been made of thermal energy in the definition of the relativistic energy given in chapter R9?

The answer is direct and surprising. Since the final object is motionless, its relativistic energy is simply equal to its mass M. But by conservation of the time component of four-momentum, we have

$$M = E_1 + E_2 = \frac{m}{\sqrt{1 - \left(\frac{3}{5}\right)^2}} + \frac{m}{\sqrt{1 - \left(\frac{3}{5}\right)^2}} = \frac{2m}{\sqrt{\frac{16}{25}}} = \frac{10}{4}m = 10 \text{ kg} \quad \text{(R10.10)}$$

which is *not* equal to $2m = 8$ kg! Conservation of four-momentum thus

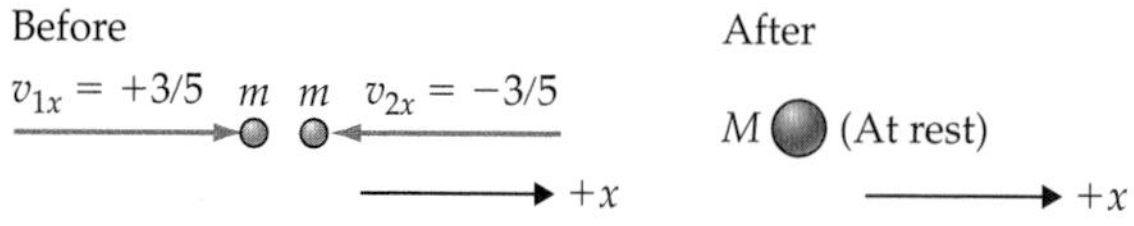

Figure R10.7

The inelastic collision of two balls of putty (each having mass $m = 4$ kg) as seen in the frame where they initially have equal speeds but opposite directions.

Before

$v_{1x} = +3/5$ m m $v_{2x} = -3/5$

After

M ⬤ (At rest)

requires that the final object have a *greater* mass than the sum of the masses that collided to form it!

We know from experience with collisions at low speeds that when two objects collide and stick together, their energy of motion gets converted to thermal energy, leaving the final object a little warmer than the original objects. (In this case, actually, the final object will be a *lot* warmer than the original objects, so much so that any *real* putty balls colliding at such speeds would vaporize instantly.) What equation R10.10 is telling us is that the final object *has* to be more massive than the original objects, and that the increased thermal energy is somehow correlated with this.

But where does this extra mass actually reside? The final object has the same number of atoms as the original objects did. Does each atom gain some extra mass somehow? This seems absurd. The increased thermal energy in the final object means that its atoms will jostle around more vigorously. Can the motion of these atoms "have mass" in some sense? This seems crazy: *individual* particles have the same mass no matter how they move. *So where is this extra mass?*

There is only one fully self-consistent answer to this question: *the extra mass is a property of the system as a whole; it does not reside in any of its parts.*

This can be vividly illustrated as follows: consider the "system" consisting of the two balls of putty *before* they collide. If they are considered separate objects, the putty balls *each* have a mass *m* of 4 kg and a relativistic energy of 5 kg, and one has an *x*-momentum of −3 kg and the other +3 kg. On the other hand, if we consider the balls to constitute a single system, the system has a total *x*-momentum of zero and a total energy of 10 kg, meaning that the system's mass $M \equiv (E_T^2 - p_T^2)^{1/2}$ is equal to 10 kg. (This is illustrated by figure R10.8.) So we see that the thermal energy produced by the collision is *not* the source of this extra mass: the extra mass was present in the "system" *before* the collision and remains the same after the collision.

So in some sense, mass is not "created" by the collision process at all: the collision simply *manifests* the mass of the *system* of two initial objects in the mass of a *single* final object. If we focus on the masses of the individual objects in the system before and after the collision, we think of mass being created. But if we focus on the *system* before and after the collision, we see that its mass remains the same.

It is possible to get unnecessarily hung up on the difference between the mass of a system and the masses of its parts. The reason this seems screwy is that we are *used* to treating mass as if it were additive: the mass of a jar of beans is the sum of the masses of the individual beans plus the mass of the

The mass of a system is not the same as the sum of the masses of its parts

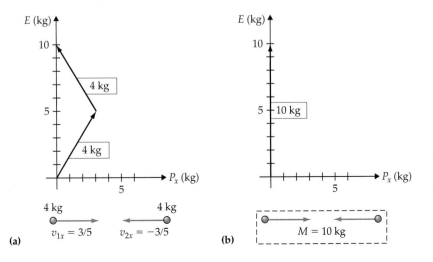

(a)

(b)

Figure R10.8
(a) Putty balls before the collision, considered as two individual objects. Each has its own mass, energy, and x-momentum.
(b) Putty balls before the collision, considered as a single system. The system's x-momentum is zero, meaning that its total energy of 10 kg is the same as the system's mass.

container, right? This is true enough if the beans travel at low velocities. But if we had common experience with a jar full of beans that bang around inside the jar with speeds close to that of light, then we would be *used* to the idea that such a jar would have a different mass than the sum of the masses of the individual beans. Mass is simply *not* additive in the way that energy and *x*-momentum are.

There are actually many examples of things in the world where the whole is greater than the sum of its parts. The meaning of a poem is not the same as the sum of the meaning of the individual letters in its words. The "life" of an organism cannot be localized in any of its parts. We simply need to start thinking about mass in the same way as we think about these things.

A system's mass can't be localized: it is a property of the system as a whole

The best way to look at this is to think of the mass of a system of particles as *a property of the system as a whole* (i.e., the magnitude of the system's total four-momentum vector) and something that simply doesn't have very much to do with the masses of its parts. The only self-consistent way to define the mass of a *system* is as the magnitude of the system's total four-momentum; and if this definition leads to the mass of a system being greater or less than the masses of its parts, well, that's the way it is!

Exercise R10X.4

Let's look at the system described in figure R10.8 in a frame moving with the left-hand ball. Use the Einstein velocity transformation to show that the other ball's *x*-velocity is $v'_{2x} = -\frac{15}{17}$. Find the *system's* four-momentum components E'_T and P'_{Tx} in this frame, and show that the system's mass [found using $M' = (E'^2_T - P'^2_{Tx})^{1/2}$] is still 10 kg. (A system's mass may not equal the sum of the masses in the system, but it *is* frame-independent.)

Exercise R10X.5

The molecules in a balloon at rest that is filled with air have an average kinetic energy of $K = 6.2 \times 10^{-21}$ J. If we convert this to kilograms, this is the additional energy that each molecule has (on average) above its rest energy (which is $m \approx 2.4 \times 10^{-26}$ kg on average). By about what fraction is the total mass of the air in a balloon larger than the sum of the masses of the individual air molecules? Is this very significant?

R10.4   The Four-Momentum of Light

We all have experienced the fact that light carries energy: we have felt sunlight warm our skin, or seen an electric motor powered by solar cells, or learned that plants convert the energy in sunlight to chemical energy. Since we have seen that energy is the time component of four-momentum, it follows that light should have an associated four-momentum vector. What does the four-momentum of light look like?

A light flash as a "particle" of light

Previously, we have explored the four-momenta of objects (rocks or putty balls or the like) that could be considered to be *particles* that have a well-defined position in space and thus a well-defined worldline through spacetime. The analogous thing in the case of light would be a "flash" or "burst" of light energy that is similarly localized in space. We can consider a continuous beam of light to be composed of a sequence of closely spaced flashes, much as we might imagine a stream of water to be a sequence of closely spaced drops.

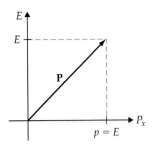

Figure R10.9
An energy-momentum diagram showing the four-momentum of a flash of light moving in the +x direction. If the four-momentum is to be parallel to the flash's worldline, then it must be drawn with slope +1 on the energy-momentum diagram. This implies that a light flash's relativistic momentum p must have the same value (in SR units) as its energy: p = E.

So what might the four-momentum of a flash of light look like? Arguably, the most basic feature of any object's four-momentum is that it is parallel to that object's worldline. If this is true for a flash of light, then it follows that the four-momentum vector for a light flash must have a slope of ± 1 on an energy-momentum diagram. The four-momentum vector of a flash with a given energy E moving in the $+x$ direction will thus look as shown in figure R10.9.

A light-flash four-momentum vector has slope ± 1

You can see from this diagram that if the flash's four-momentum vector is to have such a slope, it must have a spatial relativistic momentum p that is equal to its relativistic energy E. This is in fact consistent with equation R9.30, which in the case of light tells us that

$$\frac{p}{E} = v = 1 \quad \Rightarrow \quad p = E \qquad (R10.11)$$

The relativistic momentum of a light flash is equal to its energy

One immediate implication of this important formula is that *light must carry momentum* (as well as energy). Light bouncing off a mirror will thus transfer momentum to the mirror (causing it to recoil) in much the same way as a ball bouncing off an object transfers momentum to the object and causes it to recoil. This has been experimentally verified,[†] and it is now known that the pressure exerted by light due to its momentum plays an important part in the evolution of stars, the physics of the early universe, and a number of other astrophysical processes.

Another immediate consequence is that a "flash" of light has *zero* mass. We have defined the mass of an object in special relativity to be the invariant magnitude of its four-momentum. According to equation R10.11, the flash's mass is

$$m^2 = E^2 - p^2 = 0 \qquad (R10.12)$$

The mass of a light flash is zero

This is actually a good thing. If a flash of light were to have some *nonzero* mass m, then its relativistic energy $E = m/(1 - v^2)^{1/2}$ and relativistic momentum $p = mv/(1 - v^2)^{1/2}$ would both have to be infinite, since $v = 1$ for light, which makes the denominator zero in each expression. But since $m = 0$, these equations actually read $E = 0/0$ and $p = 0/0$. The ratio $0/0$ is technically *undefined* instead of being infinite, so instead of yielding absurdities, these equations simply don't tell us anything useful about the four-momentum of light.

If a light flash is moving in the $-x$ direction instead of the $+x$ direction, the slope of its four-momentum arrow on an energy-momentum diagram is -1 instead of $+1$, and its x-momentum is negative: $P_x = -p = -E$ (note that both p and E are positive by definition), as shown in figure R10.10.

[†]Maxwell's theory of electromagnetic waves also predicted (before relativity did) that light should carry momentum of this magnitude. Experiments performed in 1903 by Nichols and Hull in the United States and Lebedev in Russia confirmed this experimental prediction. See G. E. Henry, "Radiation Pressure," *Sci. Am.*, June 1957, p. 99.

Figure R10.10

An energy-momentum diagram showing the four-momentum of a flash of light moving in the $-x$ direction. In this case, the flash's x-momentum is negative: $P_x = -p$, where p is equal to the *magnitude* of the flash's spatial momentum ($p = |P_x|$ in this case). Note that we still have $p = E$: this is a general relation for light independent of the flash's direction of travel.

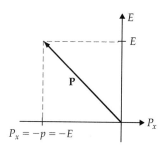

Example R10.3: A Matter-Antimatter Rocket

Problem   When a particle is brought into contact with its corresponding antiparticle, they annihilate each other, converting their rest energy entirely to light energy. A perfect rocket engine might mix antimatter with an equal amount of matter of the same type, and direct the resulting light in a tight beam out of the engine nozzle. No other kind of exhaust could possibly carry more momentum out the rear of the rocket per unit energy expended than light can. Imagine a rocket of original mass $M = 90{,}000$ kg sitting at rest in some frame in deep space (M includes the mass of the matter-antimatter fuel). Imagine that it fires its engines, emitting a burst of light having a total (unknown) energy E_L. If after this the ship's final speed is $v = \frac{4}{5}$, what is its final mass m? (See figure R10.11.)

Model   The law of conservation of four-momentum tells us that the total four-momentum of an isolated system of interacting objects will be conserved. The system's initial four-momentum vector in this case is that of a mass M at rest, which has components $P_t = M$ and $P_x = P_y = P_z = 0$. After the engines have fired, the system consists of a light flash and the somewhat lighter ship. Let the ship's final direction of motion define the $+x$ direction, as shown in figure R10.11. According to equation R10.11, the flash's relativistic momentum p_L will be equal to its energy E_L, and since the flash is moving in the $-x$ direction, the light's spatial four-momentum components are $P_x = -p_L = -E_L$ and $P_y = P_z = 0$. We don't know the final ship mass m, but we do know that it is moving with $v = \frac{4}{5}$ in the $+x$ direction; so by equations R9.10, we know that the ship's final four-momentum vector has components $P_t = m/(1 - v^2)^{1/2} = m/(1 - \frac{16}{25})^{1/2} = m/(\frac{9}{25})^{1/2} = 5m/3$, $P_x = +p = Ev = 4m/3$, $P_y = 0$, and $P_z = 0$.

Solution   The law of conservation of four-momentum therefore implies that

$$\begin{bmatrix} M \\ 0 \\ 0 \\ 0 \end{bmatrix} = \begin{bmatrix} E_L \\ -E_L \\ 0 \\ 0 \end{bmatrix} + \begin{bmatrix} \frac{5}{3}m \\ +\frac{4}{3}m \\ 0 \\ 0 \end{bmatrix} \quad \Rightarrow \quad \begin{array}{l} M = E_L + \frac{5}{3}m \\ E_L = \frac{4}{3}m \\ 0 = 0 \\ 0 = 0 \end{array} \qquad \text{(R10.13)}$$

The bottom two equations don't tell us much, but the top two represent two equations in the unknowns m and E_L. Plugging the second into the first, we get

$$M = \frac{4}{3}m + \frac{5}{3}m = \frac{9}{3}m = 3m \quad \Rightarrow \quad m = \frac{1}{3}M = 30{,}000 \text{ kg} \quad \text{(R10.14)}$$

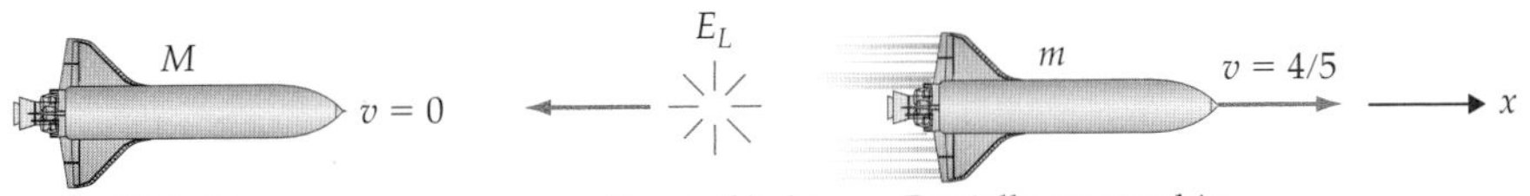

Figure R10.11
The situation discussed in example R10.3.

Evaluation   So even this perfect rocket has to use 60,000 kg of matter-antimatter fuel to boost the remaining 30,000 kg to a speed of 0.8.

Exercise R10X.6

Solve example R10.3 by using an energy-momentum diagram. (*Hint:* Since we do not know the light flash's energy, we do not know how long to draw its four-momentum vector. But we know that its slope has to be -1, and that the slope of the rocket's final four-momentum vector has to have slope $+\frac{4}{5}$.) Where do these lines intersect on the diagram?

R10.5   Applications to Particle Physics

In spite of our imaginative talk in this book about relativistic trains, spaceships, runners, and so on, special relativity has few genuine practical applications other than in the realm of subatomic particle physics. Subatomic particles (such as electrons, protons, and neutrons) have small enough masses that common processes can give them relativistic speeds. Virtually all the experimental tests of relativity theory involve such particles.

Particle physics as the most important practical application of relativity

The kilogram is an inappropriately large unit of mass or energy when one deals with elementary particles. Elementary particle physicists more commonly express the mass, energy, and momentum of such particles in terms of the energy unit of **electron volts,** where 1 eV is the energy an electron gains by going through a 1-V battery. In terms of more common units,

Appropriate units for doing particle physics

$$1 \text{ eV} = 1.602 \times 10^{-19} \text{ J} = 1.782 \times 10^{-36} \text{ kg} \qquad (R10.15)$$

So, for example, an electron (whose mass is 0.511 MeV) moving at a speed of $\frac{4}{5}$ has an energy $E = m/(1 - v^2)^{1/2} = (5/3)m = 0.852$ MeV, a kinetic energy of $K = E - m = 0.341$ MeV, and a relativistic momentum of $p = Ev = 0.682$ MeV.

Example R10.4 shows how conservation of four-momentum can be applied to a very real problem in subatomic physics.

Example R10.4: Kaon Decay

Problem   The most stable version of the subatomic particle called the K^0 meson or *kaon* (which has a mass $M = 498$ MeV) decays with a half-life of about 36 ns to two identical π^0 mesons or *pions* (which have a mass $m = 135$ MeV). If the original kaon is at rest, what are the speeds of the pions after the decay?

Translation   Let us number the pions 1 and 2, and choose the orientation of our reference frame so that the first pion moves in the $+x$ direction.

Model: Conservation of four-momentum then implies that

$$\text{Kaon four-momentum} = \begin{bmatrix} M \\ 0 \\ 0 \\ 0 \end{bmatrix} = \begin{bmatrix} E_1 \\ +p_1 \\ 0 \\ 0 \end{bmatrix} + \begin{bmatrix} E_2 \\ P_{2x} \\ P_{2y} \\ P_{2z} \end{bmatrix}$$

$$= \text{sum of pion four-momenta} \qquad \text{(R10.16)}$$

(Note that since the kaon is at rest, its spatial four-momentum components are zero and its energy is just its rest energy, which is its mass.) The bottom two lines of this equation tell us that $P_{2y} = P_{2z} = 0$, meaning that the second pion moves along the x axis. The second line tells us that $P_{2x} = -p_1$, which says that the second pion moves in the $-x$ direction with the same *magnitude* of relativistic momentum as the first: $p_2 = |P_{2x}| = p_1$. Since the pions have the same mass m, this means that the relativistic *energies* of the two pions are the same:

$$E_2 = \sqrt{m^2 + p_2^2} = \sqrt{m^2 + p_1^2} = E_1 \qquad \text{(R10.17)}$$

where the first step here follows from $m^2 = E^2 - p^2$ (equation R9.29).

Solution   The first line of equation R10.16 then tells us that $M = 2E_1$, or $E_1 = \frac{1}{2}M = 249$ MeV. If we plug this into $m^2 = E^2 - p^2$, we can solve for p_1:

$$p_1 = \sqrt{E_1^2 - m^2} = \sqrt{(249 \text{ MeV})^2 - (135 \text{ MeV})^2} = 209 \text{ MeV} \quad \text{(R10.18)}$$

Finally, we can find the pion speed by using equation R9.30:

$$v_1 = \frac{p_1}{E_1} = \frac{209 \text{ MeV}}{249 \text{ MeV}} = 0.839 \qquad v_2 = \frac{p_2}{E_2} = \frac{p_1}{E_1} = v_1 \qquad \text{(R10.19)}$$

Evaluation   These speeds end up being unitless and smaller than that of light, which are both good signs. (Note that this process "converts" kaon mass energy to pion kinetic energy.)

The cover of this book shows a colorized photograph of particle tracks in a detector called a *bubble chamber*. These tracks are curved because the detector is placed in a magnetic field which causes charged particles to follow a circular path whose radius of curvature is equal to the particle's relativistic momenta. Graduate students in the 1960s would examine such photographs and measure the curvatures by hand to determine the four-momenta and identify the particles. In modern particle physics detectors, the trajectories of particles are measured electronically (see figure R10.12), and computers do the measuring and calculating. This allows experimenters to sift through millions of collision events to look for evidence of rare particles.

E. F. Taylor and J. A. Wheeler estimated in 1963 that the annual work of physicists analyzing particle collisions already at that time was testing the law of conservation of four-momentum more often than the annual work of surveyors in the United States was testing the laws of euclidean geometry.[†] Since then, the number of particle collisions per year that are analyzed by computers has *enormously* increased. Any one of these tests could have spotted a violation of this law, but none have. The law of conservation of four-momentum is one of the best-tested laws in all physics.

[†]E. F. Taylor and J. A. Wheeler, *Spacetime Physics*, New York: Freeman, 1963, p. 123.

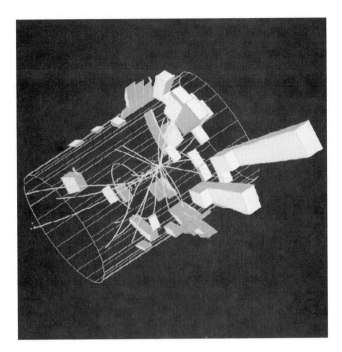

Figure R10.12
A computer reconstruction of the aftermath of a high-energy collision between a proton and an antiproton in a modern detector at the Fermilab particle accelerator facility. The bars indicate the energy deposited in various parts of the detector.

R10.6   Parting Comments

The principle of relativity is rich with fascinating implications, and this book has touched on just the most basic of these implications. I close with some suggestions as to where an interested reader can go from here.

An excellent and somewhat more advanced and detailed exploration of special relativity is found in E. F. Taylor and J. A. Wheeler's *Spacetime Physics*, 2d ed., New York: Freeman, 1992. The *American Journal of Physics* (which is found in many college libraries and is written primarily for college physics professors) is a great place to look for articles on current issues in relativity theory. Look under *special relativity* in any one of the indexes of recent volumes for a list of such articles. (Many of these articles are accessible to students.)

The next step beyond special relativity is *general* relativity. A book that one can use when beginning a study of general relativity is that by G. F. R. Ellis and R. M. Williams, *Flat and Curved Spacetimes,* New York: Oxford, 1985.

More delights await: I encourage everyone to continue the exploration!

TWO-MINUTE PROBLEMS

R10T.1 The length of the arrow representing an object's four-momentum on a energy-momentum diagram is proportional to the object's speed, true (T) or false (F)?

R10T.2 The slope of the arrow representing an object's four-momentum on an energy-momentum diagram is inversely proportional to the object's speed, T or F?

R10T.3 Which of the following processes is consistent with the law of conservation of four-momentum? (Answer C if it is consistent with the law; F if it violates it.)

a.   A particle moving eastward collides with and sticks to an identical particle at rest. The resulting single particle remains at rest but is more massive than the sum of the original particles' masses.   *(more)*

b.   Two identical particles with equal speeds, one moving east and one moving west, collide. After the collision, the same particles move north and south, respectively, with the same speeds that they had originally.

c.   Two electrons with equal speeds, one moving east and one moving west, collide. After the collision, the electrons move north and south with smaller speeds than they had originally. Nothing else is emitted.

d.   A spaceship of mass $2m$ originally at rest burns some matter-antimatter fuel until the ship's mass is m. The light energy ejected from the ship in this process has total energy m as well.

e.   A particle of mass $2m$ at rest decays to two identical particles each of mass m that move in opposite directions away from the decay position at a speed of 0.5.

R10T.4   A particle with a mass of 3.0 kg is accelerated to a speed of 0.80. The mass of this particle is now
A.   Greater than 3.0 kg

B.   Less than 3.0 kg
C.   Still 3.0 kg

R10T.5   A sealed cup of water is placed in a microwave oven. The water absorbs microwave energy, which causes its atoms to vibrate more vigorously, making the water warmer. In this process, the mass of the water in the cup
A.   Increases
B.   Decreases
C.   Does not change

R10T.6   In example R10.3, the spaceship engines convert 60,000 kg of matter-antimatter fuel to massless light. The mass of the total system (empty spaceship plus flash of light, considered as a unit) thus decreases, T or F?

R10T.7   In example R10.4, a kaon (whose mass is 498 MeV) decays to two pions (each with a mass of 135 MeV). Mass is being converted to energy in this process, T or F? The total mass of the system (considered as a whole) decreases in this process, T or F?

HOMEWORK PROBLEMS

Basic Skills

R10B.1   Draw an energy-momentum diagram for an object with a mass of 12 kg moving in the $+x$ direction at a speed of 0.80. Estimate its relativistic kinetic energy from the graph.

R10B.2   Draw an energy-momentum diagram for an object with a mass of 5 kg moving in the $-x$ direction at a speed of 0.60. Estimate its relativistic kinetic energy from the graph.

R10B.3   A particle of mass m at rest decays into two identical particles, each with mass $\frac{1}{3}m$. Conservation of spatial momentum means that the product particles have to move off in opposite directions with the same speed. What is the relativistic kinetic energy of each particle?

R10B.4   A flash of light moves in the $+x$ direction. The flash has a total energy of 2.0×10^{-18} kg. What is this in joules? What is the x component of the flash's four-momentum vector (in kilograms)? What is this in kg·m/s?

R10B.5   Two balls of putty, each of mass m, move in opposite directions toward each other with speeds of 0.95. The balls stick together, forming a single motionless ball of putty at rest. What is the mass of this ball of putty? (Express your answer as a multiple of m.)

R10B.6   A ball of putty with mass m moves in the $+x$ direction with speed $v = \frac{4}{5}$. It hits another ball of putty of mass m at rest. The balls stick together after the collision, forming a single ball. What is the x component of this final ball's four-momentum? What is its relativistic energy? (Express both as multiples of m.) What is its x-velocity?

R10B.7   A spaceship with a mass m originally at rest burns matter-antimatter fuel, ejecting light energy in the $-x$ direction until the total energy of the light that it has emitted is $E_L = \frac{1}{3}m$. What is the total relativistic energy of the partially empty spaceship now? What is the x component of its four-momentum? (Express both answers as fractions of m.) What is the ship's final speed?

Synthetic

R10S.1   A particle of mass m decays into two identical particles that move in opposite directions, each with a speed of $\frac{12}{13}$. What is the mass of each of the product particles (expressed as a fraction of m)?

R10S.2   An object with a mass m sits at rest. A light flash moving in the $+x$ direction with a total energy of $2m$ hits this object and is completely absorbed. What are the mass and the x-velocity of the object afterward?
(a)   Answer this question approximately, using an energy-momentum diagram.

(b) Solve this problem quantitatively, using four-dimensional column vectors.

R10S.3 An object with mass $m_1 = 8$ kg traveling with an x-velocity of $v_{1x} = \frac{15}{17}$ collides with an object with mass $m_2 = 12$ kg traveling with an x-velocity of $v_{2x} = -\frac{5}{13}$. After the collision, the 8-kg object is measured to have an x-velocity of $v_{3x} = -\frac{3}{5}$. Find the other object's x-velocity, and show that it has the same mass as it started with.
(a) Answer this question approximately, using an energy-momentum diagram.
(b) Solve this problem quantitatively, using four-dimensional column vectors.

R10S.4 A spaceship with rest mass m_0 is traveling with an x-velocity $v_{0x} = +\frac{4}{5}$ in the frame of the earth. It collides with a photon torpedo (an intense burst of light) moving in the $-x$ direction relative to the earth. Assume that the ship's shields totally absorb the photon torpedo.
(a) The oncoming torpedo is measured by terrified observers on the ship to have an energy of $0.75m_0$. What is the energy of the photon torpedo in the frame of the earth? (*Hint:* How do the components of four-momentum transform when we go from one inertial frame to another?)
(b) Use conservation of four-momentum to determine the final x-velocity (in the earth frame) and mass (in terms of m_0) of the damaged ship after it absorbs the torpedo. Use an energy-momentum diagram to solve this problem graphically.
(c) Find the ship's final mass and x-velocity quantitatively, using four-dimensional column vectors.

R10S.5 A spaceship of mass M is traveling through an uncharted region of deep space. Suddenly its sensors detect a black hole dead ahead. In a desperate attempt to stop the spaceship, the pilot fires the forward matter-antimatter engines. These engines convert the mass energy of matter-antimatter fuel entirely to light, which is emitted in a tight beam in the direction of the ship's motion. If the spaceship's initial speed toward the black hole is $v = \frac{4}{5}$, what fraction of its mass M must be converted to energy to bring the spaceship to rest with respect to the black hole? (*Hint:* Treat the emitted light as one big flash of light.)
(a) Answer this question approximately, using an energy-momentum diagram.
(b) Solve this problem quantitatively, using four-dimensional column vectors.

R10S.6 *Traveling to the stars, I.* As discussed in section R10.6, the most efficient possible rocket engine would take matter and antimatter fuel, combine them in a controlled way, and focus the resulting photons into a tight beam traveling away from the stern of

the spaceship. Imagine that you want to design a spaceship using such an engine that can boost a payload of mass $m = 25$ t (that is, 25 metric tons, or 25,000 kg) to a final cruising speed of 0.95. What does the ship's mass M at launch have to be? (*Hint:* The ship can essentially be considered to be a particle of mass M at rest that decays into a big flash of light and a smaller particle (the payload) of known mass m traveling at a known speed v. Use conservation of four-momentum to determine M.)

R10S.7 Consider a freely floating mirror placed initially at rest in a laser beam. Imagine the mirror to be oriented directly facing the beam so that the mirror reflects the beam back the way it came. Each flash of light that rebounds from the mirror has undergone a change in its momentum as a result of the change in direction of its velocity. This means that the mirror must recoil a bit from each rebound to conserve four-momentum. Use conservation of the spatial components of four-momentum to estimate how much power the laser beam would have to have to accelerate a perfect 1-g mirror at a rate of 1 cm/s^2. (You can assume that the mirror's momentum is essentially newtonian.) Express your final answer in watts ($1 \text{ W} = 1 \text{ J/s}$).

R10S.8 A π^- pion (mass 140 MeV) normally decays to a μ^- muon (mass 106 MeV) and a neutrino. The neutrino is a particle that is so light you can treat its mass as being essentially zero (like a flash of light). If the pion is at rest, find the speed of the emitted muon.

R10S.9 Quantum physics tells us that in certain circumstances, light can be considered a stream of massless particles called *photons*. Imagine that a photon with energy E_0 is traveling in the $+x$ direction and hits an electron of mass m at rest. The photon scatters from the electron and travels in the $-x$ direction after the collision. Find a formula for the final energy of the photon E in terms of E_0 and m. (Physicists call this process *Compton scattering*: the fact that the formula correctly describes the behavior of light scattering from electrons is one of the most important pieces of evidence supporting the photon model of light.)

Rich-Context

R10R.1 *Traveling to the stars, II.* (It is recommended, though not essential, that you do problem R10S.6 first.) Consider the rocket design discussed in problem R10S.6. (The answer to that problem is $M = 6.24m$.)
(a) Assume that you can find astronauts who are willing to travel for up to 50 years (as measured by their watches) on a round trip to the stars. In about how many light-years could the ship go out and return (assuming that the ship spends a negligible time accelerating)? Is this very far compared to the galaxy as a whole?

(b) The takeoff mass calculated in problem R10S.6 only included sufficient fuel to boost the payload to the cruising speed. For a complete round trip, one would have to carry enough fuel to boost the payload to the cruising speed, decelerate it to rest at the destination, boost it to cruising speed again for the return trip home, and decelerate it upon reaching earth. How much fuel is required for a complete round trip? Express your answer as a multiple of the payload mass m. (*Hint:* The answer is *not* 4 times the fuel required to boost the payload alone to the cruising speed!)

(c) Comment on the practicality of visiting distant stars by using a rocket which must carry its own fuel.

R10R.2 One method for getting around the difficulties discussed in problems R10S.6 and R10R.1 is to use light pressure to accelerate a payload (see problem R10S.7). Imagine that you attach a perfect mirror to the back of your payload, and then you accelerate the payload by bouncing a powerful laser beam off the mirror. (This has the big advantage that you don't have to carry the mass of the fuel or the rocket engine!) The lasers producing the beam could be massive things powered by solar energy; so not the size or mass or power of these driving lasers is a limitation (at least in principle). Imagine that you wish to accelerate a 2000-kg scientific payload outward from the earth's orbit around the sun at a rate of 1 m/s^2 (at this rate, it would take about a year to reach 10% of the speed of light). Assume that the payload has been already delivered at rest to a point far enough from the earth that the earth's gravity is negligible (but *don't* ignore the gravity of the sun). How many watts of light would the driving laser have to produce?

Advanced

R10A.1 Rework the Compton scattering problem (problem R10S.9) to find the energy of the scattered photon if its trajectory after scattering from the electron makes an angle of θ with its original direction. (This is not easy!)

ANSWERS TO EXERCISES

R10X.1 Doing the calculation, we find that

$$\frac{1}{2}m\left(\frac{3}{5}\right)^2 = \frac{9m}{50} = 0.18m < 0.25m \qquad \text{(R10.20)}$$

R10X.2 According to equation R9.23,

$$E = \frac{m}{\sqrt{1-v^2}} = \frac{1.0 \text{ kg}}{\sqrt{1-\left(\frac{4}{5}\right)^2}} = \frac{5}{3} \text{ kg} = 1.67 \text{ kg} \qquad \text{(R10.21)}$$

So to draw the arrow representing the object's four-momentum, we draw an arrow with a slope of $1/v = 5/4$ and with a vertical projection equal to 1.67 kg. The resulting energy-momentum diagram is shown in figure R10.13. (Note that if you are using hyperbola graph paper, all you need to do is to draw the tip of the arrow where the hyperbola corresponding to mass m crosses a line with slope $1/v$.) From this, we can see that $P_x = 1.33$ kg and $K = 0.67$ kg.

R10X.3 The system's total newtonian x-momentum *before* the collision is

$$m_1v_{1x} + m_2v_{2x} = (12 \text{ kg})\left(\frac{4}{5}\right) + (28 \text{ kg})(0) = \frac{48}{5} \text{ kg} \qquad \text{(R10.22)}$$

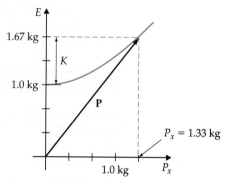

Figure R10.13
See the answer to exercise R10X.2.

After the collision, this quantity is

$$m_1v_{3x} + m_2v_{4x} = (12 \text{ kg})\left(-\frac{5}{13}\right) + (28 \text{ kg})\left(\frac{3}{5}\right)$$

$$= -\frac{60}{13} \text{ kg} + \frac{84}{5} \text{ kg}$$

$$= \frac{-300 + 1092}{65} \text{ kg}$$

$$= \frac{792}{65} \text{ kg} = \frac{60.92}{5} \text{ kg} \neq \frac{48}{5} \text{ kg} \qquad \text{(R10.23)}$$

The point is that conservation of four-momentum is different from conservation of newtonian momentum.

R10X.4 Let the Home Frame be the frame where the collision looks as shown in figure R10.8. The Other Frame in question then moves relative to the Home Frame in the $+x$ direction at $\beta = \frac{3}{5}$. According to equation R8.14a, then,

$$v'_{2x} = \frac{v_{2x} - \beta}{1 - \beta v_{2x}} = \frac{-3/5 - 3/5}{1 - (3/5)(-3/5)} = \frac{-6/5}{34/25}$$

$$= -\frac{30}{34} = -\frac{15}{17} \qquad \text{(R10.24)}$$

This means that

$$E'_2 = \frac{m}{\sqrt{1 - v_{2x}^2}} = \frac{m}{\sqrt{1 - \left(-\frac{15}{17}\right)^2}} = \frac{m}{\sqrt{1 - \frac{225}{269}}}$$

$$= \frac{m}{\sqrt{\frac{64}{269}}} = \frac{17}{8}m \qquad \text{(R10.25)}$$

$$P'_{2x} = E'_2 v'_{2x} = \left(\frac{17}{8}m\right)\left(-\frac{15}{17}\right) = -\frac{15}{8}m \qquad \text{(R10.26)}$$

The first ball is at rest in this frame, so $E'_1 = m$ and $P'_{1x} = 0$. The system's total four-momentum components thus are

$$E'_T = E'_1 + E'_2 = m + \frac{17}{8}m = \frac{25}{8}m \qquad \text{(R10.27a)}$$

$$P'_{Tx} = P'_{1x} + P'_{2x} = 0 + \frac{15}{8}m = \frac{15}{8}m \qquad \text{(R10.27b)}$$

and its total mass is

$$M' = \sqrt{E_T'^2 - P_{Tx}'^2} = \frac{m}{8}\sqrt{25^2 - 15^2} = \frac{m}{8}\sqrt{625 - 225}$$

$$= \frac{m}{8}\sqrt{400} = \frac{20}{8}m = \frac{10}{4}m \qquad \text{(R10.28)}$$

Since $m = 4$ kg, $M' = 10$ kg $= M$, as claimed.

R10X.5 Converting the value of K given to kilograms, we get

$$6.2 \times 10^{-21} \, \text{J}\left(\frac{1 \, \text{kg} \cdot \text{m}^2/\text{s}^2}{1 \, \text{J}}\right)\left(\frac{1 \, \text{s}}{3.0 \times 10^8 \, \text{m}}\right)^2$$

$$= 6.9 \times 10^{-38} \, \text{kg} \qquad \text{(R10.29)}$$

The total mass (i.e., rest energy) of a system at rest is the sum of the energies of its constituent particles. If there are N molecules in the gas, the ratio of the gas's total mass to the sum of the masses of the molecules is

$$\frac{NE}{Nm} = \frac{N(m + K)}{N(m)} = 1 + \frac{K}{m} = 1 + \frac{6.9 \times 10^{-38}}{2.4 \times 10^{-26}}$$

$$= 1 + 2.9 \times 10^{-12} \qquad \text{(R10.30)}$$

So in this case, the system's mass is different from the sum of the masses of the molecules by only about 3 parts in 1 trillion. In nonrelativistic situations like this one, the idea that mass is additive (as we have always assumed before) is an excellent approximation.

R10X.6 Since we don't know either the light flash's or the ship's final relativistic energy, we cannot immediately draw their four-momentum vectors on the diagram. But we do know that the slopes of these vectors are -1 and $\frac{4}{5}$, respectively, and that the two vectors have to add up to the ship's original four-momentum vector, which (since the ship is initially at rest) is vertical. So sketch a line with slope $\frac{4}{5}$ from the origin and another with slope -1 down from the tip of the ship's original four-momentum vector, as shown in figure R10.14a. Since the flash's and ship's final vectors have to add to the ship's original four-momentum, the vectors have to lie on these lines and stretch to their intersection, as shown in figure R10.14b. From the diagram, then, we can see that E and p for the ship after the engines fire are about $\frac{5}{9}M$ and $\frac{4}{9}M$, respectively, and using $m = (E^2 - p^2)^{1/2}$, one can show that $m = \frac{1}{3}M$.

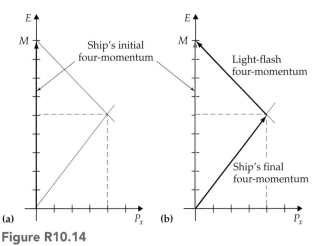

Figure R10.14

See the answer to exercise R10X.6.

Appendix **RA**
Conversion of Equations to SI Units

RA.1   Why Use SR Units?

The equations of special relativity are greatly simplified when one uses SR units to measure distance, as we have seen. But the purpose of using SR units is not merely to simplify a few equations: using such units also vividly draws one's attention to the connections that special relativity makes between quantities that were previously considered to be fundamentally distinct. For example, special relativity teaches us that energy, momentum, and mass are in fact different aspects of the same basic quantity, the four-momentum. It is not merely *convenient* to measure the time component E, the spatial components P_x, P_y, P_z, and the magnitude m of this four-vector in the same units: it is fundamentally *appropriate* as well. Similarly, the basic metaphor of spacetime geometry that lies at the root of both special and general relativity is obscured if one insists on using different units to measure time and distance.

SR units make relativistic relationships clearer

Nonetheless, this choice of units does lead to complications when one tries to apply the ideas presented in this book to practical situations, since practicing physicists in their everyday work still use traditional units to describe quantities. It is important to be able to use the simple and beautiful equations in this book in situations where the quantities in question are expressed in traditional units. Fortunately, it is straightforward to convert equations appearing in this text to equivalent equations involving quantities measured in SI units. The purpose of this appendix is to describe an easy method for doing this.

But SI units are used in most practical applications

RA.2   Conversion of Basic Quantities

SR units, as defined in chapter R2, differ from SI units only in the substitution of the *second* for the *meter* as the basic unit of distance. Mass and time thus have the same units in both unit systems and need no conversion. The most important quantities that are affected by the shift in units as one changes systems are *distance*, *velocity*, *energy*, and *momentum*. We also should consider what happens to values of universal constants such as c and Planck's constant h. (Indeed, we will treat h in what follows as an example of how we can handle such constants in general.)

How to convert basic quantities between the SR and SI units

To help keep things straight in what follows, let me denote quantities measured in SR units with an "(SR)" subscript; for example, the speed of an object in SR units will be written in this appendix as $v_{(SR)}$, the energy of an object in the SR unit of joules will be written $E_{(SR)}$, and so on. Quantities without this subscript are assumed to be measured in SI units.

In chapter R2, we saw that the general rule for converting SI quantities to SR quantities was to multiply the SI quantity by the appropriate power of c that leads to the correct SR dimensions. Let us apply this rule to the quantities of interest listed above.

Distance in SR units is measured in seconds. Distance in SI units is measured in meters. To convert an SI distance x to an SR distance $x_{(SR)}$, we must divide x by one factor of c (in meters per second): $x_{(SR)} = x/c$. The SR unit of

Table RA.1   SI equivalents for SR quantities

Quantity	SR Symbol	SI Equivalent
Time coordinate	$t_{(SR)}$	t
Spatial coordinate	$x_{(SR)}$	x/c
Speed (of an object)	$v_{(SR)}$	v/c
Speed (of a reference frame)	$\beta_{(SR)}$	β/c
Mass	$m_{(SR)}$	m
Momentum	$p_{(SR)}$	p/c
Energy	$E_{(SR)}$	E/c^2
Speed of light	1	c
Planck's constant	$h_{(SR)}$	h/c^2

energy is the *kilogram,* but the SI unit is the joule, where $1\ \text{J} = \text{kg·m}^2/\text{s}^2$. To convert from E (in joules) to $E_{(SR)}$ (in kilograms), we must divide E by two powers of c: $E_{(SR)} = E/c^2$. Planck's constant h has units of energy multiplied by time. In SR units, energy is measured in kilograms instead of joules, so again we have to divide the SI version of Planck's constant by two powers of c to get the correct SR units: that is, $h_{(SR)} = h/c^2$. Conversion equations involving velocity and momentum can be derived in a similar manner. The results are summarized in table RA.1.

RA.3   Converting SR Unit Equations to SI Unit Equations

To convert equations: take the SR equation, substitute SI quantities from table RA.1

The trick for converting equations from SR to SI units is now very simple: you simply replace the SR quantities in an equation by the SI equivalents given in table RA.1. For example, consider the metric equation

$$\Delta s_{(SR)}^2 = \Delta t_{(SR)}^2 - \Delta x_{(SR)}^2 - \Delta y_{(SR)}^2 - \Delta z_{(SR)}^2 \qquad \text{(RA.1)}$$

Both $\Delta s_{(SR)}$ and $\Delta t_{(SR)}$ have units of seconds and so have the same value in both systems. But the SI units of Δx, Δy, and Δz are meters; therefore, $\Delta x_{(SR)} = \Delta x/c$, $\Delta y_{(SR)} = \Delta y/c$, and $\Delta z_{(SR)} = \Delta z/c$. The metric equation in SI units is thus

$$\Delta s^2 = \Delta t^2 - \left(\frac{\Delta x}{c}\right)^2 - \left(\frac{\Delta y}{c}\right)^2 - \left(\frac{\Delta z}{c}\right)^2 \qquad \text{(RA.2)}$$

As another example, in unit Q we will study equations that give the energy of a light photon in terms of the frequency f or the wavelength λ of the light involved. In SI units, this equation is $E_{(SR)} = h_{(SR)} f_{(SR)} = h_{(SR)}/\lambda_{(SR)}$. Both energy and Planck's constant gain a factor of c^{-2} when we switch from SR to SI units, so this factor divides out in these equations above. The wavelength λ, on the other hand, has SI units of meters but SR units of seconds, so $\lambda_{(SR)} = \lambda/c$. The second equation thus becomes $E = hc/\lambda$ in SI units.

In some cases, equations can be made prettier by multiplying through by c^n

Table RA.2 lists some of the important equations in the text and their SI equivalents. In many of the cases described there, the SI equations are simply found by substituting the SI unit equivalents from table RA.1 for the SR unit quantities in the equation from the text. However, in many cases, the SI unit equations have been further simplified by dividing out common factors of c. For example, the SR unit version of the equation giving the magnitude of the relativistic momentum of a particle in terms of its speed reads

$$p_{(SR)} = \frac{m_{(SR)} v_{(SR)}}{\sqrt{1 - (v_{(SR)})^2}} \qquad \text{(RA.3)}$$

Table RA.2   Some important equations and their SI equivalents

Equation	SR Version	SI Equivalent
Metric	$\Delta s^2 = \Delta t^2 - \Delta x^2 - \Delta y^2 - \Delta z^2$	$\Delta s^2 = \Delta t^2 - \dfrac{\Delta x^2 + \Delta y^2 + \Delta z^2}{c^2}$
Proper time	$d\tau = dt\sqrt{1 - v^2}$	$d\tau = dt\sqrt{1 - (v/c)^2}$
Lorentz transformations (t and x)	$\gamma \equiv 1/\sqrt{1 - \beta^2}$ $t' = \gamma(t - \beta x)$ $x' = \gamma(-\beta t + x)$	$\gamma \equiv 1/\sqrt{1 - (\beta/c)^2}$ $t' = \gamma(t - \beta x/c^2)$ $x' = \gamma(-\beta t + x)$
Lorentz contraction	$L = L_R\sqrt{1 - v^2}$	$L = L_R\sqrt{1 - (v/c)^2}$
Transformation for x-velocity	$v'_x = \dfrac{v_x - \beta}{1 - \beta v_x}$	$v'_x = \dfrac{v_x - \beta}{1 - \beta v_x/c^2}$
Energy in terms of speed	$E = \dfrac{m}{\sqrt{1 - v^2}}$	$E = \dfrac{mc^2}{\sqrt{1 - (v/c)^2}}$
Relativistic momentum	$p = \dfrac{mv}{\sqrt{1 - v^2}}$	$p = \dfrac{mv}{\sqrt{1 - (v/c)^2}}$
Mass in terms of E, p	$m^2 = E^2 - p^2$	$(mc^2)^2 = E^2 - (pc)^2$
Speed in terms of E, p	$v = \dfrac{p}{E}$	$\dfrac{v}{c} = \dfrac{pc}{E}$
Photon energy in terms of λ	$E = hf = \dfrac{h}{\lambda}$	$E = hf = \dfrac{hc}{\lambda}$

If we simply perform the substitutions called for in table RA.1, we get

$$\frac{p}{c} = \frac{m(v/c)}{\sqrt{1 - (v/c)^2}} \tag{RA.4}$$

The equation can be made prettier, however, by multiplying through by c

$$p = \frac{mv}{\sqrt{1 - (v/c)^2}} \tag{RA.5}$$

This is the simplified equation given in table RA.2. Many of the equations in the right-hand column of the table have been simplified in this manner.

RA.4   Energy-Based SR Units

Most of the practical applications of special relativity are in nuclear and particle physics. Physicists in these fields typically focus on *energy* as the most important dynamic quantity; they usually modify SR units so that the preferred unit for energy, momentum, and mass is an *energy* unit (typically the electronvolt) instead of the kilogram (see section R10.5). Let us call a unit system where four-momentum quantities are measured in units of energy-based SR units (ESR). Note that in the SR equations dealing with four-momentum quantities (i.e., the last five equations in table RA.2), it doesn't really matter what units one uses to express the quantities m, p, and E as long as one uses the *same* units for these quantities. Note that mc^2, pc, and E all have SI units of energy, and thus they are the quantities that most directly correspond to the quantities m, p, and E in energy-based SR units.

In energy-based SR units, we express E, p, and m in energy units instead of mass units

Planck's constant is the same in SI units and energy-based SR units

In regard to the last equation in the table, if you use energy units instead of mass units to express quantities related to four-momentum, you should note that Planck's constant h (which has SI units of energy multiplied by time) has the same value in ESR units: $h_{(ESR)} = h$, whereas $h_{(SR)} = h/c^2$ in ordinary SR units. Other constants involving mass or energy will also (of course) be different in energy-based and ordinary SR units.

RA.5   Exercises for Practice

Exercise RAX.1

Check that $p_{(SR)} = p/c$, as claimed in table RA.1.

Exercise RAX.2

Convert the equation describing the relativistic kinetic energy of a particle (equation R9.36 in the text) to its equivalent in SI units.

Exercise RAX.3

The spacetime separation $\Delta\sigma$ between two events is given by the equation

$$\Delta\sigma^2_{(SR)} = \Delta d^2_{(SR)} - \Delta t^2_{(SR)} \tag{RA.5}$$

Since $\Delta\sigma$ is actually directly measured using a measuring stick as opposed to a clock, it would make sense to express its value in meters instead of seconds if one were going to use the SI units. With this in mind, what would be the equivalent of this equation expressed in terms of SI units?

Exercise RAX.4

What are the SR units of the universal gravitational constant $G_{(SR)}$? (*Hint:* The SI units of G can be inferred from Newton's law of universal gravitation: $F_g = Gm_1m_2/r^2$.) Derive an equation expressing $G_{(SR)}$ in terms of G in SI units.

SHORT ANSWERS TO EXERCISES

RAX.1   (Answer is given.)

RAX.2 $K = E - mc^2$

RAX.3 $\Delta\sigma^2 = \Delta d^2 - c^2\,\Delta t^2$

RAX.4   The SR units of $G_{(SR)}$ are s/kg, $G_{(SR)} = G/c^3$.

Appendix **RB**
The Relativistic Doppler Effect

RB.1   Introduction to the Doppler Effect[†]

An important application of the metric equation is the computation of the shift in the wavelength of light emitted by a relativistic moving source. You may already know that light emitted by a source moving with respect to an observer is measured by that observer to be red-shifted toward longer wavelengths (if the object is departing from the observer) or blue-shifted toward shorter wavelengths (if the object is approaching the observer). This shift in wavelength is called a **Doppler shift** and the general effect the **Doppler effect** (after Christian Johann Doppler, the 19th-century physicist who first described the effect for light). This effect has numerous and important applications in astrophysics and biophysics (as well as many other fields), and it forms the basis of such technologies as Doppler weather radars (that can detect and study tornadoes and other forms of severe weather) and the radar guns that police use to detect speeders.

Introduction to the concept of the Doppler effect

In this appendix, we will examine this effect in some detail, using the tools and concepts we have developed through chapter R5. In section RB.2, we will use a simple model of the emission/detection process and a spacetime diagram to find a formula for the magnitude of the shift in wavelength. Section RB.3 looks at an example application, and section RB.4 explores the nonrelativistic limit of our formula. Finally, section RB.5 looks at the slightly different situation involved in Doppler shift radar.

Overview of the appendix

RB.2   Deriving the Doppler Shift Formula

To simplify matters, let us consider a source that moves directly toward or away from the observer in question, and let us take the x axis of the observer's frame (the Home Frame) to coincide with the line connecting the source and the observer. Let us also assume that the observer is located at $x = 0$ in that frame.

Some basic assumptions

First consider a clock that emits brief flashes of light (we will generalize this to continuous light waves shortly). Let the event of the emission of any one flash be event A, and the emission of the next flash be event B. Since the clock is present at both events, it measures a proper time between these flashes. If the time between flashes happens also to be so short that the clock follows an essentially straight worldline between the events, then the relationship between the proper time $d\tau_{AB}$ measured in the emitting clock's frame and the coordinate time dt_{AB} measured in the observer's frame is given by equation R5.5:

A simplified model

$$d\tau_{AB} = \sqrt{1 - v^2}\, dt_{AB} = \sqrt{1 - v_x^2}\, dt_{AB} \qquad (RB.1)$$

where v is the emitting clock's speed as measured in the observer's frame. (The last step follows because if the emitting clock is moving directly toward

---

[†]This appendix follows an approach to the Doppler shift formula suggested as an exercise by E. F. Taylor and J. A. Wheeler in *Spacetime Physics*, 2d ed., San Francisco: Freeman, 1992, p. 82.

Implication of relativistic time
dilation in this case

or away from the observer, it is moving along the x axis, so $v = |v_x|$.) Therefore,

$$dt_{AB} = \frac{d\tau_{AB}}{\sqrt{1 - v_x^2}} \qquad \text{(RB.2)}$$

Now, dt_{AB} is the time our observer would measure between the *emission* of the flashes. But how much time dt_R passes between the observer's *reception* of those flashes? The times dt_{AB} and dt_R are not necessarily the same! Since both flashes travel at a speed of 1 back to the observer, the time (in the observer's frame) that it takes a flash to get from its emission event back to the observer is equal to the *distance* the emission event is from the origin (as measured in the observer's frame). If the emitting clock were at rest in the observer's frame, then these distances would be the same for both flashes, and therefore the time between reception would be the same as the time between emission: $dt_R = dt_{AB}$. But if the source is moving relative to the observer, then the two events will *not* happen at the same place in the observer's frame, meaning that the light travel times for the two flashes will not be the same, implying that $dt_R \neq dt_{AB}$.

Different flashes take different times to return to the observer

Exactly how should we correct for this effect? We can answer this question fairly easily with the help of a spacetime diagram. Figure RB.1 shows the worldlines of the light flashes in question as they travel back to the observer from a clock that happens to be moving in the $+x$ direction. In the time dt_{AB} between the emission events (as measured in the observer's frame), the emitting clock moves a distance $v_x\, dt_{AB}$ away from the observer. This means (since the speed of light is 1 in the observer's frame) that it takes $v_x\, dt_{AB}$ *more* time for the second flash to make it back to the observer than it took for the first flash. Therefore, the time between the observer's reception of the flashes is simply the time between their emission dt_{AB} (as measured in the observer's frame) plus the extra light travel time $v_x\, dt_{AB}$ required for the second flash to reach the observer:

$$dt_R = dt_{AB} + v_x\, dt_{AB} = dt_{AB}\, (1 + v_x) \qquad \text{(RB.3)}$$

Combining this with the result given by equation RB.2, we get

$$dt_R = \frac{d\tau_{AB}}{\sqrt{1 - v_x^2}}(1 + v_x) = d\tau_{AB}\, \frac{\sqrt{1 + v_x}\sqrt{1 + v_x}}{\sqrt{1 + v_x}\sqrt{1 - v_x}} = d\tau_{AB}\, \sqrt{\frac{1 + v_x}{1 - v_x}} \quad \text{(RB.4)}$$

The link to wavelength of a continuous light wave

Now what has all this to do with the wavelength of a continuous beam of light waves? Consider! Each crest of a continuous light wave moves at the speed of light, just as a flash of light would. Therefore, for the purposes of

Figure RB.1

The time between the reception of the flashes dt_R is not the same as the coordinate time dt_{AB} between their emission because the light from flash event B has to travel an extra distance $v_x dt_{AB}$ to get back to the observer compared to the flash traveling from event A.

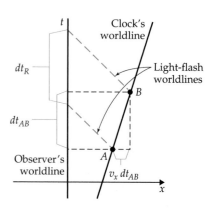

this calculation, a wave crest is analogous to a light flash, and there is a direct analogy between a sequence of light flashes and a continuous series of light wave crests. Therefore equation RB.4 should also apply to light *waves* if we interpret $d\tau_{AB}$ as the time between emission of light wave crests in the emitter's frame and dt_R as the time between the reception of light wave crests in the observer's frame.

Moreover, the **wavelength** λ of a light wave is defined to be the distance between successive crests in the wave. Since the crests also move at a speed of 1, this means that the distance λ between crests is equal to the time dt that it takes two successive crests to pass a given point in space (or emerge from the emitter): $\lambda = dt$. Plugging this into equation RB.4, we find that

$$\frac{\lambda_R}{\lambda_E} = \sqrt{\frac{1 + v_x}{1 - v_x}} \qquad \text{(RB.5)}$$

The relativistic Doppler shift formula

where λ_R is the wavelength of the light in the observer's frame and λ_E is its wavelength as measured in the emitter's frame.

Equation RB.5 is the **relativistic Doppler shift formula.** Although figure RB.1 (and the argument following) assumes that the emitter is moving *away from* the observer ($v_x > 0$), the formula also applies if the emitter is moving *toward* the observer ($v_x < 0$).

Exercise RBX.1

Verify the claim that equation RB.5 is correct even when $v_x < 0$.

Note also that if $v_x > 0$, then $\lambda_R > \lambda_E$, meaning that the received light has a longer (*red-shifted*) wavelength than its wavelength observed in the emitter frame. On the other hand, if $v_x < 0$, then $\lambda_R > \lambda_E$: the received light has a shorter (*blue-shifted*) wavelength than the wavelength observed in the emitter frame. This coincides with what you have probably heard before.

Exercise RBX.2

We have assumed in this derivation that the time between emission events (or successive crests of the light wave) is small compared with the time required for significant changes in the emitter's motion (so that it can be assumed to follow an approximately straight worldline during the interval in question). Is this approximation likely to be valid in the case of light waves (which have a wavelength ≈ 600 nm)? Justify your answer.

RB.3   Applications in Astronomy

Since excited atoms emit light having a characteristic set of wavelengths (in their own frame), equation RB.5 is commonly used by astronomers to compute the radial velocity of astrophysical objects relative to the earth. This equation technically only applies to an emitter that we know is moving directly toward or away from an observer (a somewhat more complicated formula applies when the emitter has a tangential velocity as well), but it is a

A standard application of this formula in astronomy

useful approximation in most cases (since tangential velocities are rarely large enough to matter much). The following is a typical example of such an astronomical application.

An example calculation

Problem   Light from excited atoms in a certain quasar is received by observers on earth. The wavelength of a certain spectral line of this light is measured by those observers to be 1.12 times longer than it would be if the atoms were at rest in the laboratory (i.e., the light has been *red-shifted* by about 12%). What is the quasar's speed relative to earth (assuming that it is moving directly away from earth)?

Solution   We are told that $\lambda_R/\lambda_E = 1.12$. Equation RB.5 then implies that

$$\frac{\lambda_R}{\lambda_E} = \sqrt{\frac{1 + v_x}{1 - v_x}} = 1.12 \qquad \text{(RB.6)}$$

To solve this for v_x, let us define $u \equiv \lambda_R/\lambda_E = 1.12$ and square both sides of equation RB.6. We get

$$u^2 = \frac{1 + v_x}{1 - v_x}$$

$$\Rightarrow \quad (1 - v_x)u^2 = 1 + v_x$$

$$\Rightarrow \quad u^2 - u^2 v_x = 1 + v_x$$

$$\Rightarrow \quad u^2 - 1 = v_x + u^2 v_x = (u^2 + 1)v_x$$

$$\Rightarrow \quad v_x = \frac{u^2 - 1}{u^2 + 1} = \frac{(1.12)^2 - 1}{(1.12)^2 + 1} = 0.11 \qquad \text{(RB.7)}$$

The quasar's speed relative to the earth is thus 11% of the speed of light.

Exercise RBX.3

How should we correct equation RB.5 if the emitter does not move directly toward or away from the observer? Assume that we still choose the observer's x axis to go through the emitter's position at the time it emits the light flashes we observe. (*Hints:* Where did we assume in section RB.2 that the motion was along the x axis? Argue that the emitter's motion in the y or z direction during a small time interval will not significantly affect the distance between the emitter and the observer.)

RB.4   The Nonrelativistic Limit

The derivation of the relativistic Doppler shift formula given by equation RB.5 actually combines two effects: (1) the relativistic distinction between the emitter's proper time and the observer's coordinate time between successive pulses (see equation RB.2) and (2) the delay of successive received pulses due to the changing separation between emitter and observer (see equation RB.3). The first effect arises only in the theory of relativity; the latter would apply even if time were absolute (as long as light moves with a speed of roughly 1 in the observer's frame). We might therefore suspect that when $v_x \ll 1$, we can ignore the relativistic effect in comparison to the other, so that equation RB.5 becomes

$$dt_R = (1 + v_x)\, dt_{AB} \approx (1 + v_x)\, d\tau_{AB} \qquad \Rightarrow \qquad \frac{\lambda_R}{\lambda_E} \approx 1 + v_x \qquad \text{for } v_x \ll 1$$

$$(\text{RB.8})$$

Exercise RBX.4

Verify that equation RB.8 is indeed correct by using the binomial approximation to rewrite both the numerator and the denominator of equation RB.5 in the limit that $v_x \ll 1$.

RB.5   Doppler Radar

Doppler radar (used by weather observers and police) involves a somewhat different situation. In this case, the emitter and observer are typically in the same frame, and the emitted waves are reflected by an object moving relative to both. You can use the spacetime diagram shown in figure RB.2 (and an argument analogous to that given in section RB.2) to show that in this situation

$$\frac{\lambda_R}{\lambda_E} = \frac{1 + v_x}{1 - v_x} \qquad (\text{RB.9})$$

Doppler shift formula when the emitter and observer are in the same frame but waves are reflected from a moving object

Exercise RBX.5

Verify this, using figure RB.2. (*Hint:* Note that all the times shown on the diagram are coordinate times measured in the observer's frame. In this case, the time between events A and B measured by a clock traveling with the reflector is irrelevant.)

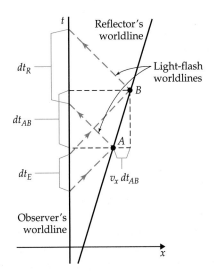

Figure RB.2

A spacetime diagram showing the relationship between the time dt_E between two light flashes emitted by an observer and the time dt_R between their reception after being reflected by a moving object at events A and B.

HOMEWORK PROBLEMS

BASIC SKILLS

RBB.1 A spaceship moves directly away from earth at a speed of 0.5. By what factor is the spaceship's tail-light red-shifted?

RBB.2 The hydrogen fusion flare from an alien ship's engine is detected from earth. If the light emitted from the hydrogen atoms is observed to be blue-shifted by 35% compared to normal, how rapidly is the ship approaching?

SYNTHETIC/RICH-CONTEXT

RBS.1 (Adapted from Taylor and Wheeler, *Spacetime Physics*, 2d ed., pp. 263–264.) A physicist brought to court for running a red light argues that the light *looked* green and thus must have been Doppler-shifted. The judge changes the charge to speeding and fines the physicist a penny for every mile per hour the physicist was traveling over the speed limit of 45 mi/h. What was the fine (approximately)? (*Hints:* Red light has a wavelength of $\approx$ 650 nm; green light has a wavelength of $\approx$ 530 nm. Also 2.2 mi/h = 1.0 m/s.)

ANSWERS TO EXERCISES

RBX.1 If the emitter is approaching, the time between reception events is $dt_R = dt_{AB} - |v_x| dt_{AB} = dt_{AB} + v_x dt_{AB}$, where the last step follows because $v_x = -|v_x|$ here. The rest of the derivation follows as before.

RBX.2 If the distance between visible light crests is about 600 nm, the time (in SI units) between crest emission events will be the time that it takes the first crest to travel 600 nm from the emitter at the speed of light, or about $(6.0 \times 10^{-7}$ m$)/(3.0 \times 10^8$ m/s$) = 2.0 \times 10^{-15}$ s. It is hard to imagine something accelerating so violently that its motion would change much in such a time interval!

RBX.3 The corrected equation is

$$\frac{\lambda_R}{\lambda_E} = \frac{1 + v_x}{\sqrt{1 - v^2}} \qquad \text{(RB.10)}$$

RBX.4 The binomial approximation tells us that

$$(1 + v_x)^{1/2} \approx 1 + \tfrac{1}{2}v_x$$

$$\text{and} \quad (1 - v_x)^{-1/2} \approx 1 + \tfrac{1}{2}v_x \qquad \text{(RB.11)}$$

in the limit that $v_x \ll 1$. Multiplying these factors together and dropping the v_x^2 term (which will be *very* small), we get equation RB.8.

RBX.5 Note that $dt_{AB} = dt_E + v_x dt_{AB}$. Combine this with equation RB.3 to get equation RB.9.

Glossary

binomial approximation: the very useful approximation that states that $(1 + x)^a \approx 1 + ax$ if $x \ll 1$. (Section R5.3)

causal influence: any effect (particle, object, wave, or message) produced by one event that can cause another event to occur. (Section R8.1)

causally connected (events): a pair of events connected by a causal influence in such a way that one event causes the other. (Section R8.1)

coordinate time Δt: the time interval between two events as measured in an inertial reference frame by computing the difference between the time coordinates of the events registered by two synchronized clocks in the frame (one present at each event). If the two events happen to occur at the same place in the frame, then one clock suffices to measure the coordinate time. (Section R3.2)

cosmic speed limit: the speed of light (in a vacuum). The speed of light is the fastest that anything can travel because anything (even a message) traveling faster than light would violate causality. (Section R8.1)

diagram t' **axis:** the line on a spacetime diagram that connects all events that occur at $x' = 0$ (spatial origin) of a given inertial reference frame. This line is the worldline of the spatial origin of that frame in the spacetime diagram. (This description applies to the Home Frame t axis as well as the Other Frame t' axis: simply substitute t for t' and x for x' in the definition given above.) (Section R6.3)

diagram x' **axis:** the line on a spacetime diagram that connects all events that occur at $t' = 0$ (that is, that are simultaneous with the origin event) in a given inertial reference frame. This is to be sharply distinguished from the *spatial x' axis*, which is the line in physical space connecting all points in space having coordinates $y' = z' = 0$. (This description applies to the Home Frame diagram x axis as well as the Other Frame diagram x' axis: simply substitute t, x, y, z for t', x', y', z' in the definition given above.) (Section R6.4)

direct Lorentz transformation equations: the equations (equations R6.11) that describe mathematically how to compute the coordinates t', x', y', z' of an event in the Other Frame given that same event's coordinates t, x, y, z in the Home Frame (or the similar equation R6.12 which does the same for coordinate differences between two events). See the entry for **Lorentz transformation equations** for a fuller discussion. (Section R6.6)

Doppler effect: a term describing fact that when one measures the properties of a wave in a frame that is in relative motion to the wave's source, one observes a change (**Doppler shift**) in the wave's wavelength. (Section RB.1)

Doppler shift: the shift in the wavelength or frequency of a wave whose properties are measured in a frame that is moving relative to the wave's source. (Section RB.1)

dynamics: the study of how the interactions between objects affect and determine their motion. (Section R9.1)

Einstein velocity transformation equations: the set of equations that allow us to convert a particle's velocity in one inertial frame into its velocity in another (assuming that we know the relative velocity and orientation of the two frames). The **direct** transformation equations (equations R8.14) yield Other Frame velocity components in terms of Home Frame components; the **inverse** versions of the equations (equations R8.8) yield Home Frame velocity components in terms of Other Frame components. (Section R8.4)

electronvolt (eV): a unit of energy, equivalent to the energy gained by one electron as it moves through a 1-V battery. Particle physicists typically use the electronvolt in place of the kilogram for mass energy, and momentum, since the kilogram is an inappropriately large unit. (In SR units, we can use any unit that we like for mass, energy, and momentum, as long as we use the *same* unit for all.) Conversion factors to other units are $1 \text{ eV} = 1.602 \times 10^{-19}$ J $= 1.782 \times 10^{-36}$ kg. (Section R10.5)

energy-momentum diagram: a diagram for illustrating four-momentum vectors of particles moving in one spatial dimension. A particle's four-momentum vector is represented on such a diagram as an arrow whose projection on the horizontal axis is the x component of the particle's four-momentum and whose projection on the vertical axis is the particle's relativistic energy. An energy-momentum diagram is to relativistic dynamics what a spacetime diagram is to relativistic kinematics. (Section R10.1)

ether: a hypothetical substance that was thought at the turn of the century to provide the medium that light waves moved through. The ether was also thought to define the reference frame where the speed of light had the value c predicted by Maxwell's equations. (Section R2.1)

event: a physical occurrence or process that happens at a well-defined position in space and instant of time. (Section R1.2)

first-law detector: a device that continually examines the trajectories of isolated objects in its vicinity for evidence of nonconstant velocities. (Section R1.4)

four-magnitude (of a four-vector): the quantity found by subtracting the squares of the four-vector's spatial components from the square of its time component and taking the square root: $|\mathbf{A}| \equiv [A_t^2 - A_x^2 - A_y^2 - A_z^2]^{1/2}$. The four-magnitude of a four-displacement between two events is the spacetime interval between them. The four-magnitude of a particle's four-momentum is its frame-independent mass m. (Section R9.4)

four-momentum (of a particle) $\mathbf{P}$: the four-vector defined by $\mathbf{P} \equiv m(d\mathbf{R}/d\tau)$, where m is the particle's invariant mass, $d\mathbf{R}$ is an infinitesimal four-displacement along the particle's worldline, and $d\tau$ is the proper time for that displacement measured along the particle's worldline. The spatial components $[P_x, P_y, P_z]$ of the four-momentum become essentially equivalent to the components of newtonian momentum at low speeds, but conservation of four-momentum *is* consistent with the principle of relativity and in fact is amply supported by experiment. (Section R9.3)

four-vector: a vector $\mathbf{A} = [A_t, A_x, A_y, A_z]$ with four components that (when one changes inertial frame) transform according to the Lorentz transformation equations, that is, like the components of the **four-displacement** four-vector $\Delta\mathbf{R} = [\Delta t, \Delta x, \Delta y, \Delta z]$. (We will always use bold capital letters for four-vectors in this text.) (Section R9.3)

frame-independent: a quantity whose value does not depend on one's choice of reference frame. For example, a particle's electric charge q is frame-independent, but its velocity $\vec{v}$ is not. In special relativity, the coordinate time Δt between two events is *not* frame-independent. (Section R3.5)

future (of an event P): the set of all events having a timelike or lightlike spacetime interval with P and occurring *after* P in all inertial frames. These events could be caused by P. (Section R8.3)

galilean transformation equations: the four equations $t' = t$, $x' = x - \beta t$, $y' = y$, and $z' = z$ that specify the spacetime coordinates $[t', x', y', z']$ of an event in the Other Frame in terms of the coordinates $[t, x, y, z]$ of the same event in the Home Frame. These equations are true (1) if the frames are in standard orientation and (2) if the newtonian assumption that time is universal and absolute is true. (Section R1.6)

galilean velocity transformation equations: the equations $v_x' = v_x - \beta$, $v_y' = v_y$, $v_z' = v_z$ that specify the velocity components $[v_x', v_y', v_z']$ of an object in the Other Frame in terms of the object's velocity components $[v_x, v_y, v_z]$ in the Home Frame. These equations are true if (1) the frames are in standard orientation and (2) if the newtonian assumption that time is universal and absolute is true. (Section R1.6)

Home Frame and **Other Frame:** convenient names for any pair of inertial reference frames we are comparing. When two inertial reference frames are in standard orientation, we conventionally define the Other Frame so that this frame moves in the $+x$ direction with respect to the Home Frame. (Equivalently, the Home Frame is the frame that moves in the

$-x$ direction relative to the Other Frame.) The coordinates of the Other Frame are conventionally written with **primes** (say, t', x'). (Sections R1.4 and R6.2)

inertial clock: a clock that moves at a constant velocity in an inertial reference frame. Alternatively, we can attach a first-law detector to a clock to test whether it is inertial. If the detector registers no violations of the first law, the clock is inertial. (Section R3.5)

inertial reference frame: a reference frame in which an isolated object is always and everywhere observed to move with a constant velocity. (Section R1.4)

inverse Lorentz transformation equations: the equations (equations R6.10) that describe mathematically how to compute the coordinates t, x, y, z of an event in the Home Frame, given that same event's coordinates t', x', y', z' in the Other Frame (or the similar equations R6.13 which do the same for coordinate differences between two events). Note that these equations represent the inverse of the coordinate transformation described by the direct LTEs. See the entry for **Lorentz transformation equations** for a fuller discussion. (Section R6.6)

kinematics: the study of how the motion of objects is to be measured and described mathematically. (Section R9.1)

length (of a moving object in a given inertial frame) L: the distance (measured in that frame) between two events that occur simultaneously (in that frame) at opposite ends of the object. (Section R7.1)

light clock: an imaginary device that uses a flash of light bouncing back and forth between two mirrors a fixed distance L apart as the reference for measuring time intervals. The time displayed by the clock is determined by counting complete cycles of the bouncing flash: each cycle lasts $2L$ of time (if L is measured in seconds) by definition. (Section R4.2)

light cone (of an event P): the frame-independent structure of spacetime around P. When drawn on a two-observer spacetime diagram with two spatial axes, the **past** and **future** of P look like two infinite cones whose points touch at P. The boundaries of these cones are the set of events touched by rings of light that converge on or expand away from P. All observers will agree which events fall where in this structure. (Section R8.3)

lightlike (spacetime interval): a spacetime interval Δs between two events such that $\Delta s^2 = 0$. If a light flash can connect the events, the interval between them is lightlike. (Section R8.2)

Lorentz contraction: the term describing the fact that an object's length is always observed to be *smaller* in a frame where the object is moving than in a frame where the object is at rest: we say that a moving object's length in the direction of motion is *Lorentz-contracted*. The equation linking the object's rest length L_R with its length L observed in a frame where it is moving at a speed β is given by $L = L_R(1 - \beta^2)^{1/2}$. (Section R7.2)

Lorentz transformation equations (LTEs): the equations that describe mathematically how to compute the coordinates $[t', x', y', z']$ of an event in the Other Frame given that same event's coordinates $[t, x, y, z]$ (or coordinate difference between the same two events) in the Home Frame, or vice versa. The equations that give $[t', x', y', z']$ in terms of $[t, x, y, z]$ are called the **direct Lorentz transformation equations;** the equations that give $[t, x, y, z]$ in terms of $[t', x', y', z']$ are called the **inverse Lorentz transformation equations.** The term is also applied to the very similar equations that convert the coordinate differences $[\Delta t, \Delta x, \Delta y, \Delta z]$ or the components $[A_t, A_x, A_y, A_z]$ of a four-vector from one frame to another. These equations are named for the Dutch physicist H. A. Lorentz, who stated them in a paper published in 1904, the year before Einstein published the theory of relativity. Lorentz is not given credit for the theory of relativity, because he gave them a very different interpretation than Einstein did. (Section R6.6)

Maxwell's equations: a set of equations that summarize the laws of electromagnetism. These equations predict the existence of electromagnetic waves that travel with a certain completely determined speed c. (If these waves have the appropriate wavelength, we perceive them as light.) (Section R2.1)

metric equation: the equation $\Delta s^2 = \Delta t^2 - \Delta x^2 - \Delta y^2 - \Delta z^2$ that links the frame-independent spacetime interval Δs between two events with the frame-*dependent* coordinate separations $\Delta t, \Delta x, \Delta y, \Delta z$ measured between those two events in any inertial reference frame. This equation, which is to spacetime what the pythagorean theorem is to plane geometry, is one of the most important equations of relativity. (Section R4.2)

muon: an elementary particle of the lepton family that is essentially a massive version of the electron (about 206 times as massive as the electron). Muons are unstable and decay with a half-life of $1.52 \, \mu$s to an electron and a pair of neutrinos. (Section R4.4)

newtonian momentum $\vec{p}$: the three-dimensional momentum vector we defined in unit C: $\vec{p} = m\vec{v} = m(d\vec{r}/dt)$. This kind of momentum is not consistent with the principle of relativity; because if a collision conserves newtonian momentum in one inertial frame, it will not in another. (Section R9.1)

noninertial reference frame: a reference frame in which an isolated object is observed at least at some place and some time to move with a nonconstant velocity. (Section R1.4)

observer: a (possibly hypothetical) person who interprets measurements made in a reference frame. (Section R1.3)

operational definition: a definition of a quantity that describes the quantity by specifying a method for *measuring* it.

origin event: an event that defines the origin of time ($t = 0$) and the origin of space ($x = y = z = 0$) in a given reference frame.

past (of an event P): the set of all events having a timelike or lightlike spacetime interval with P and occurring *before* P in all inertial frames. These events could conceivably cause P. (Section R8.3)

principle of relativity: the statement that *the laws of physics are the same in all inertial reference frames.* This principle is the foundation of the special theory of relativity, which is little more than an unfolding of the logical consequences of this statement. (Section R1.1)

proper length: another term for **rest length.** (Section R7.3)

proper time $\Delta \tau$: the time interval between two events measured by any clock that is physically present at both events. The value of this quantity is frame-independent (since one does not use a reference frame to measure it), but does depend on the nature of the worldline that the clock follows between the two events. The proper time between two events is analogous to the *pathlength* between two points in space. (The adjective *proper* here should be thought of as meaning *proprietary*, not *correct*.) (Section R3.5)

radar method: a method of assigning spacetime coordinates to events by using a single clock and light flashes rather than a clock lattice synchronized according to the Einstein method. (Both methods are equivalent, but sometimes one is more useful than the other for explaining something.) (Section R2.6)

reference frame: by definition, a rigid cubical lattice with appropriately synchronized clocks at every lattice location (or its functional equivalent). (Section R1.3)

relative (quantity): a quantity whose value depends on one's choice of inertial reference frame. For example, the coordinate time Δt between two events is *relative*, but the space-time interval Δs between those events is not. (Section R3.2)

relativistic Doppler shift formula: the equation (equation RB.5) that specifies the ratio of the received and emitted wavelengths of a light wave whose source moves relative to the observer with an x-velocity of v_x. (Section RB.2)

relativistic dynamics: the study of how the interactions between objects affect their motion in cases where relativistic effects are important (see **dynamics**). (Section R9.1)

relativistic energy (of a particle) E: the time component of the particle's four-momentum (which must be conserved along with the spatial components if the principle of relativity is to be consistent with conservation of four-momentum): $E \equiv P_t = m/(1 - v^2)^{1/2}$. (Section R9.6)

relativistic kinematics: the study of how the motion of objects is to be measured and described mathematically in situations where relativistic effects are important (see **kinematics**). (Section R9.1)

relativistic kinetic energy K: the difference between a particle's relativistic energy E and its rest energy: $K \equiv E - m$ (this is the part of the particle's relativistic energy that is due to its motion). At low speeds, $K \approx \frac{1}{2}mv^2$. (Section R9.6)

relativistic momentum p: the magnitude of the spatial components of the particle's four-momentum vector: $p \equiv [P_x^2 + P_y^2 + P_z^2]^{1/2}$. (Section R9.7)

relativistic three-momentum: the spatial components $[P_x, P_y, P_z]$ of a particle's four-momentum vector. These components define a three-dimensional vector that reduces to the particle's newtonian momentum at low velocities. (Section R9.6)

rest energy (also called **mass energy**): the relativistic energy $E_{\text{rest}} = m$ that a particle has when at rest. (Section R9.6)

rest length L_R: the length of an object measured in the frame in which it is at rest. (Since the ends of an object are not going anywhere in the object's rest frame, the events that mark out the object's ends do not *necessarily* have to be simultaneous in the rest frame to accurately mark out the object's length in that frame.) (Section R7.2)

second (of distance): the distance that light travels in 1 s of time (also known as a *light-second*). By international convention, 1 s (of distance) = 299,792,458 m. (Section R2.3)

second law of thermodynamics: a physical law that states that the entropy of an isolated system cannot decrease (see unit T). This law implies that some sets of events must occur in a certain temporal order in all inertial frames, and lies behind the more colloquial idea of "causality." (Section R8.1)

spacelike (spacetime interval): a spacetime interval between two events such that $\Delta s^2 < 0$. The **spacetime separation** (see next) corresponding to this interval can be measured by using a ruler to determine the distance between the events in an inertial reference frame where they happen at the same time. (Section R8.2)

spacetime coordinates (of a given event in a given reference frame): an ordered set of four numbers, the first specifying the *time* of the event as registered by the nearest clock in the reference frame lattice, followed by three that specify the spatial coordinates of that clock in the lattice. (Note that these coordinates depend not only on one's choice of reference frame but also on one's choice of origin for the spatial axes and origin of time.) (Section R1.3)

spacetime diagram: a graph where we plot the time coordinates of events versus their position coordinates. Spacetime diagrams are very helpful for displaying the relationships between events. An event in spacetime is plotted as a point in space. (Section R2.4)

spacetime interval Δs: the time interval between two events measured by an *inertial* clock that is physically present at both events. The value of this quantity is frame-independent (since one does not use a reference frame to measure it); and since there is only one unique worldline that takes a clock along a straight line from one event to the other at a constant velocity, this quantity has a unique value for a given pair of events. The spacetime interval between two events is analogous to the *distance* between two points in space. (Section R3.5)

spacetime separation $\Delta\sigma$: a way of expressing the magnitude of a spacetime interval between two events that is a real number when that spacetime interval is spacelike: $\Delta\sigma \equiv (-\Delta s^2)^{1/2}$. (Section R8.2)

spatial x' **axis**: the line in physical space connecting all points in space having coordinates $y' = z' = 0$. This is distinct from the **diagram** x' **axis**, which is the line on a two-observer spacetime diagram that connects all events occurring along the spatial x' axis that also happen at $t' = 0$. (Section R6.3)

special theory of relativity: the model of mechanics that follows from the principle of relativity. This theory is a special case of the *general theory of relativity*, which also handles noninertial reference frames and is our currently accepted theory of gravitation. (Section R1.1)

squared spacetime interval Δs^2: the frame-independent quantity $\Delta s^2 = \Delta t^2 - \Delta x^2 - \Delta y^2 - \Delta z^2 = \Delta t^2 - \Delta d^2$. Depending on the two events in question, this quantity can be positive, negative, or zero, but its value (including its sign) is frame-independent. (Section R8.2)

SR units: the same unit system as SI units, except that we measure distance in seconds instead of meters. In this unit system, velocity is unitless and mass, momentum, and energy are all measured in kilograms. [We can convert quantities in one unit system to the other by multiplying or dividing by an appropriate power of the conversion factor $1 = 3.00 \times 10^8$ m/(1 s).] (Section R2.3)

standard orientation (for two inertial frames): the orientation of a Home Frame and an Other Frame when (1) their coordinate axes point in the same directions in space, (2) their relative velocity is directed along their common x axis, and (3) the Other Frame moves in the $+x$ direction relative to the Home Frame. (Section R1.6)

synchronized (according to Einstein): a property of two clocks in an inertial reference frame that have been adjusted so the time interval registered by the clocks for a light flash to travel between them is equal to their separation divided by c (that is, if the speed of a light flash moving between them is measured by them to have speed c). (Section R2.2)

temporal order: the order in which two events occur in time in a given reference frame. For causality to make sense, the temporal order of *causally connected* events must be the same in all reference frames (an effect should never be observed to precede its cause). (Section R8.1)

time axis: the line on a spacetime diagram connecting all events that occur at $x = 0$ in a given inertial frame. See **diagram t' axis** for more discussion. (Section R6.3)

timelike (spacetime interval): a spacetime interval between two events such that $\Delta s^2 > 0$. This interval can be measured by the clock present at both events in an inertial frame where the events occur at the same place. (Section R8.2)

twin (or **clock**) **paradox:** a famous problem that has puzzled students of relativity almost since the theory was first stated. The situation involves a pair of twins, one of whom travels to a distant star and back, while the other stays at home. A naïve application of the the proper time formula leads each twin to think that the other is younger, while a naïve application of the principle of relativity itself seems to imply that the twins should have the same age. The apparent paradox dissolves when we recognize that (at least) the traveling twin is *not* in an inertial reference frame. Each of the twins follows his or her own worldline and thus registers his or her own time between the departure and arrival events. (Section R5.6)

two-observer spacetime diagram: a spacetime diagram that displays the t and x coordinate axes for two observers in different inertial reference frames. (Section R6.1)

wavelength: the distance between successive crests of a wave. (Section RB.2)

worldline: the continuous set of events that represents a particle's passage through space and time. On a spacetime diagram, a worldline is represented by a line or curve. (Section R2.4)

world-region: the region on a spacetime diagram that is the set of worldlines of all particles in an extended object. (Section R7.2)

Index

Entries followed by *t* and *f* refer to tables and figures, respectively.

Periodic Table of the Elements

Key:

Atomic number	1
Symbol	H
Atomic mass	1.008

1 1A																	18 8A
1 H 1.008	2 2A											13 3A	14 4A	15 5A	16 6A	17 7A	2 He 4.003
3 Li 6.941	4 Be 9.012											5 B 10.81	6 C 12.01	7 N 14.01	8 O 16.00	9 F 19.00	10 Ne 20.18
11 Na 22.99	12 Mg 24.31	3 3B	4 4B	5 5B	6 6B	7 7B	8	9 8B	10	11 1B	12 2B	13 Al 26.98	14 Si 28.09	15 P 30.97	16 S 32.07	17 Cl 35.45	18 Ar 39.95
19 K 39.10	20 Ca 40.08	21 Sc 44.96	22 Ti 47.88	23 V 50.94	24 Cr 52.00	25 Mn 54.94	26 Fe 55.85	27 Co 58.93	28 Ni 58.69	29 Cu 63.55	30 Zn 65.39	31 Ga 69.72	32 Ge 72.59	33 As 74.92	34 Se 78.96	35 Br 79.90	36 Kr 83.80
37 Rb 85.47	38 Sr 87.62	39 Y 88.91	40 Zr 91.22	41 Nb 92.91	42 Mo 95.94	43 Tc (98)	44 Ru 101.1	45 Rh 102.9	46 Pd 106.4	47 Ag 107.9	48 Cd 112.4	49 In 114.8	50 Sn 118.7	51 Sb 121.8	52 Te 127.6	53 I 126.9	54 Xe 131.3
55 Cs 132.9	56 Ba 137.3	57 La 138.9	72 Hf 178.5	73 Ta 180.9	74 W 183.9	75 Re 186.2	76 Os 190.2	77 Ir 192.2	78 Pt 195.1	79 Au 197.0	80 Hg 200.6	81 Tl 204.4	82 Pb 207.2	83 Bi 209.0	84 Po (210)	85 At (210)	86 Rn (222)
87 Fr (223)	88 Ra (226)	89 Ac (227)	104 Rf (257)	105 Db (260)	106 Sg (263)	107 Bh (262)	108 Hs (265)	109 Mt (266)	110	111	112	(113)	114	(115)	116	(117)	

58 Ce 140.1	59 Pr 140.9	60 Nd 144.2	61 Pm (147)	62 Sm 150.4	63 Eu 152.0	64 Gd 157.3	65 Tb 158.9	66 Dy 162.5	67 Ho 164.9	68 Er 167.3	69 Tm 168.9	70 Yb 173.0	71 Lu 175.0
90 Th 232.0	91 Pa (231)	92 U 238.0	93 Np (237)	94 Pu (242)	95 Am (243)	96 Cm (247)	97 Bk (247)	98 Cf (249)	99 Es (254)	100 Fm (253)	101 Md (256)	102 No (254)	103 Lr (257)